Repositioning Pedagogical Content Knowledge in Teachers' Knowledge for Teaching Science

Anne Hume · Rebecca Cooper ·
Andreas Borowski
Editors

Repositioning Pedagogical Content Knowledge in Teachers' Knowledge for Teaching Science

Springer

Editors
Anne Hume
University of Waikato
Hamilton, New Zealand

Rebecca Cooper
Monash University
Melbourne, VIC, Australia

Andreas Borowski
University of Potsdam
Potsdam, Germany

ISBN 978-981-13-5897-5 ISBN 978-981-13-5898-2 (eBook)
https://doi.org/10.1007/978-981-13-5898-2

Library of Congress Control Number: 2018966439

This Springer imprint is published by the registered company Springer Nature Singapore Pte Ltd.
The registered company address is: 152 Beach Road, #21-01/04 Gateway East, Singapore 189721, Singapore

Foreword

This foreword is designed to set the scene for this book and is based on our experience as the facilitators of the Second (2nd) PCK Summit. Through a brief reflection on that role, we hope to provide some context for the work and learning that has culminated in this book and give a real sense of the collaborative learning experience that comprised the 2nd Summit.

The 2nd PCK Summit came about as a consequence of the work conducted at the First (1st) Summit, which was held in Colorado Springs in October 2012. At that Summit, the participants (mostly in the field of science education and some from mathematics) spent a week discussing research into pedagogical content knowledge (PCK)—much of which they had conducted and led—in an effort to develop some form of consensus and shared understanding around the construct of PCK. That Summit resulted in the book *Re-examining Pedagogical Content Knowledge in Science* (Berry, Friedrichsen, & Loughran, 2015) that highlighted our learning at the Summit, but also introduced what became known as the Consensus Model (CM) of Teacher Professional Knowledge and Skill.

The success of the 1st Summit led to ongoing discussions, especially among science PCK researchers internationally, about the CM as the notion of an agreed way of viewing the construct began to resonate with others. As a consequence, in December 2016, 24 PCK researchers in science education met in Leiden to continue the discussion about PCK and to push our learning further; hence, the 2nd PCK Summit was born.

Some of the participants of the 1st Summit were in attendance at the 2nd Summit, but there were also a number of new active PCK researchers and thinkers in attendance. The focus of the 2nd PCK Summit was largely on data and analysis. In so doing, it offered an opportunity for participants to tease out the intricacies of collecting data through a variety of instruments, and to seriously examine approaches and techniques of analysis in identifying, capturing and portraying science PCK. Participants were asked to write 'outlines' of their current PCK research in science education, and these outlines were shared and read by all prior to the Summit. As the outlines focused on data collection and analysis, they offered a deep dive into instruments, processes, practices and procedures used in identifying aspects of science PCK.

The programme for the Summit was designed in such a way as to allow participants to work in small groups interested and/or experienced in the use of similar data collection instruments and analytic processes. This arrangement fostered in-depth discussion that was often focused on a specific task set by the Summit facilitators, such as which aspect/s of PCK does a particular instrument best articulate, or what criteria might best be used for identifying PCK using a given instrument? The small group discussions were reported back to the whole group in a variety of ways with the purpose of generating further discussion designed to challenge our thinking and generate questions to help us make meaningful progress in better understanding and portraying science PCK.

As the facilitators of the Summit, our role was to keep the learning moving forward, to not let things get too bogged down and to ensure that contributions were fully worked through at each level (small group and whole group) in order to build an agenda for ongoing development and understanding of PCK. Sometimes this role meant participating in a small group, sometimes it meant sitting back and observing, and at other times it meant offering a thought-provoking or challenging question. We were privileged to be able to participate in this way as it allowed us to use our observations to better facilitate the whole group discussions; the pedagogical purpose is to use the emerging learning to shape and reshape upcoming activities while still keeping an eye on the big picture.

Participant engagement across the activities helped to build a real sense of common purpose and, as this book illustrates, led to new insights and shared understandings of PCK in science education. It was a most demanding and enjoyable experience, and we were certainly very grateful for the opportunity to be involved and work with such a fine team of scholars.

As the Summit progressed, it was clear to all of us that that the workshops offered challenges and opportunities to move well beyond our individual ideas and views. The small group sessions created powerful agendas for the whole group and ideas started to come together in ways that led to a common expectation of conceptual coherence as a concrete outcome. As the Summit progressed, Julie Gess-Newsome helped to bring that coherence to our work by inviting us to revisit the CM from the 1st Summit. She offered her insights into the model and her observations about how it had been taken up in science education research and interpreted by others in the years since that Summit. Not surprisingly, it quickly became apparent that the model was rich with opportunity for us, as a group, to bring together our ideas and all our learning through the 2nd Summit in order to revisit and refine the CM. On the last day of the Summit, guided by Janet Carlson and Kirsten Daehler, the group refined the CM in a highly engaging and constructive whole group session. This book introduces the Refined Consensus Model (RCM) of science PCK and begins to unpack the possibilities it offers for the further development of PCK research.

Both of the PCK Summits have been invaluable experiences for all involved. They have enriched the work of many PCK researchers through the opportunity to publicly interrogate their work and to do so in a research community with a common interest and concern. This book captures the progress made during the 2nd Summit, introduces the RCM of PCK in science and invites all PCK researchers to become part of the conversation.

Clayton, Australia

John J. Loughran
Faculty of Education, Monash University

Rebecca Cooper
Faculty of Education, Monash University

Reference

Berry, A., Friedrichsen, P., & Loughran, J. (2015). *Re-examining pedagogical content knowledge.* London: Routledge press.

Preface

Background

Lee Shulman introduced the construct of pedagogical content knowledge (PCK) into the literature in the mid-1980s largely via articles in journals such as the Education Researcher (e.g. Shulman, 1987). Through PCK Shulman sought to acknowledge and represent a specialised form of professional knowledge, possessed by teachers, that sets teachers aside from other professionals. This knowledge typically grows with classroom experience and underpins how effective teachers are able to teach their subject matter in ways that support student understanding. The idea of a specialised form of professional knowledge crucial to expertise in teaching resonated well with academics, so PCK was quickly explored, adopted and adapted in a diversity of ways by researchers in the field across different domains, particularly in science and mathematics. Many researchers went on to focus on understanding of how such knowledge develops and how its development might be successfully supported. The first major book about PCK was the Gess-Newsome and Lederman (1999) publication *Examining Pedagogical Content Knowledge* (Springer). Not surprisingly perhaps, the burgeoning research produced a plethora of interpretations and uses of the original concept, which became problematic as inconsistencies and/or vague applications of the PCK concept began to emerge in the literature. In a review of PCK research, largely in the science education field, Abell (2007) identified this problem and she urged researchers 'to use PCK more explicitly and coherently to frame their studies' (p. 1407) to bring greater clarity and rigour into the research around PCK. However, to achieve clear and consistent practice, science and mathematics education researchers realised they first needed consensus in the field about the very meaning and understanding of the PCK construct itself.

The First PCK Summit

In 2012, the challenge to revisit PCK was taken up by more than 30 lead researchers (mostly in science education and some from mathematics education) from five different continents. They gathered as a forum to share their work and discuss issues and challenges in PCK research and the implications for policy and practice. In a week-long workshop, known as 'the PCK Summit', the researchers grappled with three broad questions:

(1) What are the attributes of PCK?
(2) What are the tools for measurement and analysis of PCK?
(3) What are, and how can we explain, the complementary and contradictory nature of research results of PCK?

Among the many outcomes of the Summit was a 'consensus model' (CM) of PCK. This generic model featuring PCK within teachers' wider professional knowledge and other outcomes of the Summit were presented at a range of (inter)national research conferences and in the spring of 2015, the book *Re-examining Pedagogical Content Knowledge* (Berry, Friedrichsen, & Loughran, 2015) was published by Routledge. This book showcased the work of the PCK Summit at Colorado in 2012 and comprised 17 chapters written by Summit participants. The book was well received, particularly in the science education PCK community, and the CM quickly became part of the international rhetoric in science education PCK research and a potential conceptual framework for research agendas in science teacher education.

By early 2016 though, it became evident within the science teacher education research community that PCK researchers—including those who participated in the Summit—were appearing to interpret and operationalise the CM differently. For example, many variants of PCK (such as dynamic PCK, canonical PCK, static PCK, enacted PCK, PCK-in-action, topic-specific PCK and domain PCK to name a few) were still being cited, along with various components of PCK. Consequently, PCK in science education continued to be assessed in very diverse ways, including the use of a wide range of instruments that attempted to measure or capture PCK, which in turn generated different kinds of qualitative and quantitative data. In part, these differences could be attributed to different goals of the researchers. Some science education studies, for example, aimed to describe the content and structure of PCK in a specific context, whereas other studies sought to assess the quality of science teachers' PCK, often times in relation to other variables. So once again, an international group of lead researchers in PCK (comprised of many from the first PCK Summit) gathered in 2016 to address these different and sometimes implicit interpretations and operationalisations of PCK, centring this time on science education for pragmatic reasons related to common purpose and focus, in a second PCK Summit that was organised as a Lorentz Workshop@Snellius in Leiden.

The Second PCK Summit in Science Teacher Education

Through invitation the organisers of the Second (2nd) PCK Summit brought together 24 researchers to attend the Lorentz Workshop. The group members, including the organising committee, were invited on the basis of their expertise and strength (both proven and potential) as researchers in the science education PCK field, their particular specialised contributions to the international research literature on science education PCK and their availability. As a result, researchers came to the 2nd Summit having used a variety of research lenses and instruments to identify and capture PCK in the fields of biology, chemistry, physics and science across primary, secondary and tertiary levels of education. With over half of the 2nd Summit group comprising original Summit members, this expert composite group also brought the advantages of continuity in thinking combined with the 'fresh eyes' of newcomers to the evaluation of progress in the field. In the workshop, the researchers were asked to revisit: the roots of their work; the data they had collected; the instruments used to collect these data; and the procedures used to infer PCK in science education from these data. The strengths and weaknesses of different instruments and procedures of data collection and analysis were discussed, and the potential of multimethod study designs were considered, in relation to the purposes of studying PCK in science education. As different kinds of instruments and their associated data were presented and discussed, participants gained insight into each other's data (and data analysis) and arrived at a more shared understanding of PCK in science education. Out of these discussions also evolved a more honed understanding of PCK in science education, enabling the group at the end of the 2nd Summit to arrive at a refined definition of PCK in the form of a refined consensus model of PCK in science education.

This Book

This current book reports the findings of the researchers participating in the 2nd PCK Summit in 2016 as they pursue the ideas and research agendas that have developed rapidly and enthusiastically out of that first 2012 PCK Summit in the science education field. It contains an evaluation of the consensus PCK model from the First (1st) PCK Summit and introduces the new model, known as the Refined Consensus Model (RCM) of PCK in science education as a refinement of the earlier model. Our book seeks not only to introduce the RCM but also to clarify and demonstrate its use in research and teacher education and practice. After an initial chapter that provides a rationale for the new model via a literature review of PCK research in science education over the last decade, a whole chapter is dedicated to the RCM, where it is presented in diagrammatic form and explanatory text. We strongly recommend that all readers of the book take extra time to carefully read and digest this second chapter before reading any of the following chapters, i.e. Chaps. 3–14. Familiarity with the RCM will allow readers to more fully appreciate the research work of contributing authors to this book. Subsequent chapters show how this new consensus model of PCK in science education is strongly connected with empirical data of varying nature, contains a tailored language to describe the nature of PCK in science education, and can be used as a framework for illuminating past studies and informing the design of future PCK studies in science education. Specifically the book informs and enhances our knowledge of science teachers' professional knowledge (especially important in these times when standards and other measures are being used to 'define' the knowledge, skills and abilities of teachers); illustrates how the PCK research agenda in science education can make a difference to science teachers' practice and students' learning of science; and makes research and knowledge about the construct more useable and applicable to the work of teachers through the RCM. It arguably contains the most relevant, recent and internationally strong collection of studies on PCK in science education available. While this book is available only three years out from the last major PCK publication, we believe it will have great appeal for researchers and science educators alike as the need and interest in deepening understanding of PCK gathers momentum internationally. As PCK grows as an attractive field of research across the globe, this book is positioned to offer an up-to-date, international perspective on the evolving nature of PCK in science education, how it is shaping the science education research agenda and how it can inform science teaching and learning.

Finally, this book is not a restatement of what already exists; it is about the ways in which the PCK construct in science education is being better understood, used and measured. It provides leverage for advancing future PCK research in science education by: repositioning PCK within teachers' professional knowledge (as depicted in the RCM of PCK in science education); providing a shared language of PCK in science education; and showcasing new methodologies for more effectively capturing, measuring and representing aspects of PCK for science teaching.

We hope you enjoy reading this book as much as we all enjoyed conceiving and writing it!

(The Editorial team was given the honour of compiling this book on behalf of our colleagues participating in the 2nd PCK Summit.)

Hamilton, New Zealand — Anne Hume
Melbourne, Australia — Rebecca Cooper
Potsdam, Germany — Andreas Borowski

References

Abell, S. K. (2007). Research on science teacher knowledge. In S. K. Abell & N. Lederman (Eds.), *Handbook of research on science education* (pp. 1105–1149). Mahwah, NJ: Lawrence Erlbaum Associates.

Berry, A., Friedrichsen, P., & Loughran, J. (2015). *Re-examining pedagogical content knowledge*. London: Routledge press.

Gess-Newsome, J., & Lederman, N. G. (1999). *Examining pedagogical content knowledge. The construct and its implications for science education*. Dordrecht, The Netherlands: Kluwer.

Shulman, L. S. (1987). Knowledge and teaching: Foundations of the new reform. *Harvard Educational Review, 57*(1), 1–22.

The original version of this book was revised with the changes of authorship for Chapter 2 and changed incorrect references citations and references for Chapter 12 and Backmatter. A correction to this book can be found at https://doi.org/10.1007/978-981-13-5898-2_15

Acknowledgements

The editors would like to acknowledge the work undertaken by the organisers of the 2nd PCK Summit, without whom the Summit and this book would not have materialised. Their vision and effort brought together 24 international researchers in science teacher education to work collaboratively and with common purpose towards greater clarity around the PCK construct and research agendas. The organisers' achievements included a successful funding bid, the coordination and monitoring of multiple research projects contributing to the Summit, and the organisation and facilitation of the face-to-face Summit workshop—the outcomes of which have informed the contents of this book. The organisers of the 2nd PCK Summit were:

Sophie Kirschner, Justus-Liebig University, Germany
Jan van Driel, Leiden University, the Netherlands
Amanda Berry, RMIT, Australia
Andreas Borowski, University of Potsdam, Germany
Janet Carlson, Stanford University, USA

The editors would also like to acknowledge the funding and facilities provided by the Lorentz Center to support the 2nd PCK Summit workshop at Leiden, The Netherlands, in December 2016. The Lorentz Center is an international centre for scientific workshops, based on the philosophy that science thrives on interaction between creative researchers. Lorentz Center workshops focus on new collaborations and interactions between researchers from different countries and fields, and with varying seniority. The Lorentz Center receives structural financial support by Leiden University, the Dutch research councils and the Netherlands Organisation for Scientific Research NWO. It also is granted funding for specific workshops by the Lorentz Fund, Netherlands eScience Center, Centre Européen de Calcul Atomique et Moléculaire CECAM and the Royal Netherlands Academy of Science KNAW.

Contents

Part I Introduction to the Refined Consensus Model of Pedagogical Content Knowledge in Science Education

1 **Towards a Consensus Model: Literature Review of How Science Teachers' Pedagogical Content Knowledge Is Investigated in Empirical Studies** 3
Kennedy Kam Ho Chan and Anne Hume

2 **The Refined Consensus Model of Pedagogical Content Knowledge in Science Education** 77
Janet Carlson, Kirsten R. Daehler, Alicia C. Alonzo, Erik Barendsen, Amanda Berry, Andreas Borowski, Jared Carpendale, Kennedy Kam Ho Chan, Rebecca Cooper, Patricia Friedrichsen, Julie Gess-Newsome, Ineke Henze-Rietveld, Anne Hume, Sophie Kirschner, Sven Liepertz, John Loughran, Elizabeth Mavhunga, Knut Neumann, Pernilla Nilsson, Soonhye Park, Marissa Rollnick, Aaron Sickel, Rebecca M. Schneider, Jee Kjung Suh, Jan van Driel and Christopher D. Wilson

3 **Vignettes Illustrating Practitioners' and Researchers' Applications of the Refined Consensus Model of Pedagogical Content Knowledge** 95
Janet Carlson, Rebecca Cooper, Kirsten R. Daehler, Patricia J. Friedrichsen, Joan I. Heller, Sophie Kirschner, Nicole Lusiani Elliott, Karen Marangio and Nicole Wong

Part II Reimagining the Existing

4 **Reconciliation Between the Refined Consensus Model of PCK and Extant PCK Models for Advancing PCK Research in Science** 119
Soonhye Park

5 **Exposing Pathways for Developing Teacher Pedagogical Content Knowledge at the Topic Level in Science** 131
Elizabeth Mavhunga

6 **The Development of Science Teachers' Professional Competence** 151
Stefan Sorge, Anita Stender and Knut Neumann

7 **Illustrating and Developing Science Teachers' Pedagogical Content Knowledge Through Learning Study** 167
Rebecca M. Schneider

8 **The PCK Map Approach to Capturing the Complexity of Enacted PCK (ePCK) and Pedagogical Reasoning in Science Teaching** 187
Soonhye Park and Jee Kyung Suh

Part III New Approaches to PCK Research in Science

9 **Unravelling Student Science Teachers' pPCK Development and the Influence of Personal Factors Using Authentic Data Sources** 203
Ineke Henze and Erik Barendsen

10 **Investigating Practising Science Teachers' pPCK and ePCK Development as a Result of Collaborative CoRe Design** 225
Jared Carpendale and Anne Hume

11 **A Grand Rubric for Measuring Science Teachers' Pedagogical Content Knowledge** 253
Kennedy Kam Ho Chan, Marissa Rollnick and Julie Gess-Newsome

12 **Unpacking the Complexity of Science Teachers' PCK in Action: Enacted and Personal PCK** 273
Alicia C. Alonzo, Amanda Berry and Pernilla Nilsson

Part IV So What and What Next?

13 **Perspectives on the Future of PCK Research in Science Education and Beyond** 291
Christopher D. Wilson, Andreas Borowski and Jan van Driel

14 **Developing Research on PCK as a Community** 303
Rebecca Cooper and Jan van Driel

Correction to: Repositioning Pedagogical Content Knowledge in Teachers' Knowledge for Teaching Science C1

Postscript: Considerations from an External Perspective 317
Oliver Tepner and Elke Sumfleth

Editors and Contributors

About the Editors

Anne Hume is a retired Honorary Lecturer in the Faculty of Education at the University of Waikato, Hamilton, New Zealand. She was formerly a member of the TEMS Education Research Centre from 2005 to 2016, leading the Science Research Group, and lecturing in science education at undergraduate, graduate and postgraduate levels including doctoral supervision. In her last year, she was Director of the TEMS Centre. Previously Anne had been an evaluator with the New Zealand Education Review Office for 4 years, following 25 years of secondary science and chemistry teaching. She also had extensive experience in the wider science education scene in New Zealand, including a term as President of the New Zealand Science Teachers' Association (1993–1994) and positions on national curriculum and qualifications development groups. Her key research interest is science PCK development for pre-service and in-service teachers and teacher educators, using interventions such as reflective writing, simulation of classroom practice and Content Representation (CoRe design). She was Associate Editor for the *Research in Science Education (RISE)* journal and Member of the *International Journal of Science Education (IJSE)* editorial board. She continues to supervise doctoral students in science education and to do some writing.

Rebecca Cooper is a Senior Lecturer in the Faculty of Education, Monash University, Australia. She works predominantly with pre-service and in-service science teachers and her research interests include: considering how science teachers and science teacher educators develop pedagogical knowledge and pedagogical content knowledge throughout their career, improving the quality of science teaching to increase student engagement, and working with teachers on promoting values in their science teaching in an effort to better understand the development of scientific literacy with students. She has also worked with other lecturers at Monash to better understand and develop their teaching (School of Mathematics and Faculty of Medicine) and has recently been part of programme for aspiring principals in government schools in Victoria to unlock their potential. She has established and maintained partnerships between the Faculty of Education and many secondary schools in Victoria to provide richer experiences for pre-service teachers and their mentor teachers while on placement, and she continues to be inspired by the great work that so many teachers do to support students' learning. Prior to working at Monash University, she taught physics, science and mathematics in secondary schools for many years.

Andreas Borowski is a Professor in the Faculty of Mathematics and Science, University of Potsdam, Germany, where he is also Director of the Teacher Education and Education Research Center (ZeLB). His research interest is the professional knowledge of pre-service and in-service physics teachers. With pre-service teachers, he investigates the connections between pedagogical knowledge, pedagogical content knowledge and content knowledge, while in in-service education he studies the influence of the professional knowledge of physics teachers through videotaped classroom performance and the learning gain as well as the motivation of the students. He also investigates the development of content knowledge and pedagogical content knowledge during teacher training at the university and analyses the connection to performance situations like preparing or reflecting on a lesson as well as explaining physics phenomena. As Director of the Teacher Education and Education Research Center, he is interested in developing a connection between professional knowledge and the experience of practice during the internship of student teachers as well as connections between lecturers of the university and teachers at school. He establishes and maintains partnerships between the University of Education and many schools in the Brandenburg State. He has recently become the scientific head of the statewide teacher training in grade 5 and 6 science for all schools. Prior to working at University of Potsdam, he worked as Professor at the RWTH Aachen University and as Lecturer at the University of Duisburg-Essen. He also taught physics and mathematics in secondary schools for many years.

Contributors

Alicia C. Alonzo Michigan State University, East Lansing, USA

Erik Barendsen Radboud University, Nijmegen, Netherlands; Open University of the Netherlands, Heerlen, Netherlands

Amanda Berry Monash University, Melbourne, Australia

Andreas Borowski University of Potsdam, Potsdam, Germany

Janet Carlson Center to Support Excellence in Teaching, Stanford University, Palo Alto, USA

Jared Carpendale University of Waikato, Hamilton, New Zealand

Kennedy Kam Ho Chan The University of Hong Kong, Hong Kong, Hong Kong

Rebecca Cooper Faculty of Education, Monash University, Melbourne, Australia

Kirsten R. Daehler WestEd, San Francisco, USA

Nicole Lusiani Elliott Center to Support Excellence in Teaching, Stanford University, Palo Alto, USA

Patricia J. Friedrichsen University of Missouri, Columbia, USA

Julie Gess-Newsome Oregon State University-Cascades, Bend, USA

Joan I. Heller Heller Research Associates, Oakland, USA

Ineke Henze Delft University of Technology, Delft, Netherlands

Anne Hume University of Waikato, Hamilton, New Zealand

Sophie Kirschner Justus Liebig University, Giessen, Germany

Karen Marangio Faculty of Education, Monash University, Melbourne, Australia

Elizabeth Mavhunga Science Education Division, School of Education, University of the Witwatersrand, Johannesburg, South Africa

Knut Neumann Leibniz Institute for Science and Mathematics Education, Kiel University, Kiel, Germany

Pernilla Nilsson Halmstad University, Halmstad, Sweden

Soonhye Park Department of STEM Education, College of Education, North Carolina State University, Raleigh, NC, USA

Marissa Rollnick University of Witwatersrand, Johannesburg, South Africa

Rebecca M. Schneider University of Toledo, Toledo, OH, USA

Stefan Sorge Leibniz Institute for Science and Mathematics Education, Kiel University, Kiel, Germany

Anita Stender Department of Physics Education, University Duisburg-Essen, Duisburg, Germany

Jee Kyung Suh Curriculum and Instruction, The University of Alabama, Tuscaloosa, AL, USA

Jan van Driel Melbourne Graduate School of Education, The University of Melbourne, Melbourne, Australia

Christopher D. Wilson Biological Sciences Curriculum Study, Colorado Springs, CO, USA

Nicole Wong Heller Research Associates, Oakland, USA

Part I
Introduction to the Refined Consensus Model of Pedagogical Content Knowledge in Science Education

Janet Carlson

Overview

This part of the book introduces the Refined Consensus Model (RCM) of PCK in science teacher education and research. This introduction begins with a review of the literature base of models and studies of science teacher PCK in Chap. 1, setting the stage for learning about the RCM, which is the focus of Chap. 2. Then, Chap. 3 unpacks the very centre of the model. Finally, Chap. 4 provides illustrations of the RCM in science teacher education and in research through a set of vignettes describing the use of the RCM in very specific settings. Our goal in this part, as well as the entire book, is to continue to articulate this complex model and to show how it is being used and can be applied across multiple contexts so that we can continue to make progress towards a model that can be used by many.

The ideas represented in this part had their origin at the 2016 PCK Summit in Leiden, Netherlands, where 24 invited researchers from the science education field shared their work in PCK using a variety of research lenses and instruments to identify and capture the PCK of science teachers. As the strengths and weaknesses of different instruments and procedures of data collection and analysis were discussed, and the potential of multi-method study designs were considered, in relation to the purposes of studying PCK, it became clear that an updated model for PCK in science teacher education was needed.

As mentioned above, Chap. 1—*Towards a Consensus Model: Literature Review of How Science Teachers' Pedagogical Content Knowledge is Investigated in Empirical Studies*— provides a sophisticated analysis of the literature base related to studying PCK in science teachers. In particular, the chapter presents a systematic review of the methodologies used for investigating PCK. For each of the reviewed studies, the authors identified (1) the research sample of the investigation; (2) the major purpose of the study; (3) the conceptualisation of PCK in the study; (4) the data source(s) used; and (5) the approach(es) used to determine teachers' PCK. As expected, the review reveals that these researchers are conceptualising, operationalising, and researching PCK differently. This chapter reveals gaps in the extant PCK literature and highlights several points of divergence in thinking around the PCK concepts within the PCK research community in the field of science education.

Next, Chap. 2—*The Refined Consensus Model of PCK in Science Education*—chronicles the developmental journey of the RCM of PCK in science teacher education and research and seeks to introduce the new model by describing the different components of the model in both diagrammatic form and explanatory text. Science education researchers and practitioners from around the world contributed to this model. As such, it is intended to withstand scrutiny in different countries, be relevant across different policy environments, be useful for different research paradigms, and inform a wide range of teacher preparation and professional learning programmes.

The very centre of the RCM diagram represents enacted PCK (ePCK), which is a particularly complicated aspect of the model.

Finally, Chap. 3—*Vignettes Illustrating Practitioners' and Researchers' Applications of the Refined Consensus Model of Pedagogical Content Knowledge*—provides examples of how the RCM moves from an abstract visualisation to a tool used by educators and researchers alike. The chapter is comprised of short vignettes from a variety of science education settings and perspectives. The vignettes bring the RCM to life by illustrating how the model is being used in teacher education or research. The first and second vignettes provide pre-service perspectives from two different countries. The third vignette shares insights into the use of the RCM during a teacher professional learning course, while the fourth and fifth vignettes offer a researcher lens with examples from large-scale studies. Each vignette follows a similar structure with a description of context, explicit connections to the RCM, and author reflections on the role of the model.

Chapter 1
Towards a Consensus Model: Literature Review of How Science Teachers' Pedagogical Content Knowledge Is Investigated in Empirical Studies

Kennedy Kam Ho Chan and Anne Hume

Abstract This chapter presents a systematic review of the science education literature to identify how researchers investigate science teachers' pedagogical content knowledge (PCK). Specifically, we focus on empirical studies of individual science teachers' PCK published in peer-reviewed science education and teacher education journals since 2008. For each of the reviewed studies, we identify (1) the research context of the investigation; (2) the major purpose of the study; (3) the conceptualisation of PCK in the study; (4) the data sources used to investigate teachers' PCK; and (5) the approaches used to determine teachers' PCK. Using this collated information, we provide an overview of how the PCK concept is used, interpreted and investigated within the science education community. The review reveals that researchers conceptualise and operationalise PCK differently. Consequently, they investigate PCK in highly diverse ways and use a wide range of data sources and approaches to capture and determine teachers' PCK, which in turn generates different kinds of qualitative and quantitative data. Collectively, our findings reveal gaps in the PCK literature and highlight several points of divergence in thinking around the PCK concept within the PCK research community in the field of science education. The findings also provide evidence from the literature supporting the need to build upon and further refine the Consensus Model (CM) that emerged from the first (1st) PCK Summit in 2012 to further science education research.

Introduction

For many decades, what teachers know and how they make use of their knowledge to accomplish the work of teaching has been a subject of interest for education researchers, teacher educators and educational policy-makers (Guerriero, 2017). In

K. K. H. Chan (✉)
The University of Hong Kong, Hong Kong, Hong Kong
e-mail: kennedyckh@hku.hk

A. Hume
University of Waikato, Hamilton, New Zealand
e-mail: annehume@waikato.ac.nz

A. Hume et al. (eds.), *Repositioning Pedagogical Content Knowledge in Teachers' Knowledge for Teaching Science*, https://doi.org/10.1007/978-981-13-5898-2_1

his seminal articles, Shulman (1986, 1987) introduced a new way of thinking about the content of teacher knowledge. Shulman (1987) proposed that teachers' knowledge base comprises at least seven categories of knowledge: content knowledge; pedagogical knowledge; curriculum knowledge; knowledge of learners; knowledge of contexts; knowledge of educational ends, purposes and values; and pedagogical content knowledge (PCK). Of these categories, PCK has generated particular interest as it represents the unique province of knowledge *for* teaching that distinguishes teachers from content specialists, and the idea has spawned many empirical studies on teachers' knowledge, particularly in the domains of science and mathematics. Yet, despite its potential to move the field forward, the diverse understanding and interpretation of PCK that has occurred since its inception have greatly limited its utility in research, teacher education and policy (Settlage, 2013).

The purpose of this chapter is to review and synthesise the empirical literature to specifically identify how PCK researchers investigate science teachers' PCK. Although there are a number of review articles on science teachers' PCK (e.g., Berry, Depaepe, & van Driel, 2016; Kind, 2009; Miller, 2007; Schneider & Plasman, 2011; van Driel & Berry, 2010, 2017) and science teachers' professional knowledge (e.g., Abell, 2007; Fischer, Borowski, & Tepner, 2012; van Driel, Berry, & Meirink, 2014), these reviews do not provide an in-depth analysis of the research methodologies used. Recent reviews of how researchers investigate mathematics teachers' PCK (Depaepe, Verschaffel, & Kelchtermans, 2013) and teachers' technological pedagogical content knowledge (TPACK) (Abbitt, 2011; Matthew, Tae Seob, & Punya, 2012; Willermark, 2017) can be found, but the most recent review focusing on how science PCK is investigated in empirical studies was published about 20 years ago (Baxter & Lederman, 1999). Our chapter not only complements the extant reviews but also presents a departure from the current literature, with its particular focus on PCK conceptualisation and methodologies for investigating PCK in recent studies. We believe that this systematic review reveals the diverse ways in which the PCK concept is used, interpreted and investigated within the science education research community. Such information is important for advancing the field by revealing the convergence and divergence in thinking around the PCK concept and identifying gaps in the current state of knowledge about the investigation of science teachers' PCK.

Literature Review

The Pedagogical Content Knowledge (PCK) Concept

In his presidential address for the American Education Research Association, Shulman (1986) first introduced the academic construct of PCK to attend to what he and his colleagues referred to as the missing paradigm or 'blind spot with respect to content that now characterises most research on teaching' (pp. 7–8). Although its introduction served to turn research attention away from a sole focus on teaching procedures to include subject matter knowledge, Shulman argued that PCK 'goes beyond the knowledge of subject matter per se to the dimension of subject matter *for teaching*' (p. 9). For Shulman, PCK encompasses the unique idea that teachers have knowledge:

> for the most regularly taught topics in one's subject area, the most useful forms of representation of those ideas, the most powerful analogies, illustrations, examples, explanations and demonstrations – in a word the ways of representing and formulating the subject that make it comprehensible to others. … [and] an understanding of what makes the learning of specific topics easy or difficult: the conceptions and preconceptions that students of different ages and backgrounds bring with them to the learning of those most frequently taught topics and lessons. (p. 9)

Such a distinct province of knowledge is particularly appealing to science teacher educators and researchers alike, as it is well known that student (pre-)conceptions can heavily influence science learning (Driver, Asoko, Leach, Scott, & Mortimer, 1994), requiring teachers to make use of specific instructional strategies and representations to aid student learning. Since the 1980s, the idea of PCK has heavily pervaded the science education literature. It is widely used, for example, as a theoretical lens for researching the professional knowledge of science teachers (Abell, 2007). Policy documents (e.g., National Research Council, 1996), professional science standards (e.g., National Council for Accreditation of Teacher Education, 2008) and the international literature (Kind, 2009; van Driel & Abell, 2010) unequivocally regard a good grasp of PCK as essential for high-quality science teaching.

Scholarly Debates About the PCK Concept

Although the science education community has embraced the idea of PCK, various debates regarding its nature and content have arisen as researchers and educators have endeavoured to use the construct in their work. Notable debates revolve around the following questions:

(1) Is PCK a 'stand-alone' and distinct body of knowledge? How is PCK then related to the professional knowledge base for teaching?
Shulman (1987) regarded PCK as one of the seven categories of teacher knowledge within the teacher knowledge base and believed that PCK 'emerges and grows as teachers transform their content knowledge for the purpose of teaching' (Wilson, Shulman, & Richert, 1987, p. 118). Critics of his proposal doubted whether PCK can be distinguished from other knowledge categories, such as content knowledge (Marks, 1990). Others followed Shulman's original proposal and argued that viewing PCK as a distinct category of knowledge highlights that PCK is more than the sum of the knowledge categories from which PCK is synthesised (Magnusson, Krajcik, & Borko, 1999). Abd-El-Khalick (2006) added that taking this position would make explicit a transformative mechanism by

which teachers develop their PCK (i.e., how teachers 'convert' their subject matter to PCK).

(2) Is PCK a knowledge form, a skill set, a disposition or some combination thereof? Shulman (1986) originally regarded PCK as a form of professional knowledge needed to successfully teach a certain content to particular groups of students. He identified two subsets of knowledge that together comprised PCK, that is, knowledge of appropriate topic-specific instructional strategies and representation and understanding of students' learning difficulties and (pre)conceptions. However, a range of epistemological views about what counts as 'knowledge' began to emerge and the debate arising from such divergent views added complexity to the nature of PCK. For example, some researchers believe that PCK should also embrace non-cognitive attributes, hence the addition of affective components to their PCK models such as teachers' conceptions of purpose for teaching subject matter (Grossman, 1990), teacher efficacy (Park & Oliver, 2008b), emotions (Zembylas, 2007) and orientations to teaching science (Magnusson et al., 1999). Similarly, some researchers take a more static view of knowledge, viewing 'what is known' as something that a person possesses in his/her mind, whereas others take a more dynamic view that regards *knowing* as part of action (Cook & Brown, 1999). Researchers in the latter camp strive to capture and portray PCK *in the act of* teaching particular content (Loughran, Milroy, Berry, Gunstone, & Mulhall, 2001).

(3) If PCK is a stand-alone body of knowledge, what components should be included?
Most, if not all, researchers have expanded upon the two components advanced in Shulman's (1986) PCK model. Apart from affective components, other commonly added components include curricular knowledge (i.e., knowledge about the goals and scope of the curriculum) (Grossman, 1990) and assessment knowledge (i.e., knowledge about what and how to assess) (Tamir, 1988). Some take a narrower and more parsimonious view of what PCK comprises and limit the PCK components to a few components in their models, while others adopt a more encompassing view of PCK (see Kirschner, Borowski, Fischer, Gess-Newsome and von Aufschnaiter (2016) for a comparison of the composition of different PCK models).

(4) Is PCK context-specific? Can PCK be investigated out of the classroom context? Some researchers contend that PCK can be investigated using standardised paper and pencil tests that are devoid of classroom contexts, whereas others claim that a teacher's PCK does not exist in a vacuum but is situated in *specific* classroom contexts. Depaepe et al. (2013) distinguished two perspectives for investigating PCK—a cognitive and a situated perspective. Those who take a cognitive perspective typically investigate PCK independently outside of the classroom context, whereas researchers who adopt a situated perspective examine how PCK is enacted in a *specific* classroom context. Paradigmatically, the former group sees knowledge as located 'in the head' of a teacher, whereas the latter views knowledge as a 'social asset' that is meaningful only in the context of its application (p. 22). van Driel et al. (2014) suggest that the former camp

regards PCK as 'knowledge *for* teachers'—knowledge that is rather static and prescriptive in nature. The latter group sees PCK as the 'knowledge *of* teachers', which teachers develop from their formal learning opportunities and their own professional practices.

(5) Is PCK individual or collective?
There is some debate as to whether PCK exists at the individual or the group level. Some scholars underscore the personal and private nature of PCK (Hashweh, 2005). Others take the view that despite the idiosyncratic nature of individuals' PCK, commonalities can be identified across the PCK of a group of teachers to generate a more general form of PCK (van Driel, Verloop, & de Vos, 1998). Some even go a step further by introducing the idea of indispensable PCK (Park, Suh, & Seo, 2017) or canonical PCK (Smith & Banilower, 2015). Canonical PCK, for example, refers to 'PCK that is widely agreed upon and formed through research and/or collective expert wisdom of practice' (Smith & Banilower, 2015, p. 90). This perspective suggests that PCK is normatively defined by researchers and experts.

(6) What are appropriate levels or grain sizes of PCK?
Some researchers highlight the generic nature of PCK (Fernández-Balboa & Stiehl, 1995), while others contend that the value of PCK lies in its topic specificity (van Driel et al., 1998). Still others believe that apart from topic-specific PCK, teachers also require PCK for disciplinary practices, which encompasses knowledge 'to help students understand the authentic activities of a discipline, the ways knowledge is developed in a particular field, and the beliefs that represent a sophisticated understanding of how the field works' (Davis & Krajcik, 2005, p. 5). Veal and MaKinster (1999) further propose that PCK exists at three different levels: general PCK (also called subject-specific PCK); domain-specific PCK; and topic-specific PCK.

These divergent views about PCK prompted Abell (2007), in her comprehensive review of studies on science teacher knowledge, to label PCK research as 'pre-science'. She deplored the inconsistent and incoherent use of terms and methodologies within the PCK field and called for a more coherent conceptualisation that would enable researchers to build their findings on those of others in the field and create a common body of the literature.

Progress Towards a Consensus View About PCK in Science

To move the field forward, Abell (2008) attempted to identify consensual views about the nature of PCK through her critical analysis of the articles in a special issue of the *International Journal of Science Education*. She identified common ground among researchers on four key aspects of PCK:

- PCK is a *distinct* category of knowledge that involves 'the transformation of other types of knowledge' (p. 1407) and has inherent close links to other knowledge

categories (e.g., subject matter knowledge, pedagogical knowledge and knowledge of context).
- PCK comprises discrete knowledge components. However, when applied in teaching practice, these knowledge components are integrated and blended together.
- A teacher's PCK can develop over time as a result of different experiences, such as teacher preparation programmes and a teacher's *own* teaching and learning experiences. Importantly, this perspective recognises teachers' *own* classroom experience as a source of development of PCK in their *own* context.
- Content is regarded as central to PCK. Teachers develop and use PCK for teaching *specific* science topics.

Another breakthrough in reaching a consensus about PCK occurred at a meeting of researchers in science PCK, including a few in mathematics education research, known as the PCK Summit (Berry, Friedrichsen, & Loughran, 2015) held in Colorado in 2012. Importantly, this meeting led to the creation of a 'consensus' model for professional teaching knowledge and skills, including PCK (Gess-Newsome, 2015), known as the Consensus Model (CM). The nature of PCK is made more explicit in this model, such that:

- PCK is reaffirmed as a '*separate*' category of knowledge within the overall knowledge bases pertinent to teaching professionals. The close links between PCK and other knowledge categories are emphasised by the many arrows within the diagrammatic representation of the CM. The model also underscores the multiple sources of PCK, its position relative to other forms of professional knowledge and the iterative nature of its development via the use of feedback loops.
- PCK is defined as *both* knowledge and skills.
  - Within the CM, the affective components (e.g., teachers' orientations and beliefs) are removed from the PCK construct itself and included as filters and/or amplifiers. Their repositioning recognises them as factors that influence the content and nature of a teacher's PCK.
  - The model further delineates two variants of PCK: personal PCK and personal PCK and skills (PCK&S). The former is related to 'teachers' knowledge of, reasoning behind, and planning', while the latter pertains to 'the act of teaching a particular topic in a particular way for a particular purpose to particular students for enhanced student outcomes' (Gess-Newsome, 2015, p. 36). Epistemologically, PCK embraces both *knowledge* (i.e., what a teacher knows about teaching a certain content and applies in planning) and *knowing* (i.e., what a teacher does, meaning his/her actions *during* the act of teaching in the classroom). In other words, teachers draw on their existing personal PCK to inform their lesson planning, enactment of teaching and reflection, and their personal PCK&S becomes evident in the teaching artefacts that teachers create, through their articulation of their pedagogical decisions and the use of pedagogical moves in their teaching.
  - A teacher's reasoning is considered as part of a teacher's PCK. This perspective reinforces the idea that 'PCK is constituted by what a teacher knows, what a

teacher does, and the reasons for the teacher's actions' (Baxter & Lederman, 1999, p. 158).

- PCK is specific to particular subject matter. It is regarded as *topic-specific* teacher knowledge.
- PCK is considered to be highly *personal* and *idiosyncratic*. In acknowledgement of these qualities, PCK is relabelled as *personal PCK* in the CM and is distinguished from the more canonical topic-specific professional knowledge (TSPK) held by the field. The model also highlights the role of context in influencing the content and nature of a teacher's PCK.
- Finally, the model also explicitly explains the reciprocal links between teachers' PCK and student outcomes (e.g., student achievement, motivation).

Although the CM was created through the collective ideas and efforts of leading scholars who had done extensive PCK research in science and mathematics education, the model was *not* informed by empirical data. It remains unclear whether the ideas articulated in the consensus model accurately represent the prevalent thinking around the PCK concepts within the science education community.

Conflicting Findings Between PCK Studies in Science Education Research

Although the CM delineates a link between teachers' PCK and student outcomes, this relationship is theoretically assumed rather than empirical. Different or even conflicting findings have been reported in the few empirical studies that have investigated this relationship (e.g., Alonzo, Kobarg, & Seidel, 2012; Davidowitz & Potgieter, 2016; Förtsch, Werner, von Kotzebue, & Neuhaus, 2016; Gess-Newsome et al., 2017; Kanter & Konstantopoulos, 2010; Keller, Neumann, & Fischer, 2017; Mahler, Großschedl, & Harms, 2017; Sadler, Sonnert, Coyle, Cook-Smith, & Miller, 2013). For example, while some studies pointed to a positive relationship between science teachers' PCK and student achievements (Sadler et al., 2013), some found that PCK did not predict student achievements (Gess-Newsome et al., 2017). One possible reason may be the different conceptualisations of PCK adopted in the studies. Another reason may be related to the different methodologies used to investigate science teachers' PCK. Some measured PCK by scoring teachers' responses to paper and pencil tests (Keller et al., 2017), while others analysed teachers' (written) comments on videos (Kanter & Konstantopoulos, 2010; Roth et al., 2011). Still, others measured PCK using rubrics for analysing teachers' (oral and written) reflections and videos of classroom teaching (Gess-Newsome et al., 2017). In some studies, researchers focused on measuring a *specific* PCK component (Sadler et al., 2013), whereas others measured how PCK, as a whole, is applied in practice and informs the teaching action (Gess-Newsome et al., 2017). This brief survey of the literature suggests that even though researchers may share the same goal of measuring teachers' PCK, what they are actually measuring appears to be very different. These findings signal that a closer look at the different ways researchers investigate science teachers' PCK—including how they conceptualise it, the data sources they use and the approaches they take—would allow readers to more critically compare and contrast the findings of different studies.

Summary

To summarise, while the idea of PCK pervades the science education literature, the different interpretations of the concept have limited our ability to generalise from the research findings to inform future research, teacher education programmes and policy. Although a 'consensus' model has been created by a group of leading PCK researchers, whether the ideas articulated in the CM accurately reflect the prevalent views of the larger science PCK researcher community is currently unknown. Hence, this review chapter sets out to organise, integrate and synthesise empirical investigations of science teachers' PCK to provide an overview of how the PCK concept is being used, interpreted and investigated within the science education community. We focus on studies that investigate individual science teachers' PCK because this form of PCK is arguably the one most directly related to students' learning experiences in the classroom. This argument is reflected in the PCK studies included in this review, most of which investigated individual teachers' PCK, while relatively few looked at teachers' collective PCK, for example (see exceptions Daehler & Shinohara, 2001; Falk, 2012). Hence, we are of the view that a systematic review of studies with this focus has the most potential to move the field forward. The research question and subquestions guiding the present review are:

- How is individual science teachers' PCK investigated in empirical science educational research?

    1. What is the research context in the studies?
    2. What are the major research foci of the studies?
    3. How do the studies conceptualise individual science teachers' PCK?
    4. What types of data sources are used to investigate individual science teachers' PCK?
    5. What approaches are used to determine individual science teachers' PCK?

Methods

In an extensive review of science teacher knowledge studies, Abell (2007) called for greater consensus on the conceptualisation of and methodologies for investigating science teachers' PCK. As an initial response to this call for greater unanimity in approaches to empirical studies in the field, this review conducted a broad search of the literature from 2008 to the present. Our intent was to comprehensively explore how science teachers' PCK has been investigated since the extensive review by Abell (2007). We followed the systematic review process outlined by Bennett, Lubben and Hogarth (2007), which involves the following sequential stages: (1) identifying the review topic areas; (2) identifying the review research questions; (3) developing inclusion and exclusion criteria; (4) systematically searching electronic databases; (5) coding studies; (6) producing an overview/systematic map of studies; and (7) reviewing in depth the studies. We first identified our review focus (i.e., methodologies for investigating individual science teachers' PCK) and then formulated the research questions (see above). The following sections describe in greater detail the subsequent stages of the review process i.e., stages 3–7.

In the third stage, we formulated the following selection criteria. First, all articles had to report an empirical study or studies that investigated science teachers' PCK (i.e., they were not solely conceptual). Second, as our focus was on individual science teachers' PCK, we excluded studies investigating (1) science teachers' TPACK and the PCK of mathematics teachers, science teacher educators, university science teaching assistants, university science instructors and early childhood teachers; (2) science teachers' collective PCK; (3) as PCK concerns teacher knowledge for teaching specific subject matter, we excluded studies investigating PCK for disciplinary practices or argumentation; and moreover, (4) we discarded studies that *only* focused on science teachers' orientations to teaching science (Friedrichsen, van Driel, & Abell, 2011), an often-included component of existing PCK models (Magnusson et al., 1999; Park & Oliver, 2008b). We made this decision because orientations have not been considered part of PCK since the development of the CM in the first PCK Summit (Gess-Newsome, 2015), but rather an influence on PCK (i.e., amplifiers and filters). Finally, the data analysis part of the article should include sufficient description of how the teachers' PCK was investigated. In instances where the method section of the article referred to a prior publication reporting the instrument used for investigating the teachers' PCK, we included the instrument that was referred to in the analysis. These criteria for exclusion are listed in Table 1.1.

Bearing in mind these criteria, we then conducted searches in 14 peer-reviewed high-ranking ISI-listed journals in the field of science education and teacher education, using the keyword 'pedagogical content knowledge'. The journals were as follows: *Journal of Research in Science Teaching*; *Science Education*; *International Journal of Science Education*; *Research in Science Education*; *Journal of Science Teacher Education*; *Science & Education*; *International Journal of Science and Mathematics Education*; *Chemistry Education Research and Practice*; *Eurasia Journal of Mathematics, Science and Technology Education*; *Research in Science and Technological Education*; *Teaching and Teacher Education*; *Journal of Teacher Education;* and *Teaching and Teachers: Theory and Practice*. This initial search resulted in 1261 articles (as of 31 December 2017)[1] (see Table 1.2). The first author then scanned the retrieved articles to identify the studies to be included in the review

[1]Articles that were accepted and published online, even if not assigned to an issue, before 31 December 2017 were included in this review.

Table 1.1 Inclusion and exclusion criteria for this review

Criteria		Description of excluded studies	Examples of articles excluded
1	Empirical studies	• Non-empirical in nature (i.e., conceptual works)	van Dijk (2014), van Driel and Berry (2012)
2	Individual PCK of science teachers	• Mathematics teachers' PCK or mathematics knowledge for teaching (MKT)	Charalambous (2016), Kaiser et al. (2017)
		• Science teachers' TPACK	Cetin-Dindar, Boz, Yildiran Sonmez, and Demirci Celep (2018), Wang, Tsai, and Wei (2015)
		• Science teacher educators' PCK	Abell, Rogers, Hanuscin, Lee, and Gagnon (2008), Faikhamta and Clarke (2012)
		• University science teaching assistants' PCK/university science instructors' PCK	Fraser (2016), Seung (2013)
		• Early childhood teachers	Nilsson and Elm (2017)
		• Teachers' collective PCK	Akerson, Pongsanon, Park Rogers, Carter, and Galindo (2017), Nilsson (2014)
3	PCK for teaching specific subject matter	• Science teachers' discipline/domain-specific PCK (e.g., PCK for argumentation, PCK for scientific practices)	Avraamidou and Zembal-Saul (2005), McNeill, González-Howard, Katsh-Singer, and Loper (2016)
4	Not solely investigating teachers' orientations	• Teachers' orientation to teaching science (e.g., teachers' views about the purposes of teaching science)	Boesdorfer and Lorsbach (2014), Mellado, Bermejo, Blanco, and Ruiz (2007)
5	Sufficient description on analysis of PCK	• Insufficient methodological details	Khourey-Bowers and Fenk (2009)

using the selection criteria. Next, 99 articles (see Appendix) were read and analysed in detail. The following parameters comprised the analytical framework for coding the studies.

(1) *The research context.* The analysis included the sample size, the location of the PCK studies, the subject domain, the grade levels of the teachers and the topic(s) under investigation.

Table 1.2 Total number of hits in the databases using the keywords 'pedagogical content knowledge' and the number of articles selected for review

Journal name		Total number of hits	Number of articles included in the review
		After 2008	
Science education journals	*Chemistry Education Research and Practice*	52	11
	Eurasia Journal of Mathematics, Science and Technology Education	37	3
	Journal of Research in Science Teaching	97	12
	Journal of Science Teacher Education	161	7
	Science & Education	44	0
	Science Education	65	3
	International Journal of Science Education	153	28
	International Journal of Science and Mathematics Education	141	6
	Research in Science Education	101	17
	Research in Science and Technological Education	27	3
Teacher education journals	*Educational Researcher*	27	0
	Journal of Teacher Education	75	1
	Teaching and Teacher Education	226	7
	Teachers and Teaching: Theory and Practice	55	1
Total		1261	99

(2) *The major research focus of the study.* To identify the major research focus of the reported study, we opted to use the lines or categories of PCK research reported in the review of mathematics PCK research by Depaepe et al. (2013) (e.g., investigating the nature of PCK, exploring the development of teachers' PCK) to guide our analysis. In addition to this deductive analysis of the data, we adopted an inductive approach and noted the emergence of several new categories, i.e., reporting the development of a PCK measurement instrument and investigating changes in PCK following or during an intervention. Then, within each category that was deductively or inductively derived, the reviewed studies were further classified into groups based on the research approaches adopted in the studies (i.e., qualitative, quantitative or mixed methods). Because some studies addressed more than one line of research, these studies were placed in more than one category of PCK research.

(3) *The conceptualisation of PCK in the study.* To characterise how science education researchers conceptualised PCK in their studies, we identified the PCK models that the authors used to frame their studies and the PCK component(s)/knowledge category/categories[2] that they investigated. The definitions of the different knowledge categories were drawn mainly from the CM (Gess-Newsome, 2015) and the definition of knowledge categories by Grossman (1990), with slight modifications. As the CM did not unpack the composition of personal PCK (i.e., PCK components), we drew on the often-cited PCK model of Magnusson et al. (1999) in our analysis. We slightly modified the original definitions of the PCK components to better fit the data (see Table 1.3).

(4) *The types of data source(s) used by researchers to investigate science teachers' PCK.* To analyse in a more nuanced manner how researchers investigate and determine PCK, we decided to identify the different types of data sources used in the research studies. In our analysis, we drew on the categories devised by Matthew et al. (2012) in their analysis of measurement of teachers' TPACK, but slightly modified them to provide a better fit to our analysis. In brief, the categories included: (1) written questionnaires, surveys and tests; (2) interviews; (3) artefacts from teaching tasks; and (4) lesson observations. Table 1.4 shows the definitions of our final categories with examples alongside that typify each of the data sources.

(5) *The approaches researchers use to determine science teachers' PCK*. Based on the results of RQ#4, we assigned each study to one or more of the following categories: (1) investigation of PCK via *self-reports* and (2) investigation of PCK via *performance* in teaching tasks. The studies in the first category relied on teachers' self-reports to determine their PCK, and these articles were further divided into three subcategories: (1) use of PCK tests in which teachers' responses were numerically scored; (2) use of questionnaires and surveys; and (3) self-reports of actual experience. The studies in the second category, which investigated PCK via *performance* in teaching tasks, involved the use or collec-

---

[2]In line with Shulman (1987), we use the term category to refer to a *distinct* domain of knowledge. PCK *components* refer to the knowledge components of PCK.

Table 1.3 Definitions of the different knowledge categories/PCK components

		Description
Knowledge category	Assessment knowledge (AK)	Encompasses teachers' knowledge of how to design formative and summative assessments, and their knowledge of interpretation and action-taking based on assessment data (Gess-Newsome, 2015)
	Content knowledge (CK)	The part of teachers' subject matter knowledge that is pertinent to the teaching task (Cochran & Jones, 1998). It refers to teachers' knowledge of the key ideas and concepts for teaching and their relationships (Cochran & Jones, 1998; Grossman, 1990).
	Contextual knowledge (C*x*K)	Refers to teachers' knowledge of particular teaching contexts. It includes teachers' knowledge of their school setting and the districts in which they are working (e.g., the expectations, constraints and the 'culture') (Grossman, 1990)
	Curricular knowledge (C*u*K)	Concerns teacher knowledge of the goals of a curriculum, its structures, scope and sequence (Gess-Newsome, 2015)
	Knowledge of students (KS)	Entails teachers' knowledge of students' cognitive development and variations in their approaches to learning and general characteristics (Gess-Newsome, 2015)
	Pedagogical knowledge (PK)	Includes teachers' general, not subject-specific, knowledge and skills related to teaching. It includes, for example, teachers' knowledge and skills about learning theories, instructional principles and classroom management (Grossman, 1990) and the related strategies (Gess-Newsome, 2015)

(continued)

Table 1.3 (continued)

		Description
PCK components	Knowledge of assessment (KA)	Refers to teachers' knowledge of the dimensions of science learning that are important to assess (i.e., what to assess) and the methods by which that learning can be assessed (i.e., how to assess) (Magnusson et al., 1999)
	Knowledge of curriculum (KC)	Refers to teachers' knowledge of the goals and objectives for students in the subject(s) they are teaching, their knowledge about the relevant instructional materials and resources as well as their knowledge about the horizontal and vertical curricula (Magnusson et al., 1999). It also encompasses teachers' knowledge of the importance of topics relative to the whole curriculum, which enables them to identify core concepts and big ideas and eliminate trivial facts (Park & Oliver, 2008b). This aspect is referred to as curricular saliency (Geddis, Onslow, Beynon, & Oesch, 1993) in some PCK models (e.g., Mavhunga & Rollnick, 2013)
	Knowledge of instructional strategies and representations (KISR)	Refers to teachers' knowledge of specific strategies, including activities and representations for helping students comprehend a science topic (Magnusson et al., 1999)
	Knowledge of students' understanding (KSU)	Covers teachers' knowledge about the science concepts or topics that students find difficult to learn, the prerequisite knowledge for learning specific scientific knowledge, as well as variations in students' approaches to learning as they relate to the development of knowledge within specific science topic areas (Magnusson et al., 1999)

(continued)

Table 1.3 (continued)

		Description
	Orientations to teaching science (OTS)	Refer to a set of beliefs encompassing (1) teachers' goals and the purposes of science teaching, (2) teachers' views of science and (3) teachers' beliefs about science teaching and learning (Friedrichsen et al., 2011)

tion of teaching artefacts related to the teaching cycle, i.e., the pre-active phase (planning), interactive phase (enactment) and post-active phase (reflection), or lesson observations. Two subcategories were identified, namely (1) studies that examined the artefacts/teaching actions *only* and (2) studies that *also* investigated teachers' instructional decisions around teaching tasks. The studies in the latter subcategory typically made use of open-ended questions in interviews and/or written tasks to probe into the teachers' decision-making process underpinning their judgements and actions. The studies in the second subcategory were further classified into two subgroups: (1) those that used (a) simulated teaching task(s) and (2) those that investigated real-life teaching in situ. Simulated teaching tasks were designed to represent and approximate authentic, real-life science teaching tasks, e.g., asking the teachers to complete a unit, topic or lesson planning template such as CoRes,[3] or to comment on authentic student responses. The second subgroup entailed investigating the teaching cycle in authentic real-life settings. For each subgroup, we further sorted the studies into the following subcategories based on the phase of the teaching cycle they investigated (Jackson, 1986): (1) the pre-active phase (i.e., the planning phase involving *reflection on action*); (2) the interactive phase (i.e., the enactment phase of teaching involving *reflection in action*); (3) the post-active phase[4] (i.e., the reflection phase involving *reflection on action*); and (4) the whole teaching cycle. A rather strict criterion was applied for coding studies in categories 2 and 4, involving the interactive phase of teaching. A study was only coded in this way if it reported lesson observations or investigated teachers' comments on authentic classroom video clips.

This classification scheme enables the different *forms* of PCK under investigation to be identified. Shavelson, Ruiz-Primo and Wiley (2005) distinguished four types

[3]CoRe stands for content representation—a portrayal of PCK structured by big ideas related to a topic with responses to key prompts (Loughran, Mulhall, & Berry, 2004).

[4]Despite the cyclical nature of the teaching process, it can be difficult to determine whether a particular activity belongs to the pre-active or the post-active phase of the teaching cycle. As such, analysis of student work, predicting student responses in assessment tasks, classifying assessment questions and reflecting on one's *own* teaching actions were assigned to the post-active phase of teaching.

Table 1.4 Types of data sources used to determine teachers' PCK

Category	Description	Representative studies
Written questionnaires/surveys/tests	• Participants provide written responses to a set of prompts created by researchers (e.g., a pedagogical scenario) or questions and statements in a survey, questionnaire or test	Schmelzing et al. (2013), Sorge, Kröger, Petersen, and Neumann (2017)
Artefacts from teaching tasks	• Artefacts related to the teaching cycle (i.e., lesson planning, enactment and reflection) are collected. Examples of artefacts include unit/topic/lesson plans, classroom videos[a], student work, teaching documents (e.g., worksheets, handouts) and teachers' written reflections on enacted lesson(s)[b]	Cohen and Yarden (2009), Friedrichsen et al. (2009), Roth et al. (2011)
Interviews	• Participants provide responses in an oral interview, which is recorded on videotape or audio tape or in notes, and later transcribed for analysis	Henze, van Driel, and Verloop (2008), Luft et al. (2011)
Lesson observations	• Participants' classes or sessions are observed or recorded on audio tape or videotape or in field notes. All or part of the recording(s) is (are) later transcribed for analysis	Alonzo et al. (2012), Marshall, Smart, and Alston (2016)

[a]If the lesson was video-recorded for analysis in the study, the data source was regarded as lesson observations. It was counted as a teaching artefact only if the classroom video was used for the purpose of stimulated recall

[b]Reflection papers that were not on specific enacted lessons were not included in this category

of knowledge as learning goals for teaching and learning: (1) declarative knowledge (factual, conceptual knowledge) or 'knowing *that*'; (2) procedural knowledge (step-by-step or condition–action knowledge) or 'knowing *how*'; (3) schematic knowledge or 'knowing *why*'; and (4) strategic knowledge (knowledge of *when*, *where* and *how* knowledge applies). Investigating a teacher's PCK via *self-report* allows investigation of '*what a teacher knows*' (i.e., knowing '*that*') and '*his/her reasons for his/her judgements/action*' (i.e., knowing '*why*') if the teacher's reasoning is elicited. Investigating a teacher's *performance* in teaching tasks allows the investigation of '*what a teacher knows*' and '*what a teacher does*'. It enables researchers to investigate how a teacher *applies* his/her knowledge when carrying out teaching tasks (i.e., *enacted knowledge/skills*) and his/her actual *teaching acts* in the classroom (i.e., *knowledge embedded in practice/skills*) when conducting lesson observation. Examining *only* the artefacts or teaching actions allows a teacher's procedural knowledge and strategic knowledge to be investigated. If a teacher's instructional decisions and reasoning around teaching tasks are also investigated, researchers can gain access to the teacher's capacity to reason (i.e., knowing '*why*') and/or to the declarative knowledge the teacher does not utilise in the teaching tasks. Finally, distinguishing between simulated teaching tasks and real-life teaching differentiates between performance in '*performance tasks*' and in '*real-life*' teaching. See Fig. 1.1 for a summary.

After the systematic coding and analysis, we produced summary tables and charts for each of the analysis foci above (see Tables 1.5, 1.6, 1.7, 1.8 and 1.9 and Fig. 1.2). We then selected studies that were representative of each of the categories for more detailed reviews.

This chapter relied on investigator triangulation (Denzin, 1989) to ensure the trustworthiness of the findings. The first author coded and analysed all of the articles. Both authors then discussed cases that did not easily fit the categories to reach a consensus on the coding. The authors also shared and discussed the emergent findings in more than ten Skype meetings and many email exchanges.

Findings

We organise our findings by first describing the research context of the studies and the major research lines. We then discuss the various conceptualisations of PCK before turning our attention to the different data sources and approaches used to investigate PCK. A brief overview of the 99 articles can be found in the Appendix at the end of this chapter.

Research Context

Location of PCK Studies As shown in Table 1.5, the PCK studies were conducted in diverse locations, spanning six continents, suggesting that PCK is a popular theoretical lens for researching science teachers' knowledge all over the world. Most studies were conducted in Europe ($N = 35$; 35.4% of all studies), followed by North America ($N = 28$; 28.3%) and Asia ($N = 26$; 26.3%). The USA contributed the most studies to the investigation of teachers' PCK ($N = 27$). Note that only three studies involved teachers from more than one European country (see Table 1.5), while hardly any studies could be found that compared PCK of teachers from different *continents*. These findings are somewhat surprising given that comparative studies of PCK of

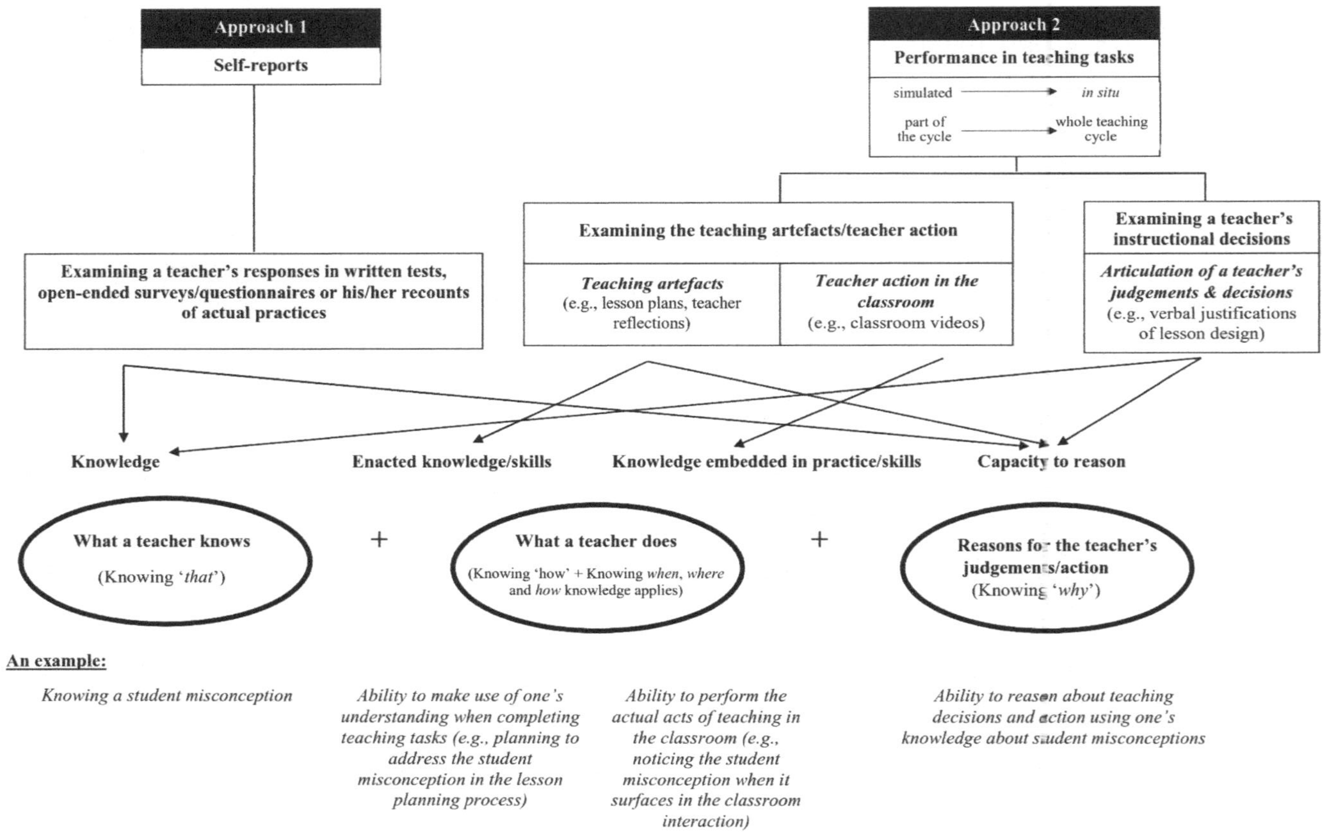

Fig. 1.1 Different approaches for determining science teachers' PCK and their relationship with different forms of PCK

Table 1.5 Countries in which the PCK studies were conducted

Continent	Country	No. of studies	% of studies	Studies
Africa	Lesotho	1	8.1	Qhobela and Kolitsoe Moru (2014)
	South Africa	6		Davidowitz and Potgieter (2016), Mavhunga (2016), Mavhunga and Rollnick (2016), Moodley and Gaigher (2017), Rollnick (2017), Rollnick, Bennett, Rhemtula, Dharsey, and Ndlovu (2008)
	Swaziland	1		Mthethwa-Kunene, Onwu, and de Villiers (2015)
Asia	Cambodia	2	26.3	Melo-Niño, Cañada, and Mellado (2017a, 2017b)
	China	2		Chen and Wei (2015), Zhou, Wang, and Zhang (2016)
	Hong Kong	2		Chan and Yung (2015, 2017)
	Israel	1		Cohen and Yarden (2009)
	Korea	1		Oh and Kim (2013)
	Lebanon	1		Salloum and BouJaoude (2008)
	Singapore	1		Tay and Yeo (2017)
	Taiwan	3		Lin (2016, 2017), Lin and Chiu (2010)
	Thailand	1		Supprakob, Faikhamta, and Suwanruji (2016)
	Turkey	12		Adadan and Oner (2014), Akin and Uzuntiryaki-Kondakçı (2018), Aydin and Boz (2013), Aydin, Demirdöğen, Akin, Uzuntiryaki-Kondakçı, and Tarkin (2015), Aydin et al. (2013, 2014), Bektas et al. (2013), Demirdöğen (2016), Demirdöğen, Hanuscin, Uzuntiryaki-Kondakçı, and Köseoğlu (2016), Demirdöğen and Uzuntiryaki-Kondakçı (2016), Kaya (2009), Uzuntiryaki-Kondakçı, Demirdöğen, Akin, Tarkin, and Aydın (2017)
Oceania	New Zealand	1	1.0	Donnelly and Hume (2015)

(continued)

Table 1.5 (continued)

Continent	Country	No. of studies	% of studies	Studies
Europe	Finland	1	35.4	Käpylä, Heikkinen, and Asunta (2009)
	Germany	17		Alonzo et al. (2012), Bindernagel and Eilks (2009), Förtsch et al. (2016), Großschedl, Harms, Kleickmann, and Glowinski (2015), Großschedl, Mahler, Kleickmann, and Harms (2014), Jüttner and Neuhaus (2012), Kirschner et al. (2016), Krepf et al. (2017), Mahler et al. (2017), Meschede, Fiebranz, Möller, and Steffensky (2017), Paulick, Großschedl, Harms, and Möller (2016), Rosenkränzer, Hörsch, Schuler, and Riess (2017), Scharfenberg and Bogner (2016), Schmelzing et al. (2013), Sorge et al. (2017), Stender, Brückmann, and Neumann (2017), van Dijk (2009)
	Greece	2		Piliouras, Plakitsi, Seroglou, and Papantoniou (2017), Stasinakis and Athanasiou (2016)
	The Netherlands	3		Barendsen and Henze (2017), Henze et al. (2008), Wongsopawiro, Zwart, and van Driel (2017)
	Scotland	1		Findlay and Bryce (2012)
	Sweden	7		Bergqvist and Chang Rundgren (2017), Bergqvist, Drechsler, and Chang Rundgren (2016), Kellner, Gullberg, Attorps, Thorén, and Tärneberg (2011), Nilsson and Loughran (2012), Nilsson and van Driel (2010), Nilsson and Vikström (2015), Walan, Nilsson, and Ewen (2017)
	UK	1		Kind (2017)
	More than one European country	3		Keller et al. (2017), Smit, Rietz, and Kreis (2017), Smit, Weitzel et al. (2017)

(continued)

Table 1.5 (continued)

Continent	Country	No. of studies	% of studies	Studies
North America	USA	27	28.3	Alonzo and Kim (2016), Barnett and Friedrichsen (2015), Boesdorfer and Lorsbach (2014), Brown, Friedrichsen, and Abell (2013), Burton (2013), Diezmann and Watters (2015), Friedrichsen et al. (2009), Gess-Newsome et al. (2017), Hallman-Thrasher, Connor, and Sturgill (2017), Hanuscin (2013), Hanuscin, Lee, and Akerson (2011), Heller et al. (2012), Jin, Shin, Johnson, Kim, and Anderson (2015), Kanter and Konstantopoulos (2010), Lucero, Petrosino, and Delgado (2017), Luft (2009), Luft et al. (2011), Marshall et al. (2016), Monet and Etkina (2008), Park and Chen (2012), Park, Jang, Chen, and Jung (2011), Park and Oliver (2008a, 2008b), Park et al. (2017), Roth et al. (2011), Sickel and Friedrichsen (2017), Suh and Park (2017)
	Mexico	1		Alvarado, Cañada, Garritz, and Mellado (2015)
South America	Chile	1	1.0	Bravo and Cofré (2016)

teachers from different continents, both large scale (e.g., Blömeke, Suhl, & Kaiser, 2011) and small scale (e.g., An, Kulm, & Wu, 2004), have been published in the mathematics education literature.

Research Participants and Sample Size The research participants in the PCK studies included pre-service teachers (e.g., Kind, 2017), beginning in-service teachers (e.g., Luft et al., 2011), experienced teachers with more than five years of teaching experience (e.g., Chan & Yung, 2015) and scientists (e.g., Schmelzing et al., 2013) (see Appendix). The majority of the studies ($N = 73$)[5] used in-service teachers as the research participants. Some studies ($N = 7$) investigated the PCK of both novice and more experienced, or expert, teachers (e.g., Krepf, Plöger, Scholl, & Seifert, 2017) and mentor teachers (e.g., Nilsson & van Driel, 2010). Others ($N = 2$) compared

[5] Because 9 studies involved both pre-service and in-service teachers, the total number of studies is 108, rather than 99.

Table 1.6 Research foci and research methods of the science PCK studies

Research focus	No. of articles[a]	Studies		
		Qualitative	Quantitative	Mixed methods
		65	21	13
Nature of science teachers' PCK	42	*In-service teachers (N = 29):* Akin and Uzuntiryaki-Kondakçı (2018), Alonzo and Kim (2016), Alonzo et al. (2012), Alvarado et al. (2015), Aydin and Boz (2013), Aydin et al. (2014), Barendsen and Henze (2017), Bergqvist and Chang Rundgren (2017), Bergqvist et al. (2016), Bindernagel and Eilks (2009), Boesdorfer and Lorsbach (2014), Chen and Wei (2015), Cohen and Yarden (2009), Diezmann and Watters (2015), Hanuscin et al. (2011), Lin (2016), Lin and Chiu (2010), Melo-Niño et al. (2017b), Moodley and Gaigher (2017), Mthethwa-Kunene et al. (2015), Oh and Kim (2013), Park and Chen (2012), Park and Oliver (2008b), Qhobela and Kolitsoe Moru (2014), Salloum and BouJaoude (2008), Supprakob et al. (2016), Tay and Yeo (2017), van Dijk (2009), Walan et al. (2017) *Pre-service teachers (N = 4):* Demirdöğen (2016), Kellner et al. (2011), Kind (2017), Zhou et al. (2016) *Both pre-service and in-service teachers (N = 2):* Friedrichsen et al. (2009), Lin (2017) *From pre-service education to first year of teaching (N = 1):* Hallman-Thrasher et al. (2017)	*Pre-service teachers (N = 3):* Großschedl et al. (2014, 2015), Sorge et al. (2017) *In-service teachers (N = 1):* Stasinakis and Athanasiou (2016)	*Pre-service teachers (N = 1):* Kaya (2009) *In-service teachers (N = 1):* Krepf et al. (2017)

(continued)

Table 1.6 (continued)

Research focus	No. of articles[a]	Studies		
		Qualitative	Quantitative	Mixed methods
		65	21	13
Development of science teachers' PCK	14	*Pre-service teachers (N = 5):* Adadan and Oner (2014), Barnett and Friedrichsen (2015), Bektas et al. (2013), Brown et al. (2013), Hanuscin (2013) *In-service teachers (N = 6):* Chan and Yung (2015, 2017), Henze et al. (2008), Melo-Niño et al. (2017a), Park and Oliver (2008a), Sickel and Friedrichsen (2017) *Both pre-service and in-service teachers (N = 1):* Nilsson and van Driel (2010) *From pre-service education to beginning years of teaching (N = 1):* Findlay and Bryce (2012)	*Pre-service teachers (N = 1):* Sorge et al. (2017)	/
Relationship between PCK and other variable(s) (e.g., content knowledge, instructional quality)	20	*Pre-service teachers (N = 2):* Käpylä et al. (2009), Uzuntiryaki-Kondakçı et al. (2017) *In-service teachers (N = 5):* Lucero et al. (2017), Melo-Niño et al. (2017a), Rollnick (2017), Rollnick et al. (2008), Suh and Park (2017)	*Pre-service teachers (N = 2):* Paulick et al. (2016), Smit, Weitzel, et al. (2017) *In-service teachers (N = 7):* Davidowitz and Potgieter (2016), Förtsch et al. (2016), Kanter and Konstantopoulos (2010), Keller et al. (2017), Mahler et al. (2017), Park et al. (2011), Stender et al. (2017) *Both pre-service and in-service teachers (N = 1):* Meschede et al. (2017)	*Pre-service teachers (N = 2):* Kaya (2009), Mavhunga and Rollnick (2016) *In-service teachers (N = 1):* Gess-Newsome et al. (2017)

(continued)

Table 1.6 (continued)

Research focus	No. of articles[a]	Studies		
		Qualitative	Quantitative	Mixed methods
		65	21	13
Changes in science teachers' PCK following or during an intervention/professional development programme	23	*Pre-service teachers (N = 6):* Aydin et al. (2015), Aydin et al. (2013), Demirdöğen et al. (2016), Demirdöğen and Uzuntiryaki-Kondakçı (2016), Donnelly and Hume (2015), Nilsson and Loughran (2012) *In-service teachers (N = 4):* Bravo and Cofré (2016), Nilsson and Vikström (2015), Piliouras et al. (2017), Wongsopawiro et al. (2017)	*Pre-service teachers (N = 1):* Smit, Rietz, et al. (2017) *In-service teachers (N = 3):* Heller et al. (2012), Kanter and Konstantopoulos (2010), Roth et al. (2011)	*Pre-service teachers (N = 4):* Mavhunga (2016), Mavhunga and Rollnick (2016), Rosenkränzer et al. (2017), Scharfenberg and Bogner (2016) *In-service teachers (N = 5):* Burton (2013), Gess-Newsome et al. (2017), Luft (2009) Luft et al. (2011), Monet and Etkina (2008)
Development of a PCK measurement instrument	6	–	*In-service teachers (N = 2):* Jin et al. (2015), Marshall et al. (2016) *Both pre-service and in-service teachers (N = 2):* Kirschner et al. (2016), Schmelzing et al. (2013)	*In-service teachers (N = 1):* Park et al. (2017) *Both pre-service and in-service teachers (N = 1):* Jüttner and Neuhaus (2012)

[a]The total number of article in this column does not add up to 99 as 6 articles (Gess-Newsome et al., 2017; Kanter & Konstantopoulos, 2010; Kaya, 2009; Mavhunga & Rollnick, 2016; Melo-Niño et al., 2017a; Sorge et al., 2017) were assigned to more than one line of research; hence, the total number is 105

Table 1.7 Conceptualisation of PCK

Integrative model: *In about 1/10 of the studies, PCK is conceptualised as an integration of different knowledge categories rather than as a distinct entity*

Author(s)	Model	*N*	CK	PK	C*x*K	KS	C*u*K
Krepf et al. (2017)	Shulman (1986)	1	X	X			
Findlay and Bryce (2012)	Grossman (1990)	1	X	X	X		X
Kind (2017)	Gess-Newsome (2015)	1	X	X			
Piliouras et al. (2017)	Not specified	1	X	X			
Qhobela and Kolitsoe Moru (2014)		1	X		X		
Gess-Newsome et al. (2017), Stasinakis and Athanasiou (2016)		2	X	X	X		
Mthethwa-Kunene et al. (2015)		1	X	X		X	
Tay and Yeo (2017)	Cochran, DeRuiter, and King (1993)	1	X	X	X	X	
	Total	9	9	8	5	2	1

Transformative model: *For the majority of studies, PCK is conceptualised as a distinct knowledge category*

Author(s)	Model	N	KSU	KISR	KA	KC	OTS	Others
Lucero et al. (2017), Zhou et al. (2016)	Shulman (1986)	2	X					
Oh and Kim (2013)		1		X				
Alonzo and Kim (2016)		1	X	X				

(continued)

Table 1.7 (continued)

Transformative model: *For the majority of studies, PCK is conceptualised as a distinct knowledge category*								
Author(s)	Model	N	KSU	KISR	KA	KC	OTS	Others
Großschedl et al. (2015)	Tamir (1988)	1	X	X	X	X		
van Dijk (2009)	van Dijk and Kattmann (2007)	1	X	X				X
Moodley and Gaigher (2017)	Hill, Ball, and Schilling (2008)	1	X	X				
Diezmann and Watters (2015)		1	X	X		X		
Jüttner and Neuhaus (2012)	Tepner et al. (2012)	1	X					
Förtsch et al. (2016), Kirschner et al. (2016)		2	X	X				
Rollnick et al. (2008)	Rollnick et al. (2008)	1		X	X	X		
Davidowitz and Potgieter (2016), Mavhunga (2016), Mavhunga and Rollnick (2016)	Mavhunga and Rollnick (2013)	3	X	X		X		
Mahler et al. (2017)	Magnusson et al. (1999)	1	X	X				
Brown et al. (2013), Scharfenberg and Bogner (2016)		2	X	X			X	
Aydin et al. (2013), Barendsen and Henze (2017), Boesdorfer and Lorsbach (2014), Bravo and Cofré (2016), Chan and Yung (2015, 2017), Chen and Wei (2015), Donnelly and Hume (2015), Henze et al. (2008), Kanter and Konstantopoulos (2010), Kaya (2009), Melo-Niño et al. (2017a)		12	X	X	X	X		
Bektas et al. (2013), Käpylä et al. (2009)		2	X	X	X		X	

(continued)

Table 1.7 (continued)

Transformative model: *For the majority of studies, PCK is conceptualised as a distinct knowledge category*								
Author(s)	Model	N	KSU	KISR	KA	KC	OTS	Others
Adadan and Oner (2014), Alvarado et al. (2015), Aydin et al. (2014), Barnett and Friedrichsen (2015), Cohen and Yarden (2009), Demirdöğen (2016), Demirdöğen et al. (2016), Demirdöğen and Uzuntiryaki-Kondakçı (2016), Friedrichsen et al. (2009), Hanuscin (2013), Hanuscin et al. (2011), Melo-Niño et al. (2017b), Sickel and Friedrichsen (2017), Wongsopawiro et al. (2017)		14	X	X	X	X	X	
Supprakob et al. (2016)	Hanuscin et al. (2011)	1	X	X	X	X	X	
Park et al. (2017)	Park and Oliver (2008a)	1	X	X			X	
Akin and Uzuntiryaki-Kondakçı (2018), Aydin and Boz (2013), Aydin et al. (2015), Park and Chen (2012), Park et al. (2011), Park and Oliver (2008a)		6	X	X	X	X	X	
Park and Oliver (2008b)	Park and Oliver (2008b)	1	X	X	X	X	X	X
Suh and Park (2017)	Park and Chen (2012)	1	X	X	X	X	X	
Bergqvist et al. (2016), Smit, Rietz, et al. (2017), Smit, Weitzel, et al. (2017)	Gess-Newsome (2015)	3	X	X				
Rollnick (2017), Sorge et al. (2017), Stender et al. (2017)		3	X	X	X	X		
Kellner et al. (2011), Lin (2016, 2017), Lin and Chiu (2010)	Not specified	4	X					

(continued)

Table 1.7 (continued)

Transformative model: *For the majority of studies, PCK is conceptualised as a distinct knowledge category*								
Author(s)	Model	N	KSU	KISR	KA	KC	OTS	Others
Alonzo et al. (2012), Bergqvist and Chang Rundgren (2017), Großschedl et al. (2014), Heller et al. (2012), Jin et al. (2015), Luft (2009), Luft et al. (2011), Mahler et al. (2017), Meschede et al. (2017), Nilsson and van Driel (2010), Paulick et al. (2016), Roth et al. (2011), Salloum and BouJaoude (2008), Schmelzing et al. (2013)		14	X	X				
Bindernagel and Eilks (2009), Monet and Etkina (2008), Nilsson and Vikström (2015), Rosenkränzer et al. (2017), Walan et al. (2017)		5	X	X		X		
Keller et al. (2017)		1	X			X		X
Nilsson and Loughran (2012)		1	X	X	X	X	X	
Hallman-Thrasher et al. (2017)		1	X	X		X	X	
	Total	88	86	80	43	52	30	3

Notes

1. *CK* content knowledge; *PK* pedagogical knowledge; *CxK* contextual knowledge; *CuK* curricular knowledge; *KS* knowledge of students; *KSU* knowledge of students' understanding of science; *KISR* knowledge of instructional strategies and representations; *KA* knowledge of assessment; *KC* knowledge of curriculum (refer to Table 1.3 for details); *OTS* orientations to teaching science

2. Two articles (Burton, 2013; Marshall et al., 2016) were excluded from the above table as it was not possible to classify the components the paper investigated using the analytical framework

Table 1.8 Different types of data sources used to investigate science teachers' PCK

Study/studies	*N*	Data source(s)				
		Written tests/questionnaires/surveys	Teaching artefacts	Interviews	Lesson observations	Others
Single type of data source (N = 37)						
Davidowitz and Potgieter (2016), Förtsch et al. (2016), Großschedl et al. (2015), Großschedl et al. (2014), Heller et al. (2012), Jin et al. (2015), Jüttner and Neuhaus (2012), Keller et al. (2017), Kind (2017), Kirschner et al. (2016), Lin (2016, 2017), Mahler et al. (2017), Meschede et al. (2017), Monet and Etkina (2008), Park et al. (2017), Paulick et al. (2016), Rosenkränzer et al. (2017), Scharfenberg and Bogner (2016), Schmelzing et al. (2013), Smit, Rietz, et al. (2017), Smit, Weitzel, et al. (2017), Sorge et al. (2017), Stasinakis and Athanasiou (2016), Stender et al. (2017)	25	X				
Bindernagel and Eilks (2009), Findlay and Bryce (2012), Hallman-Thrasher et al. (2017), Henze et al. (2008), Kaya (2009), Luft (2009), Luft et al. (2011), Salloum and BouJaoude (2008)	8			X		

(continued)

Table 1.8 (continued)

Study/studies	*N*	Data source(s)				
		Written tests/questionnaires/surveys	Teaching artefacts	Interviews	Lesson observations	Others
Alonzo et al. (2012), Marshall et al. (2016), Oh and Kim (2013), Tay and Yeo (2017)	4				X	
Two types of data sources (N = 19)						
Kanter and Konstantopoulos (2010)	1	X	X			
Qhobela and Kolitsoe Moru (2014), Zhou et al. (2016)	2	X		X		
Adadan and Oner (2014), Alonzo and Kim (2016), Alvarado et al. (2015), Aydin et al. (2015), Demirdöğen (2016), Friedrichsen et al. (2009), Krepf et al. (2017), Lin and Chiu (2010), Moodley and Gaigher (2017), van Dijk (2009)	10		X	X		
Akin and Uzuntiryaki-Kondakçı (2018), Aydin and Boz (2013), Aydin et al. (2014), Barendsen and Henze (2017), Park et al. (2011)	5			X	X	
Piliouras et al. (2017)	1				X	X

(continued)

Table 1.8 (continued)

Study/studies	*N*	Data source(s)				
		Written tests/questionnaires/surveys	Teaching artefacts	Interviews	Lesson observations	Others
Three or more types of data sources (N = 43)						
Käpylä et al. (2009), Kellner et al. (2011), Melo-Niño et al. (2017a)	3	X	X	X		
Mavhunga (2016), Mavhunga and Rollnick (2016)	2	X	X			X
Boesdorfer and Lorsbach (2014), Bravo and Cofré (2016), Brown et al. (2013), Chan and Yung (2015, 2017), Diezmann and Watters (2015), Gess-Newsome et al. (2017), Lucero et al. (2017), Nilsson and Vikström (2015), Park and Chen (2012), Rollnick et al. (2008), Sickel and Friedrichsen (2017), Suh and Park (2017), Supprakob et al. (2016), Uzuntiryaki-Kondakçı et al. (2017)	15		X	X	X	
Barnett and Friedrichsen (2015), Bergqvist and Chang Rundgren (2017), Bergqvist et al. (2016), Donnelly and Hume (2015), Hanuscin (2013), Walan et al. (2017), Wongsopawiro et al. (2017)	7		X	X		X

(continued)

Table 1.8 (continued)

Study/studies	*N*	Data source(s)				
		Written tests/questionnaires/surveys	Teaching artefacts	Interviews	Lesson observations	Others
Roth et al. (2011)	1	X	X		X	
Chen and Wei (2015)	1			X	X	X
Aydin et al. (2013), Cohen and Yarden (2009), Demirdöğen et al. (2016), Demirdöğen and Uzuntiryaki-Kondakçı (2016), Melo-Niño et al. (2017b), Nilsson and Loughran (2012)	6	X	X	X		X
Hanuscin et al. (2011), Nilsson and van Driel (2010), Park and Oliver (2008a, 2008b), Rollnick (2017)	5		X	X	X	X
Bektas et al. (2013), Burton (2013), Mthethwa-Kunene et al. (2015)	3	X	X	X	X	X

Notes

1. If CoRes was only used as a template for formulating interview questions and the CoRes lesson/unit plan was not collected from the teacher in the study, teaching artefact was not considered as a data source (Akin & Uzuntiryaki-Kondakçı, 2018)
2. Data sources specifically used for exploring teachers' orientations to teaching science are not included in the above table. For example, Demirdöğen (2016) used a written questionnaire specifically for this purpose. Similarly, Aydin and Boz (2013) and Aydin et al. (2014) used a card-sorting task to explore teachers' OST
3. If a particular data source was collected but not analysed, the data source was not counted. For example, Oh and Kim (2013) mentioned conducting interviews, but specified that the interview transcripts were not analysed in that study
4. For studies that involved the development of a PCK measurement instrument, the final instrument was regarded as the data source. In other words, even if the validation process involved interviews (e.g., Park et al., 2017) or student work (Jüttner & Neuhaus, 2012), these data sources were not counted
5. Studies that analysed video-recorded lessons were assigned to the category 'lesson observation'. Only if the videos were further edited and used (e.g., in stimulated recalls; Chan & Yung, 2015) were the studies also assigned to the category 'teaching artefacts'

Table 1.9 Approaches used to determine science teachers' PCK

	PCK tests ($N = 22$)	Questionnaires/surveys/interviews ($N = 19$)	Self-report of actual practices ($N = 8$)
Approach 1 Self-reports ($N = 48^a$)	Davidowitz and Potgieter (2016), Förtsch et al. (2016), Großschedl et al. (2015), Großschedl et al. (2014), Heller et al. (2012), Jin et al. (2015), Jüttner and Neuhaus (2012), Keller et al. (2017), Kirschner et al. (2016), Mahler et al. (2017), Mavhunga (2016), Mavhunga and Rollnick (2016), Meschede et al. (2017), Monet and Etkina (2008), Park et al. (2017), Paulick et al. (2016), Rosenkränzer et al. (2017), Schmelzing et al. (2013), Smit, Rietz, et al. (2017), Smit, Weitzel, et al. (2017), Sorge et al. (2017), Stender et al. (2017)	Bektas et al. (2013), Burton (2013), Cohen and Yarden (2009), Demirdöğen et al. (2016), Demirdöğen and Uzuntiryaki-Kondakçı (2016), Käpylä et al. (2009), Kaya (2009), Kellner et al. (2011), Kind (2017), Lin (2016, 2017), Melo-Niño et al. (2017a, 2017b), Mthethwa-Kunene et al. (2015), Nilsson and Loughran (2012), Qhobela and Kolitsoe Moru (2014), Salloum and BouJaoude (2008), Scharfenberg and Bogner (2016), Zhou et al. (2016)	Bindernagel and Eilks (2009), Findlay and Bryce (2012), Hallman-Thrasher et al. (2017), Henze et al. (2008), Luft (2009), Luft et al. (2011), Qhobela and Kolitsoe Moru (2014), Stasinakis and Athanasiou (2016)
Approach 2 Performance in teaching tasks ($N = 64$)[b]	Examined the teaching artefacts/actions only[c] ($N = 6$)	*Pre-active phase* ($N = 0$)	*Interactive phase* ($N = 6$)
		NA	Alonzo et al. (2012), Lucero et al. (2017), Marshall et al. (2016), Oh and Kim (2013), Roth et al. (2011), Tay and Yeo (2017)

(continued)

Table 1.9 (continued)

	PCK tests ($N = 22$)		Questionnaires/surveys/interviews ($N = 19$)		Self-report of actual practices ($N = 8$)	
	Also examined teachers' decisions and reasoning around the teaching tasks ($N = 60$)[d]	*Used simulated teaching tasks* ($N = 27$)[e]	*Pre-active phase* ($N = 18$)	*Interactive phase* ($N = 4$)	*Post-active phase* ($N = 4$)	*Whole teaching cycle* ($N = 2$)
			Adadan and Oner (2014), Alvarado et al. (2015), Aydin et al. (2015), Aydin et al. (2013), Bergqvist and Chang Rundgren (2017), Bergqvist et al. (2016), Brown et al. (2013), Demirdöğen (2016), Demirdöğen et al. (2016), Demirdöğen and Uzuntiryaki-Kondakçı (2016), Donnelly and Hume (2015), Friedrichsen et al. (2009), Käpylä et al. (2009), Kellner et al. (2011), Mavhunga (2016), Mavhunga and Rollnick (2016), Melo-Niño et al. (2017b), Nilsson and Loughran (2012)	Alonzo and Kim (2016), Krepf et al. (2017), Mavhunga (2016), Roth et al. (2011)	Cohen and Yarden (2009), Lucero et al. (2017), Moodley and Gaigher (2017), van Dijk (2009)	Bektas et al. (2013), Rollnick (2017)

(continued)

Table 1.9 (continued)

	PCK tests ($N = 22$)		Questionnaires/surveys/interviews ($N = 19$)		Self-report of actual practices ($N = 8$)	
		Investigated real-life teaching in situ ($N = 37$)	*Pre-active phase* ($N = 4$)	*Interactive phase* ($N = 1$)	*Post-active phase* ($N = 4$)	*Whole teaching cycle* ($N = 28$)
			Barnett and Friedrichsen (2015), Hanuscin (2013), Melo-Niño et al. (2017a), Walan et al. (2017)	Alonzo and Kim (2016)	Kanter and Konstantopoulos (2010), Lin and Chiu (2010), Piliouras et al. (2017), Wongsopawiro et al. (2017)	Akin and Uzuntiryaki-Kondakçı (2018), Aydin and Boz (2013), Aydin et al. (2014), Barendsen and Henze (2017), Bektas et al. (2013), Boesdorfer and Lorsbach (2014), Bravo and Cofré (2016), Brown et al. (2013), Burton (2013), Chan and Yung (2015, 2017), Chen and Wei (2015), Diezmann and Watters (2015), Gess-Newsome et al. (2017), Hanuscin et al. (2011), Mthethwa-Kunene et al. (2015), Nilsson and van Driel (2010), Nilsson and Vikström (2015), Park and Chen (2012), Park et al. (2011), Park and Oliver (2008a, 2008b), Rollnick (2017), Rollnick et al. (2008), Sickel and Friedrichsen (2017), Suh and Park (2017), Supprakob et al. (2016), Uzuntiryaki-Kondakçı et al. (2017)

(continued)

Table 1.9 (continued)

Note: The total number of studies using approaches 1 and 2 does not add up to 99, as some studies adopted both approaches

[a]The total number of studies is not the sum of the three subcategories as one study used both survey and interview to determine teachers' PCK

[b]The total number of studies is not the sum of the two subcategories as some studies contained separate tasks that allowed the examination of the teaching artefacts/actions only in one task and the teachers' decisions and reasoning around the other teaching task

[c]The 'post-active phase' and 'whole teaching cycle' subcategories were absent for two reasons. First, engaging teachers in written/verbal reflection of their lessons inevitably elicits their reasoning and teaching decisions. Second, there was not one study that only analysed teachers' written reflections to determine teachers' PCK

[d]The total number is not the sum of the studies in the two subcategories as some studies used simulated teaching tasks and also investigated real-life teaching in situ

[e]The total number is not the sum of the studies in the four subcategories as one study used both simulated tasks for pre-active phase and interactive phase of teaching to determine the teachers' PCK

teachers' PCK from their last year of teacher education to their first years in the profession (e.g., Findlay & Bryce, 2012).

Figure 1.2 shows the grade levels of the teachers and the respective sample sizes of the PCK studies. There were more studies on secondary ($N = 86$) than primary ($N = 23$) teachers.[6] In ten of the studies, both primary and secondary teachers were investigated. Regarding the subject domain, secondary chemistry teachers ($N = 27$) were the most researched, followed by secondary biology teachers ($N = 25$) (Fig. 1.2). The sample size of the reviewed studies ranged from 1 (e.g., Barendsen & Henze, 2017) to 631 teachers (Paulick et al., 2016) (see Appendix). The great majority of the studies (73 out of 99) had sample sizes with fewer than 50 teachers (see Appendix and Fig. 1.2), of which 54 comprised less than 10 teachers. Only one study had a sample size larger than 500.

Topic Under Investigation Some studies investigated different teachers' PCK for teaching the *same* topic (e.g., Friedrichsen et al., 2009). Others explored the same teacher's PCK for teaching *different* topics (e.g., Aydin, Friedrichsen, Boz, & Hanuscin, 2014). Still, others investigated different teachers' PCK in the context of teaching different topics (e.g., Nilsson & Vikström, 2015). The focal subject contents were often conceptually challenging and difficult, such as chemical equilibrium in chemistry (e.g., Mavhunga & Rollnick, 2016), photosynthesis in biology (e.g., Park & Chen, 2012) and electricity in physics (e.g., Moodley & Gaigher, 2017). Electric circuit was a commonly researched topic in studies of primary teachers' PCK (e.g., Heller, Daehler, Wong, Shinohara, & Miratrix, 2012). In some studies, the subject matter seemed to be narrower in scope (e.g., ideas about chemicals), whereas in others it was broader (e.g., organic chemistry). Some topics also appeared more generic in nature (e.g., nature of science, system thinking). In some studies, the focal topic was defined by the curriculum materials provided by the professional developer (e.g., Kanter & Konstantopoulos, 2010). Collectively, there were clear variations in the scope of the focal subject matter and how a 'topic' was defined in different studies.

Summary In summary, PCK has been used in the last decade as a theoretical framework in investigations to understand the content-specific knowledge of science teachers across grade levels and career spans in many countries around the world. Most of these studies used small samples (<50 teachers) and involved in-service secondary teachers. There were obvious differences in the scope of the subject matter and the grain size of the topics in different studies, and different researchers appeared to use different definitions of a 'topic'.

[6]Because 10 studies involved both primary and secondary school teachers, the total number of studies is 109, rather than 99.

Major Research Lines

Table 1.6 displays the major purposes of the reviewed PCK studies in science and their research methods (i.e., qualitative, quantitative or mixed methods). The majority of the PCK researchers ($N = 65$) used qualitative methods. Interestingly, quantitative or mixed methods were adopted in more recent studies, with nearly two-thirds of these studies ($N = 20$ out of 34) published in or after 2016. Nearly, half of these studies had a larger sample size larger than 100. We discerned several distinct lines of research and reviewed them in this section.

The first line of research focused on the nature of science teachers' PCK and included the great majority of the studies. Most of them were small-scale qualitative studies, typically describing the nature and content of the teachers' PCK with respect to predetermined PCK components (e.g., Bergqvist et al., 2016; Kellner et al., 2011). Several studies compared the PCK between different groups of teachers, such as expert and novice teachers (e.g., Krepf et al., 2017), or primary and secondary teachers (e.g., Lin, 2017) to illuminate the difference in PCK profiles between the teacher groups. A number of studies emphasised the nature of integration between the PCK components (e.g., Aydin & Boz, 2013; Park & Chen, 2012). Others attempted to use their empirical findings to refine concepts about PCK and/or uncover new aspects of PCK. For example, by investigating how teachers responded to unexpected student thinking captured in carefully selected video clips, Alonzo and Kim (2016) argued for the presence of a more dynamic and flexible form of PCK that underpins teachers' spontaneous instructional decision-making. Some studies assigned to this strand explored how certain factors influenced the nature and content of teachers' PCK. For example, Bergqvist and Chang Rundgren (2017) investigated the influence of textbooks on chemistry teachers' PCK, while Friedrichsen et al. (2009) explored the role of teaching experience on teachers' PCK.

The second line of research investigated the development of science teachers' PCK in naturalistic settings or in the context of initial teacher preparation. All but one of these studies were qualitative and involved small samples (<10 teachers). These PCK studies generally provided a snapshot of the teachers' PCK, measured using the same instrument(s) before and after a certain time frame. The time frame ranged from a single teaching instance (e.g., Chan & Yung, 2015) to longer periods such as the four and a half-year longitudinal study conducted by Findlay and Bryce (2012). Only two longitudinal studies (i.e., Findlay & Bryce, 2012; Henze et al., 2008) exceeded two years. Longitudinal studies spanning several years and with large sample sizes are yet to appear. Sorge et al. (2017) are a rare example of a quantitative study with a large sample size. The authors investigated the PCK development of pre-service physics teachers using a cross-sectional design, in which they compared the PCK of beginning pre-service teachers with that of teachers in their advanced year.

The third line of studies explored the relationship between PCK and other variables, such as teachers' cognition (e.g., teaching script; Stender et al., 2017), affective attributes such as attitudes, self-concept and beliefs (Mavhunga & Rollnick, 2016; Paulick et al., 2016; Smit, Weitzel, et al., 2017), knowledge (e.g., CK; Davidowitz

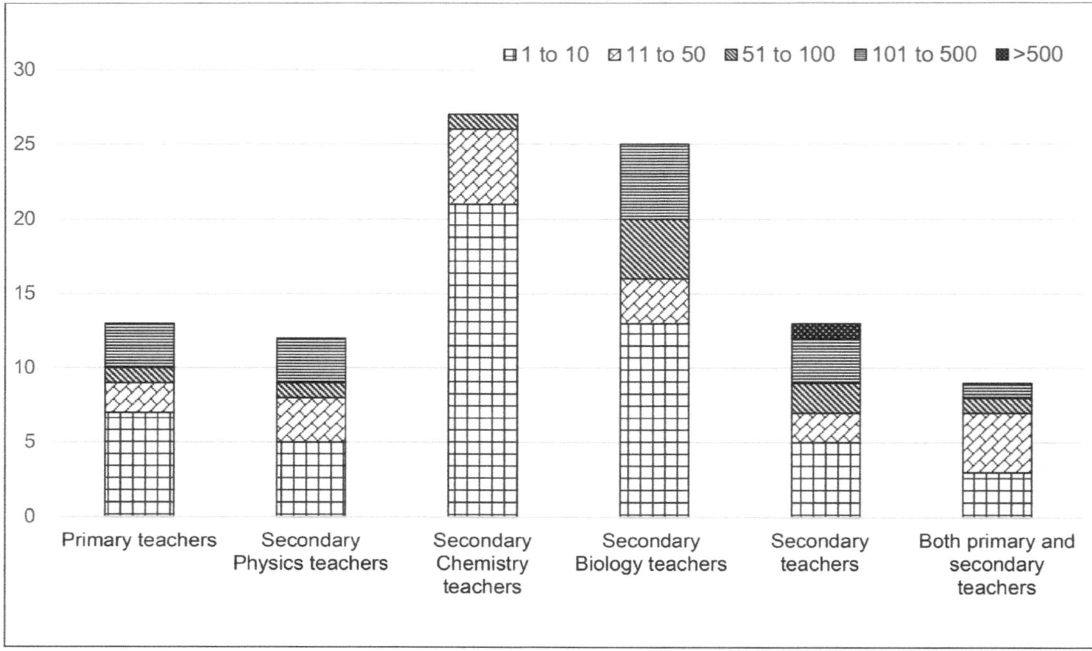

Note:

1. Teachers from grade 1 to 6 were labelled as primary teachers, whereas those that involved teachers from grade 7 to 12 were categorised as secondary teachers. Hence, middle school teachers (see *Appendix*) were assigned to the category 'both primary and secondary teachers'.

2. Studies that recruited teachers from more than one subject domain were classified as involving science teachers. The category of secondary teachers comprised studies involving science teachers and those involving science and non-science teachers.

Fig. 1.2 Grade levels of the teachers and sample sizes of the PCK studies ($N = 99$)

& Potgieter, 2016; Käpylä et al., 2009; Lucero et al., 2017; Rollnick et al., 2008), instructional practices (e.g., reformed science teaching; Park et al., 2011) cognitive activation (Förtsch et al., 2016), and student achievement (Gess-Newsome et al., 2017; Kanter & Konstantopoulos, 2010; Keller et al., 2017). These studies may be aggregated into two subgroups. The first subgroup typically involved small, qualitative studies that explored the relationship between PCK and the variable(s) of interest. The second subgroup often used standardised paper and pencil tests to investigate the PCK of a large sample of teachers. Studies within this subgroup used different statistical analysis methods to explore and verify the relationship between PCK and the variable(s) of interest. These methods included the use of doubly latent multilevel analysis (e.g., Mahler et al., 2017), multilevel structural equation modelling (e.g., Keller et al., 2017) and other similar analysis methods.

The fourth line of studies investigated changes in science teachers' PCK as a result of an intervention. These studies may be situated in the context of a method course ($N = 11$) or an in-service professional development programme ($N = 13$) and are divided into two subgroups. The first subgroup typically sought to report qualitative changes in the content and/or nature of teachers' PCK before and after the intervention. The other subgroup relied on quantitative methods to investigate the effectiveness of the intervention and often adopted quasi-experimental designs (e.g., Rosenkränzer et al., 2017; Roth et al., 2011).

The last line of research involved reporting the development and validation of an instrument for measuring science teachers' PCK. All, except the study by Marshall et al. (2016), reported the development of *written* questionnaires, surveys or tests for measuring teacher's PCK. Marshall et al. (2016) described an observational protocol for such a purpose. Studies in this line of research typically made use of statistical analysis to validate the instrument. Some studies also validated their instruments using qualitative data (e.g., think-aloud interviews by experienced teachers; Park et al., 2017).

Summary Five major lines of research were identified within the reviewed studies of science teachers' PCK. Small-scale descriptive studies aiming to capture and portray the nature and content of PCK among a small number of teachers continued to predominate. A clear trend over time is the increasing number of studies with larger samples that aim to move beyond description and explore the relationships between PCK and other variables. Alongside this trend, there has been a surge of interest in measuring PCK, as reflected in the publication of recent articles discussing how to create valid and reliable measurement instruments. The review also identified some gaps in the literature. For example, there is a lack of extended longitudinal studies using large participant samples (>50) over a period of three or more years.

Conceptualisation of PCK

The reviewed studies can be broadly classified into two groups (Table 1.7). The first group ($N = 9$; about 9.1%) adopted an *integrative* stance when conceptualising PCK (Gess-Newsome, 1999), in which it was not treated as a *distinct* category of knowledge and several knowledge categories related to PCK were investigated. The second group of studies ($N = 88$; 88.9%) adopted a *transformative* stance in conceptualising PCK (Gess-Newsome, 1999), treating it as a *distinct* category of knowledge. In these studies, PCK was often further conceptualised as comprising various PCK *components*.

All nine studies within the first group (Table 1.7) investigated the science teachers' content knowledge. Other knowledge categories that were investigated included pedagogical knowledge ($N = 8$), contextual knowledge ($N = 5$), knowledge of students ($N = 2$) and curricular knowledge ($N = 1$). Assessment knowledge, a component within teachers' professional bases in the consensus model (Gess-Newsome, 2015), was not investigated in any of the studies.

Of the 88 studies that adopted the transformative stance, it is noteworthy that about a third of them ($N = 26$) did not specify a PCK model that informed their studies. Some of these studies did not even provide a clear definition of the PCK components under investigation. Typically, these authors referred to a number of PCK models and claimed that they were looking at the component(s) commonly agreed upon in the science PCK field. They then went on to give the name(s) of their component(s) of interest without defining them. In some instances, new terminologies were introduced. Roth et al. (2011), for example, described their two PCK components as (1) knowledge about creating *a coherent science content storyline* and (2) knowledge about *eliciting, supporting and challenging students' thinking*. Similarly, Rosenkränzer et al. (2017) created a new term for their PCK component—knowledge of curriculum *and educational ends*. Although some other authors referred to Shulman's original definition of PCK, they also introduced new terms. For example, Oh and Kim (2013) introduced the term 'pedagogical transformation' to describe 'the instructional principle in which scientific ideas are simplified and reconstructed into what can be readily accessible to and understood by students without distorting the essential features of the ideas' (p. 1593).

The most commonly used PCK model across the reviewed papers was the transformative model by Magnusson et al. (1999) and its variants (i.e., Hanuscin et al., 2011; Park & Chen, 2012; Park & Oliver, 2008a, 2008b; Park et al., 2017) ($N = 41$). It was adopted by science education research groups in various geographic locations, including Asia (e.g., Chan & Yung, 2015), Oceania (e.g., Donnelly & Hume, 2015), Europe (e.g., Scharfenberg & Bogner, 2016), North America (e.g., Sickel & Friedrichsen, 2017) and South America (e.g., Bravo & Cofré, 2016). The PCK model proposed by Mavhunga and Rollnick (2013) was popular in South African science PCK studies, while the model of Tepner et al. (2012) informed several German science PCK studies. In addition, the science PCK community started to cross-reference the work of mathematics PCK communities. Several recent studies (i.e., Diezmann & Watters,

2015; Moodley & Gaigher, 2017) were conceptually grounded in a PCK model in mathematics education (i.e., the mathematics knowledge for teaching (MKT) model by Hill et al., 2008). More recently, there has been a clear trend for the increasing use of the CM for teacher professional knowledge and skills (Gess-Newsome, 2015) in PCK studies. However, while researchers generally connected their theoretical framework to the CM, a closer look at the articles suggested that the consensus thinking about PCK was not always adhered to in the studies. For example, a number of researchers equated PCK with the TSPK in the CM. Smit, Rietz, et al. (2017), and Smit, Weitzel, et al. (2017), for example, claimed that their PCK items targeted two important components of the CM (i.e., two components within TSPK). Likewise, Kind (2017) defined PCK as an amalgam of TSPK and content knowledge (CK) that is context-free and generalisable, different from the private and idiosyncratic nature of personal PCK in the CM.

The model of Magnusson et al. (1999) has proven to be a useful framework for analysing the PCK components investigated in the reviewed studies. Only a few PCK components investigated in the studies are not included in this model, including teachers' self-efficacy (Park & Oliver, 2008b), teachers' knowledge about the difficulty of tasks and content (Keller et al., 2017) and subject matter *for* teaching (van Dijk, 2009). Again, variations and deviations in the use of terminologies and meanings were apparent when describing the PCK components. Although many researchers adopted the Magnusson et al. (1999) model, they tended to use the *same* term for somewhat *different* ideas, resulting in a subtle shift in the meaning of the original definition. A case in point is the PCK component 'knowledge of curriculum', which Magnusson et al. (1999) originally conceptualised as comprising a teacher's knowledge of learning goals and instructional materials, the sequencing of instruction across particular topics (horizontal curricula) and knowledge of vertical curricula. Park and Oliver's (2008b) pentagon model aligned this PCK component with the notion of curricular saliency (Geddis et al., 1993). Friedrichsen et al. (2009) made a similar connection in defining the PCK components. At the same time, different researchers have sometimes used *different* terms for something approaching the *same* idea. For instance, although Mavhunga and Rollnick (2013) included a PCK component called curricular saliency in their PCK model, they did not connect it to the knowledge of curriculum component in the Magnusson et al. (1999) model.

As shown in Table 1.7, nearly all of the studies in the second group (86 out of 88 studies) investigated the PCK component, knowledge of students' understanding of science. Most studies (80 out of 86) studied science teachers' knowledge of instructional strategies and representations. These appear to be the two most commonly agreed-upon PCK components. About one-third (30 out of 88) of the studies included the affective attribute, orientations to teaching science, as part of their investigation. Nearly, all of these studies used the original PCK model proposed by Magnusson et al. (1999), or a version of it, as their theoretical backbone. It is also worth pointing out that nearly half (13 out of 31) of the studies that used this model ignored the orientation component. The majority of the studies ($N = 58$) did not include affective attributes as a part of PCK.

During analysis, it was found that an increasing number of studies have explored the *connections* between two or more PCK components. Recently, Park and her colleagues (Park & Chen, 2012; Suh & Park, 2017) developed the PCK mapping approach (see Chap. 9 *in this volume*) to investigate the integration between PCK components. A number of studies adopted this approach to investigate the PCK profiles of teachers (Aydin & Boz, 2013; Aydin et al., 2015). This focus on the integration between PCK components reinforces the consensus view that although PCK comprises discrete components, they are blended together and functioned synergistically when applied to solving problems in practice (Abell, 2008).

Summary Among the reviewed studies of science teachers' PCK, several prevalent views about how researchers conceptualised PCK can be identified. It seems that (1) most researchers viewed PCK as a *distinct* category of knowledge comprising various components; (2) the two most commonly agreed-upon and investigated PCK components were knowledge of students' understanding of science and knowledge of instructional strategies and representations that align with Shulman's original proposal; (3) most studies, except those adopting the Magnusson et al. (1999) model, did not consider teachers' affective attributes to be part of PCK; and (4) there was some degree of agreement about the importance of integration between PCK components.

Apart from areas of consensus, the review also highlighted issues of concern within the science PCK community. First, a significant number of studies did not explicitly define the PCK components under investigation. Second, a lack of shared terminologies about the PCK components and an inconsistent use of terms were apparent. Third, the connection to the CM was cosmetic rather than genuine, which may be an indication of the challenge of integrating the CM into extant lines of work.

Data Source(s) to Investigate PCK

Four major types of data sources for investigating science teachers' PCK were identified: (1) written tests, surveys and questionnaires; (2) interviews; (3) artefacts from teaching tasks; and (4) lesson observations. The types of data sources used in each of the reviewed PCK studies are listed in Table 1.8. About 37.4% ($N = 37$) of the studies used only a single data source, whereas nearly half ($N = 43$; 43.3%) used more than three data sources to investigate the teachers' PCK. Below, we provide examples of how these data sources were used.

(1) *Written tests/questionnaires/surveys.* These came in a variety of formats. For example, some consisted of a list of statements about teaching a particular science subject, with which teachers had to rate their level of agreement (Stasinakis & Athanasiou, 2016); some contained multiple-choice (MC) questions with a single answer (e.g., Stender et al., 2017); some combined true or false items, MC items, matching items and open-ended (short or long response) items (e.g., Sorge et al., 2017); and some contained open-ended, scenario-based items in

the form of teaching vignettes (e.g., Kind, 2017). There were also written surveys asking the participants to predict students' difficulties in learning particular science content matter (e.g., Zhou et al., 2016). These tests, questionnaires and surveys were administered on Web-based platforms (e.g., Lin, 2017) or as paper and pencil tests (i.e., time constrained) (e.g., Sorge et al., 2017).

(2) *Interviews.* Interviews were often semi-structured and carried out with individual teachers. Most researchers designed their interview protocols around the components of a PCK model (e.g., Henze et al., 2008) or the question prompts of the Content Representations (CoRes) (e.g., Walan et al., 2017). Teachers were also asked questions about a particular science lesson that they had enacted (e.g., Chan & Yung, 2017) or to describe their general 'best' science lesson in the interview (e.g., Luft et al., 2011). Alternative ways of conducting interviews included focus group interviews (e.g., Donnelly & Hume, 2015) and engaging novice interviewers such as student teachers in interviewing the participating teachers (e.g., Bindernagel & Eilks, 2009). In some studies, the interviewees were mentor teachers (e.g., Hallman-Thrasher et al., 2017) or students taught by the participating teacher(s) (e.g., Diezmann & Watters, 2015).

(3) *Artefacts from teaching tasks.* Artefacts were taken from different phases of the teaching cycle, i.e., the pre-active phase of teaching, such as lesson plans (e.g., Bergqvist et al., 2016); the interactive phase of teaching, such as teaching videos (e.g., Chan & Yung, 2015); and the post-active phase of teaching, such as students' work and teachers' written reflections on the enacted lessons (e.g., Park & Oliver, 2008a).

(4) *Lesson observations.* Teachers were observed live by the researcher(s). The lessons were often video-recorded for later analysis (e.g., Chan & Yung, 2017).

(5) *Other data sources.* These included *audio-recorded conversations between mentor and mentee* (Barnett & Friedrichsen, 2015); *audio-recorded methods' class activities and artefacts* (Mavhunga & Rollnick, 2016); *researchers' reflective journals* (Park & Oliver, 2008b); *student teachers' reflection papers* (Demirdöğen et al., 2016); *teachers' project reports* (Rollnick, 2017); *students' questionnaire responses* (Walan et al., 2017); *teacher guides* (Chen & Wei, 2015); *teacher workshop discussions* (Cohen & Yarden, 2009); and *textbooks* (Melo-Niño et al., 2017b). The purpose of these data sources often seemed to provide additional contextual information or triangulation. However, researchers rarely made explicit how the data sources were used or analysed to understand science teachers' PCK.

Summary Four major types of data sources for determining science teachers' PCK were identified. It is significant that a large number of studies ($N = 37$; 33.4%) used only a single type of data source to investigate the teachers' PCK, which allows investigation of limited aspects of a teacher's PCK. Another concern is that although researchers included various data sources (e.g., classroom artefacts, researchers' reflective journal) in the method section of their studies, they seldom made clear how these sources were used to determine teachers' PCK. Whether these data sources were only used to provide contextual information or for the purpose of triangulation in determining teachers' PCK was often not clearly explained in the articles.

Different Approaches for Determining Science Teachers' PCK

We distinguished two approaches for determining PCK. The first approach, in which teachers' PCK is revealed by teachers' *self-reports,* can be regarded as less authentic and more distant from real-life teaching contexts than the second approach, which investigates PCK via teachers' *performance* in teaching tasks. The two approaches also allow the investigation of different *forms* of PCK, i.e., what teachers know, what they do and their reasons for their judgements and action (see also Fig. 1.1).

Table 1.9 shows the approaches used in each of the studies. From the table, it can be discerned that about half ($N = 48$) adopted the self-report approach to determine teachers' PCK while about two-thirds ($N = 64$)[7] determined teachers' PCK according to their performance when carrying out teaching tasks.

Approach 1—Via Self-reports Three subgroups were identified: (1) PCK tests; (2) questionnaires, surveys and interviews; and (3) self-reports of actual practices. The following briefly reviews how PCK was determined in representative PCK studies.

(1) *PCK tests.* Teachers attempted a 'test', and their responses to the test were judged against 'correct' answers. Responses may be dichotomously scored (0 or 1 mark) or polytomously scored (0 mark, 1 mark for partially 'correct', 2 marks for 'fully correct') with respect to the 'correct' answers, as established by the researchers or experts (e.g., Rosenkränzer et al., 2017; Smit, Rietz, et al., 2017). Alternatively, the teachers' responses might be compared with those of their students (e.g., whether the student misconceptions predicted by the teachers matched with the misconceptions evident in students' interviews and tests) or with the existing literature on student misconceptions and learning difficulties as a point of reference (e.g., Schmelzing et al., 2013).
(2) *Questionnaires/surveys/interviews.* In most cases, teachers answered open-ended questions to elicit their PCK in written format (e.g., Kellner et al., 2011) or verbally in interviews (e.g., Salloum & BouJaoude, 2008). Teachers' PCK was then content-analysed to reveal its content and nature. Some studies quantified the qualitative data by first categorising the responses based on predetermined PCK components or categories that emerged from the data, and then documenting the frequencies of each category or component (Scharfenberg & Bogner, 2016). In some rare cases, teachers' self-perceived knowledge was elicited in questionnaires. Nilsson and Loughran (2012), for example, asked the participating student teachers to self-assess their own development of PCK components by asking them to rate their perceived importance of, and confidence in using, the CoRes prompts in shaping their thinking about teaching a science topic.

[7] The total number does not add up to 99 as some studies used both approaches.

(3) *Self-report of actual practices*. Teachers were asked to describe their *actual practices* in interviews or in writing. Henze et al. (2008), for instance, interviewed experienced science teachers and asked them to describe their practices of teaching the topic 'models of the solar system and the universe'. The interview transcripts were then content-analysed with respect to predetermined PCK components. A less common method was to ask teachers to self-assess and rate their level of agreement with a given statement about their teaching of a particular topic. In the study by Stasinakis and Athanasiou (2016), for instance, 181 Greek biology teachers completed a self-report questionnaire on how they taught the topic of evolution, using a Likert response scale.

Approach 2—Via Performance in Teaching Tasks Of the 64 studies that used this approach, 6 examined *only* the teaching artefacts/teacher actions, whereas 60 *also* examined the science teachers' instructional decisions[8]; 27 studies used simulated performance tasks, and 37 studied teachers' real-life teaching.[9] The following briefly reviews how PCK was determined in selected studies.

(1) Studies that *examined **only** the teaching artefacts/actions*

No study examined *only* teaching artefacts related to planning (i.e., lesson plan). All of the studies in this group examined *teacher actions*. Typically, the researchers analysed the teaching acts of the teachers using observation protocols (e.g., Marshall et al., 2016) or lesson transcripts (e.g., Oh & Kim, 2013).

(2) Studies that ***also** examined the teachers' teaching decisions and instructional reasoning*

(a) *Using simulated teaching tasks*

Most studies (18 studies) targeted the lesson planning stage, and only four investigated PCK related to the enactment phase of teaching. The following are some examples.

Pre-active phase:

(i) *Lesson preparation tasks* (e.g., Bergqvist & Chang Rundgren, 2017; Bergqvist et al., 2016; Friedrichsen et al., 2009; Käpylä et al., 2009; Kellner et al., 2011). In this method, teachers were asked to complete a science lesson plan and were then interviewed to understand the rationales underpinning their planning decisions. Alternatively, teachers were asked to complete a Content Representation (CoRe; for full explanation of a CoRe, see Loughran et al., 2004) as a means of indicating their lesson/unit plan (e.g., Adadan & Oner, 2014; Aydin et al., 2013). The lesson plan/CoRes as well as the interview transcripts were then content-analysed to reveal the teachers' PCK.

---

[8]The total number does not add up to 64 as two studies contained separate tasks that allowed the examination of the teaching artefacts/actions only in one task and the teachers' decisions and reasoning around the other teaching task.

[9]The total number does not add up to 60 as four studies used both simulated teaching tasks and investigated real life in situ.

Interactive phase:

(ii) *Video analysis task (videos from other teachers).* Roth et al. (2011) asked teachers to analyse videos of primary teachers teaching science who were unfamiliar to them and to identify the strategies used in the videos. Krepf et al. (2017) engaged both novice and experienced teachers in analysing problematic video cases. The videos used by researchers also varied in length in different studies, from several minutes (less than 1–5 min) (Alonzo & Kim, 2016) to longer video clips that captured a lesson (e.g., 15 min) (Krepf et al., 2017). Teachers' comments on the videos were further analysed to determine their PCK.

Post-active phase:

(iii) *Tasks involving researcher-created PCK elicitation probes.* Cohen and Yarden (2009) designed a tool to explore teachers' PCK based on the Magnusson et al. (1999) model. Teachers were asked to comment on illustrations representing how the topic 'cells' should be taught in the curriculum. The picture acted as a probe to elicit teachers' knowledge of the science curriculum. The teachers in the study were also asked to classify questions for the assessment of different grade levels and to identify unsuitable questions for the purpose of eliciting their knowledge of assessment. Teachers' responses to the tasks were examined to reveal their PCK.

(iv) *Students' performance prediction task.* Lucero et al. (2017) asked the biology teachers in their study to predict students' choices of particular MC items to probe their knowledge of the students. Teachers' predictions were compared with students' actual performance to determine the quality of teachers' PCK.

Whole teaching cycle:

(v) *Peer teaching sessions.* Rollnick (2017) analysed the PCK of the participants based on the lesson plan and audio-recorded the peer teaching sessions at the beginning of the course.

(b) *Investigating real-life teaching* in situ

Pre-active phase:

(i) *Planning interviews.* Chan and Yung (2015) used semi-structured interviews to investigate the teachers' planning decisions for the biology lessons they were to enact. The interview transcripts were analysed qualitatively to uncover the teachers' PCK.

(ii) *Using CoRes as a tool.* Walan et al. (2017) asked teachers to complete a CoRe to indicate their planning of the lessons that they would enact and conducted semi-structured interviews to understand their planning decisions. The CoRes and the interview transcripts were examined for teacher's PCK.

(iii) *Analysing planning conversations.* Barnett and Friedrichsen (2015), for example, captured the conversations between a student teacher and his mentor, including brief discussions between classes and more sustained discussions during planning and reflection for analysis of their PCK.

Interactive phase:

(iv) *Interactive video task (own videos).* Alonzo and Kim (2016) elicited physics teachers' reasoning in the moment by asking them to comment on short video clips selected from their own classrooms that possibly captured unexpected student thinking. The participants were forced to think on their feet and made instructional decisions in the moment *during* the interviews.

Post-active phase:

(v) *Written video analysis task (own videos).* Kanter and Konstantopoulos (2010) asked teachers to analyse their own teaching video clips by constructing episodes that documented what their students said, did and wrote in class, students' thoughts about the science concepts and their follow-up responses. The written responses were then graded using a rubric to determine the quality of teachers' PCK.

(vi) *Stimulated recall interviews using videos.* Researchers such as Brown et al. (2013) and Chan and Yung (2015) used carefully selected video excerpts during the interviews to stimulate teachers' recall and probe into their pedagogical reasoning or decisions underlying their teaching actions in the biology lessons they enacted. Nilsson and van Driel (2010) presented an alternative approach to stimulated recall interviews, which they referred to as the critical incident approach. They asked the participating teachers to pause the video and make comments when they saw 'special events' in the videos that caused them to reflect on their teaching. The PCK embedded in the teachers' responses was further analysed.

(vii) *Writing reflection on enacted lessons* (e.g., reflective journal). Several researchers analysed the knowledge embedded in science teachers' reflections on their *own* teaching practices (e.g., Park & Oliver, 2008a; Wongsopawiro et al., 2017).

Several emerging issues related to the wide variety of ways that researchers use to determine teachers' PCK warrant further discussion.

First, to interpret a study's findings it is important to take into account the approach taken to determine teachers' PCK as this decision reflects the different *form(s)* of PCK under investigation. Shulman (2015) recently reminded researchers in the field that PCK is not *only* propositional in nature, but also a type of strategic knowledge (i.e., knowledge of *when*, *where* and *how* knowledge applies). He further commented that PCK 'was not to be construed as "something" that teachers had in their heads but was a more dynamic construct that described the processes that teachers employed when confronted with the challenge of teaching particular subjects to particular learners in particular settings' (p. 9). Arguably, a teacher's PCK determined through written surveys and tests or reported verbally (i.e., knowledge, mainly declarative in nature) (*Approach 1*) may not be the same as the knowledge that the teacher *uses* for carrying out teaching tasks, such as planning, teaching in the classroom and reflection (i.e., enacted knowledge) (*Approach 2*). Given this distinction, it is of concern that few of the reviewed articles made explicit the *forms* of knowledge that they were investigating. This lack of *explicit* reference to different forms of PCK under investigation makes it virtually impossible to compare the findings of different studies in science education research.

Second, very few studies actually distinguished between the PCK related to different phases of the teaching cycle (i.e., pre-active, interactive and post-active phases of teaching) (*Approach 2*). It is clear that PCK that informs a teacher's planning decisions may not necessarily be translated into the teacher's *teaching acts* in a particular classroom. While many studies ($N = 28$) investigated a teacher's PCK associated with the *entire* teaching cycle in real-life settings, they seldom made explicit the PCK associated with each phase of the teaching cycle (see Chan and Yung, 2015, for an exception). In most cases, the researchers generated a single PCK profile for the teachers, assuming that the PCK associated with different phases of the teaching cycle was essentially the same. Moreover, as shown in Table 1.9, most of the studies (18 out of 27) that made use of simulated teaching tasks focused on the lesson planning phase, rather than the interactive phase of teaching.

Among the studies that focused on the interactive phase of teaching, another emerging issue was the diversity around how *teacher actions* in the classroom were analysed to determine teachers' PCK. Some researchers focused on the *strategies* the teachers enacted in the lessons. For example, Roth et al. (2011) analysed how teachers made use of student thinking strategies and science content storyline strategies to determine their PCK. Others focused on teachers' *specific actions* during teacher–student interactions. Lucero et al. (2017), for example, examined whether and how teachers responded to students' alternative conceptions 'on the fly' once they became apparent in real-time classroom interactions. Others constructed lesson observation protocols, but did not make explicit how each of the observed actions is related to a particular PCK component or components (e.g., Marshall et al., 2016). The pedagogical moves related to a particular PCK component or components seemed to be very different for different researchers. More problematically, some articles did not provide sufficient details on how they sought evidence of *teacher actions* in their coding as evidence for teachers' PCK. It is quite striking to find that only four PCK studies (Barendsen & Henze, 2017; Gess-Newsome et al., 2017; Lucero et al., 2017; Roth et al., 2011) designed specific instruments to capture *both* variants of PCK *separately*, i.e., what a teacher knows and the PCK manifested in the teacher's actions in the classroom. Interestingly, we noted that some studies (e.g., Förtsch et al., 2016; Luft, 2009) used the PCK lens to investigate the science teachers' *knowledge* (i.e., what the teachers knew), but did not analyse how teachers' PCK was translated into their teaching actions (i.e., what the teachers did). These researchers chose to use other theoretical constructs, such as cognitive activation (Förtsch et al., 2016) and lesson observation protocols (Luft, 2009), to characterise the teachers' teaching practices. These decisions may reflect that what PCK looks like 'in practice' (i.e., teaching acts) remains underspecified in the science education field.

Finally, and disconcertingly, all of the existing studies that involved a large sample size (i.e., >50) (Appendix, Tables 1.8 and 1.9) relied on a single PCK test or a single interview to determine teachers' PCK (i.e., *Approach 1*). Clearly, this approach can only measure part of a teacher's PCK. There is no guarantee that the knowledge represented in these tests and interviews will be *applied* to teaching tasks or *accessed* in the *teaching practices* in the classroom. Yet, there is a lack of a valid instrument for the large-scale assessment of teachers' ability to *apply* their knowledge in real classroom situations. In this regard, we can learn from our mathematics education counterparts, where various video-based instruments have been validated and used to investigate the nature and structure of teachers' professional knowledge on a large scale (e.g., Kersting, 2008; Kersting et al., 2016; Knievel, Lindmeier, & Heinze, 2015; König et al., 2014). Video clips taken from authentic classrooms are used in video-based instruments to roughly approximate real classroom situations. Video-based instruments, hence, allow researchers to investigate how teachers *apply* their knowledge in a context that is closer to the real classroom situations and actual teaching performance than paper and pencil tests. They are also less labour-intensive to administer than classroom observations. More initiatives are needed to develop similar instruments for investigating science teachers' PCK that are appropriate for large-scale studies.

Summary To summarise this section, we identified diverse ways of determining science teachers' PCK and categorised them into groups that reflect the *forms* of PCK (i.e., knowledge, enacted knowledge, knowledge in practice, pedagogical reasoning) and the phases of the teaching cycle under investigation. Several issues emerged from this analysis. First, few studies explicitly stated the forms of PCK that were investigated. Second, seldom did the studies clearly state/identify the PCK related to each phase of the teaching cycle. Third, researchers seemed to have different ideas about what PCK looks like in practice (i.e., teaching acts during the interactive phase of teaching). Finally, it became apparent that most large-scale studies relied on self-reports to determine science teachers' PCK, thus exposing a lack of valid instruments that would enable large-scale investigations of other forms of science teachers' PCK more closely related to classroom situations.

Conclusion

This chapter reviewed 99 published articles to provide an overview of the PCK research in the field of science education. Despite some limitations in the scope of the review, including the use of peer-reviewed journals in English, the timespan and the focus on individual science teachers' PCK, it nonetheless offers comprehensive coverage of 99 articles (involving more than 200 PCK researchers) published in major journals in the field of science and teacher education. Thus, it seems reasonable to assert that we have been able to capture how the PCK concept is currently being used, interpreted and investigated in science education communities.

The review synthesised and organised the studies into groups related to their research context, the major purposes of the research, the conceptualisation of PCK, the types of data sources used in the investigation and the approaches for determining teachers' PCK. In so doing, this chapter contributes to the field in two major ways. First, the analysis provides evidence that the CM that emerged from the first PCK Summit in 2012 captured some points of convergence in thinking around the PCK concept within the field. However, it also revealed some points of divergence in thinking around the PCK concept that were not fully addressed in the CM. Second, our new ways of organising and clustering the studies based on their approaches to investigating PCK not only make explicit the various methods for investigating science teacher's PCK, but also reveal how the different approaches allow the investigation of different forms of PCK (see Fig. 1.1). Collectively, the exercise brought to light several issues and knowledge gaps within the current literature related to the investigation of science teachers' PCK. Several noteworthy findings are highlighted as follows.

Issues concerning the conceptualisation and operationalisation of PCK:

- Although all of the reviewed studies concerned PCK, the scope of the focal science subject matter and how a 'topic' was defined varied considerably across different studies.
- The diverse interpretations of the PCK concept led to the development of diverse and often idiosyncratic terminology to describe PCK/PCK components. New terms were introduced, and the meaning of existing terms shifted.
- The operationalisation of PCK components remained unclear in many of the PCK studies.
- Although some studies made an explicit connection between their theoretical framework and the CM, the connection was often cosmetic and superficial rather than genuine and detailed, which may be an indication of the challenge of integrating the CM into extant lines of work.
- The categorisation of teachers' pedagogical moves and teaching acts as a representation of science teachers' PCK in action varied greatly between research groups.

Issues concerning the investigation of PCK:

- A considerable number of studies relied on a single type of data source to investigate PCK, which allows investigation of limited aspects of a science teacher's PCK.
- Studies seldom made explicit the *forms* of PCK being investigated. This omission makes the comparison between findings from different studies difficult. It speaks to the potential of a consensus model that can better unpack the content and nature of PCK, including its different forms and variants.
- Studies often generated a single profile of a science teacher's PCK rather than providing a finer distinction between the PCK associated with different phases of the teaching cycle.

Knowledge gaps within the science education field:

- Large-scale comparative studies on science teachers' PCK across countries are yet to come.
- Longitudinal studies of PCK development are lacking in the science education field.
- Valid measurement instruments that would allow large-scale investigations of how science teachers apply their PCK in authentic classroom contexts and situations are lacking.
- Most studies focused on the planning phase of the teaching cycle, whereas very few focused on the interactive phase of teaching that is most directly related to students' learning experiences in science.

This review clearly shows that science PCK researchers conceptualise and operationalise PCK differently. The diverse interpretations of the PCK concept lead to the development of increased (often idiosyncratic) terminology to describe PCK and multiple, non-aligned approaches of investigating PCK in science education. As some points of divergence in thinking around the PCK concept were not fully addressed in the CM that emerged from the PCK Summit in 2012 (e.g., the PCK components, the relationship between different forms of PCK), our review draws attention to the need to further refine and update the CM (see Chap. 2). Although Shulman (2015) pointed out that plural perspectives on PCK can be productive to the general field of PCK research, we believe that in future studies, researchers should at least make explicit their conceptualisation of PCK, the forms of PCK they investigate, the parts of the teaching cycle they focus on and the data sources and approaches they use to collect the data. Not until this explicit sharing occurs can PCK researchers build upon their findings to generate a common understanding of research agendas and outcomes.

Acknowledgements This research was supported by the Early Career Scheme of the Research Grants Council of Hong Kong [Project Number 27608717].

Appendix

An overview of the studies reviewed. Note all reviewed articles can be found in the reference list for this chapter.

	Study	Country/countries	Number of participant(s)	Pre-service teachers	In-service teachers	Grade level	Subject domain(s)	Subject matter/topic(s)
1	Adadan and Oner (2014)	Turkey	2	X		Secondary	Chemistry	Behaviour of gases
2	Akin and Uzuntiryaki-Kondakçı (2018)	Turkey	3		X (3–20 years)	Secondary	Chemistry	(1) Rate of reaction (2) Chemical equilibrium
3	Alonzo and Kim (2016)	USA	6		X (5–26 years)	Secondary	Physics	Force and motion
4	Alonzo et al. (2012)	Germany	2		X (3–4 years)	Secondary	Physics	Optics
5	Alvarado et al. (2015)	Mexico	10		X (8–39 years)	Secondary	Chemistry	Acid–base
6	Aydin and Boz (2013)	Turkey	2		X (8–15 years)	Secondary	Chemistry	(1) Redox reactions (2) Electrochemical cells
7	Aydin et al. (2015)	Turkey	3	X		Secondary	Chemistry	Rate of reaction
8	Aydin et al. (2013)		3	X		Secondary	Chemistry	Rate of reaction

(continued)

(continued)

	Study	Country/countries	Number of participant(s)	Pre-service teachers	In-service teachers	Grade level	Subject domain(s)	Subject matter/topic(s)
9	Aydin et al. (2014)	Turkey	2		X (8–15 years)	Secondary	Chemistry	(1) Nuclear reactions (2) Electrochemical cells
10	Barendsen and Henze (2017)	The Netherlands	1		X (almost 20 years)	Secondary	Chemistry	(Poly) lactic acid
11	Barnett and Friedrichsen (2015)	USA	2	X	X (25 years)	Secondary	Biology	(1) DNA/protein synthesis (2) Evolution
12	Bektas et al. (2013)	Turkey	7	X		Secondary	Chemistry	Nature of science (Particle nature of matter)
13	Bergqvist and Chang Rundgren (2017)	Sweden	10		X (3–36 years)	Secondary	Chemistry	Chemical bonding
14	Bergqvist et al. (2016)	Sweden	10		X (3–36 years)	Secondary	Chemistry	Chemical bonding
15	Bindernagel and Eilks (2009)	Germany	28		X (1 to >10 years)	Secondary	Chemistry	Particulate nature of matter
16	Boesdorfer and Lorsbach (2014)	USA	1		X (10 years)	Secondary	Chemistry	Periodic table
17	Bravo and Cofré (2016)	Chile	2		X (4, 10 years)	Secondary	Biology	Human evolution

(continued)

(continued)

	Study	Country/countries	Number of participant(s)	Pre-service teachers	In-service teachers	Grade level	Subject domain(s)	Subject matter/topic(s)
18	Brown et al. (2013)	USA	4	X		Secondary	Biology	*Lesson planning task:* Heritable variation *Interview–observation cycles:* Not specified
19	Burton (2013)	USA	17		X (3–43 years)	Secondary	Science	Nature of science
20	Chan and Yung (2015)	Hong Kong	4		X (6–23 years)	Secondary	Biology	Polymerase chain reaction
21	Chan and Yung (2017)	Hong Kong	2		X (6–14 years)	Secondary	Biology	Polymerase chain reaction
22	Chen and Wei (2015)	China	5		X (12–25 years)	Secondary	Chemistry	Multiple
23	Cohen and Yarden (2009)	Israel	Focus group: 59 Workshop: 12 Interview: 6		X (2–24 years)	Secondary	Science and non-science	Cell
24	Davidowitz and Potgieter (2016)	South Africa	89		X (1–45 years)	Secondary	Chemistry	Organic chemistry
25	Demirdöğen (2016)	Turkey	8	X		Primary and secondary (Middle)	Science	Not specified

(continued)

(continued)

	Study	Country/countries	Number of participant(s)	Pre-service teachers	In-service teachers	Grade level	Subject domain(s)	Subject matter/topic(s)
26	Demirdöğen et al. (2016)	Turkey	30	X		Secondary	Chemistry	Nature of science
27	Demirdöğen and Uzuntiryaki-Kondakçı (2016)	Turkey	30	X		Secondary	Chemistry	Nature of science
28	Diezmann and Watters (2015)		1		X (1 year)	Secondary	Biology	Microscopy
29	Donnelly and Hume (2015)	New Zealand	7	X		Secondary	Chemistry	Redox reaction
30	Findlay and Bryce (2012)	Scotland	Pre-service: 15 In-service: 6	X	X (First 3.5 years)	Secondary	Physics	Electricity
31	Förtsch et al. (2016)	Germany	39		X ($M = 6.1$ years)	Secondary	Biology	Neurobiology
32	Friedrichsen et al. (2009)	USA	4	X	X (2 years)	Secondary	Biology	Heritable variation
33	Gess-Newsome et al. (2017)	USA	35		X (1–20 years) (M = 7.4 years)	Secondary	Biology	Multiple
34	Großschedl et al. (2015)	Germany	274	X		Secondary	Biology	Multiple

(continued)

(continued)

	Study	Country/countries	Number of participant(s)	Pre-service teachers	In-service teachers	Grade level	Subject domain(s)	Subject matter/topic(s)
35	Großschedl et al. (2014)	Germany	134		X (7 months–42 years) (M = 16.4 years)	Secondary	Biology	Ecosystem (the Wadden Sea), focusing on the species *Mytilus edulis*
36	Hallman-Thrasher et al. (2017)	USA	6	X	X (First year of teaching)	Secondary	Science and non-science	Not specified
37	Hanuscin (2013)	USA	1	X		Primary	/	Nature of science
38	Hanuscin et al. (2011)	USA	3		X (2–29 years)	Primary	/	Nature of science
39	Heller et al. (2012)	USA	270		X (1–43 years)	Primary	/	Electric circuit
40	Henze et al. (2008)	The Netherlands	9		X (8–26 years)	Secondary	Science	Models of the solar system and the universe
41	Jin et al. (2015)	USA	194		X (3–40 years)	Primary and secondary	Science	Plant growth and functioning
42	Jüttner and Neuhaus (2012)	Germany	PCK test: 65 Interview: 5		X (M = 10.5 years) Interview: (M = 13.4 years)	Secondary	Biology	Human reflex arc
43	Kanter and Konstantopoulos (2010)	USA	9		X (0–13 years) (M = 6.6 years)	Primary and secondary (Middle)	Science	Middle school problem-based science curriculum topics—(1) Calorimetry (2) Body system

(continued)

(continued)

	Study	Country/countries	Number of participant(s)	Pre-service teachers	In-service teachers	Grade level	Subject domain(s)	Subject matter/topic(s)
44	Käpylä et al. (2009)	Finland	20	X		Primary and secondary	Science	Photosynthesis
45	Kaya (2009)	Turkey	75	X		Primary and secondary (Middle)	Science	Ozone
46	Keller et al. (2017)	Germany and Switzerland	77		X (1–41 years)	Secondary	Physics	Electricity
47	Kellner et al. (2011)	Sweden	32	X		Primary and secondary	Science and non-science	Multiple
48	Kind (2017)	UK	239	X		Secondary	Science	Multiple
49	Kirschner et al. (2016)	Germany	186 in-service teachers 79 pre-service teachers 9 physicists 21 non-physics teachers	X	X (not specified)	Secondary	Physics	Mechanics (PCK items also cover physics in general)
50	Krepf et al. (2017)	Germany	18		X (Experts > 15 years; novices < 15 h)	Secondary	Science and non-science	Optics: Snell's law
51	Lin (2016)	Taiwan	76		X ($M = 9.5$ years)	Primary	/	Electric circuit
52	Lin (2017)	Taiwan	171	X	X (not specified)	Primary	/	Electric circuit

(continued)

(continued)

	Study	Country/countries	Number of participant(s)	Pre-service teachers	In-service teachers	Grade level	Subject domain(s)	Subject matter/topic(s)
53	Lin and Chiu (2010)	Taiwan	1		X (6 years)	Secondary	Chemistry	Acids–bases
54	Lucero et al. (2017)	USA	4		X (1 semester–7 years)	Secondary	Biology	Evolution by natural selection
55	Luft (2009)	USA	114		X (1 year)	Secondary	Science	Any topic
56	Luft et al. (2011)	USA	98		X (1–2 years)	Secondary	Science	Any topic
57	Mahler et al. (2017)	Germany	48		X (1–20 years) ($M = 7.4$)	Secondary	Biology	The ecosystem 'Wadden Sea' and the morphology and life of *Mytilus edulis*
58	Marshall et al. (2016)	USA	37		X (not specified)	Primary and secondary	Science and non-science	Not specified
59	Mavhunga (2016)	South Africa	36	X		Secondary	Chemistry	(1) Particular nature of matter (2) Chemical equilibrium
60	Mavhunga and Rollnick (2016)	South Africa	16	X		Secondary	Chemistry	Chemical equilibrium
61	Melo-Niño et al. (2017a)	Cambodia	2		X (7–8 years)	Secondary	Physics	Electric field

(continued)

(continued)

	Study	Country/countries	Number of participant(s)	Pre-service teachers	In-service teachers	Grade level	Subject domain(s)	Subject matter/topic(s)
62	Melo-Niño et al. (2017b)	Cambodia	4		X (4–9 years)	Secondary	Physics	Electric field
63	Meschede et al. (2017)	Germany	223	X	X	Primary	/	(1) Water cycle (2) Floating and sinking
64	Monet and Etkina (2008)	USA	10		X	Primary & Secondary (Middle)	Science	Topics in Full Option Science Systems Earth History science kit
65	Moodley and Gaigher (2017)	South Africa	6		X (6–13 years)	Secondary	Science	Electricity
66	Mthethwa-Kunene et al. (2015)	Swaziland	4		X (5–22 years)	Secondary	Biology	Genetics
67	Nilsson and Loughran (2012)	Sweden	34	X		Primary	/	Air
68	Nilsson and van Driel (2010)	Sweden	4	X	X (>20 years)	Primary	/	Space and universe Water
69	Nilsson and Vikström (2015)	Sweden	6		X (Several years)	Secondary	Science	Multiple
70	Oh and Kim (2013)	Korea	5		X (6–13 years)	Primary	/	Multiple

(continued)

(continued)

	Study	Country/countries	Number of participant(s)	Pre-service teachers	In-service teachers	Grade level	Subject domain(s)	Subject matter/topic(s)
71	Park and Chen (2012)	USA	4		X (2–43 years)	Secondary	Biology	(1) Heredity (2) Photosynthesis
72	Park et al. (2011)	USA	7		X (2–43 years)	Secondary	Biology	(1) Heredity (2) Photosynthesis
73	Park and Oliver (2008a)	USA	3		X (8–21 years)	Secondary	Chemistry	Not specified
74	Park and Oliver (2008b)	USA	3		X (8–21 years)	Secondary	Chemistry	Not specified
75	Park et al. (2017)	USA	Survey: 85 Observation: 7 Interview: 6		X (Not specified)	Secondary	Biology	Photosynthesis
76	Paulick et al. (2016)	Germany	631	X		Secondary	Biology and physics	Multiple
77	Pilliouras et al. (2017)	Greece	4		X (10–29 years)	Primary	–	Nature of science
78	Qhobela and Kolitsoe Moru (2014)	Lesotho	Questionnaire: 39 Interview: 5		X (0->10 years)	Secondary	Physics	Not specified
79	Rollnick (2017)	South Africa	7		X (4–20 years)	Secondary	Science	Semiconductor

(continued)

(continued)

	Study	Country/countries	Number of participant(s)	Pre-service teachers	In-service teachers	Grade level	Subject domain(s)	Subject matter/topic(s)
80	Rollnick et al. (2008)	South Africa	2 in-service teachers 1 lecturer		X (5–10 years)	Secondary	Chemistry	Amount of substances (mole)
81	Rosenkränzer et al. (2017)	Germany	108	X		Secondary	Biology and geography	Multiple (System thinking)
82	Roth et al. (2011)	USA	48		X (>9 years)	Primary	/	Multiple
83	Salloum and BouJaoude (2008)	Lebanon	9		X (3–25 years)	Primary and secondary (Middle)	Chemistry	Ideas about chemicals
84	Scharfenberg and Bogner (2016)	Germany	92	X		Secondary	Biology	Genetics fingerprinting
85	Schmelzing et al. (2013)	Germany	93 pre-service/in-service teachers 12 biologists	X	X ($M = 12$ years)	Secondary	Biology	Blood and the human cardiovascular system
86	Sickel and Friedrichsen (2017)	USA	3		X (First 1–2 years)	Secondary	Biology	Natural selection
87	Smit, Weitzel, et al. (2017)	Germany and Switzerland	121	X		Secondary	Biology	Eye/visual perception
88	Smit, Rietz, et al. (2017)	Germany and Switzerland	118	X		Secondary	Biology	Accommodation and adaptation of the eye

(continued)

(continued)

	Study	Country/countries	Number of participant(s)	Pre-service teachers	In-service teachers	Grade level	Subject domain(s)	Subject matter/topic(s)
89	Sorge et al. (2017)	Germany	200	X		Secondary	Physics	Multiple
90	Stasinakis and Athanasiou (2016)	Greece	181		X (~56% teaches for 1–10 years)	Secondary	Biology	Evolution
91	Stender et al. (2017)	Germany	49		X (M = 13.5 years)	Secondary	Physics	Newton's law
92	Suh and Park (2017)	USA	3		X (11–30 years)	Primary	/	Force and motion
93	Supprakob et al. (2016)	Thailand	6		X (1–5 years)	Secondary	Chemistry	Nature of science
94	Tay and Yeo (2017)	Singapore	1		X (8 years)	Secondary	Physics	Newton's third law of motion
95	Uzuntiryaki-Kondakçı et al. (2017)	Turkey	5	X		Secondary	Chemistry	Gas law
96	van Dijk (2009)	Germany	9		X (6–34 years)	Secondary	Biology	Evolutionary theory
97	Walan et al. (2017)	Sweden	2		X (Several years of experience)	Primary	/	Water
98	Wongsopawiro et al. (2017)	The Netherlands	12		X (2–28 years)	Primary and secondary	Science	Multiple
99	Zhou et al. (2016)	China	143	X		Secondary	Physics	Newton's third law

Notes

1. The number of years of teaching experience (in general) of in-service teachers is given in parentheses. '*M*' denotes the mean number of teaching years of the teachers in the study.

2. Grade level refers to the grades that the teachers taught or for which they were certified. Teachers from grade 1 to 6 were labelled as primary teachers, whereas those from grade 7 to 12 were categorised as secondary teachers. Studies that involved middle school teachers were categorised as investigating both primary and secondary teachers. If the studies specified that the target participants were middle school teachers, middle school was also included in the above table.

3. If teachers from more than two subject domains were included, the subject was listed as 'science' or 'science and non-science'.

4. The subject domain of primary teachers is not shown as primary teachers are often generalists rather than subject specialists.

5. If more than two science topics were investigated in the study, the topics were labelled as 'multiple'.

6. If the topics under investigation were not clearly listed, the topic was categorised as 'not specified'.

7. Only teachers, not scientists, were included in the determination of the sample size in Kirschner et al. (2016) and Schmelzing et al. (2013).

8. As Salloum and BouJaoude (2008) adopted purposeful sampling of chemistry teachers, the subject domain was considered as 'chemistry'.

References

Abbitt, J. T. (2011). Measuring technological pedagogical content knowledge in preservice teacher education. *Journal of Research on Technology in Education, 43*(4), 281–300.

Abd-El-Khalick, F. (2006). Pre-service and experienced biology teachers' global and specific subject matter structures: Implications for conceptions of pedagogical content knowledge. *EURASIA Journal of Mathematics, Science & Technology Education, 2*(1), 1–29.

Abell, S. K. (2007). Research on science teacher knowledge. In S. K. Abell & N. Lederman (Eds.), *Handbook of research on science education* (pp. 1105–1149). Mahwah, NJ: Lawrence Erlbaum Associates.

Abell, S. K. (2008). Twenty years later: Does pedagogical content knowledge remain a useful idea? *International Journal of Science Education, 30*(10), 1405–1416.

Abell, S. K., Rogers, M. A. P., Hanuscin, D. L., Lee, M. H., & Gagnon, M. J. (2008). Preparing the next generation of science teacher educators: A model for developing PCK for teaching science teachers. *Journal of Science Teacher Education, 20*(1), 77–93.

Adadan, E., & Oner, D. (2014). Exploring the progression in preservice chemistry teachers' pedagogical content knowledge representations: The case of "Behavior of gases". *Research in Science Education, 44*(6), 829–858.

Akerson, V. L., Pongsanon, K., Park Rogers, M. A., Carter, I., & Galindo, E. (2017). Exploring the use of lesson study to develop elementary preservice teachers' pedagogical content knowledge for teaching nature of science. *International Journal of Science and Mathematics Education, 15*(2), 293–312.

Akin, F. N., & Uzuntiryaki-Kondakçı, E. (2018). The nature of the interplay among components of pedagogical content knowledge in reaction rate and chemical equilibrium topics of novice and experienced chemistry teachers. *Chemistry Education Research and Practice*.

Alonzo, A. C., & Kim, J. (2016). Declarative and dynamic pedagogical content knowledge as elicited through two video-based interview methods. *Journal of Research in Science Teaching, 53*(8), 1259–1286.

Alonzo, A. C., Kobarg, M., & Seidel, T. (2012). Pedagogical content knowledge as reflected in teacher–student interactions: Analysis of two video cases. *Journal of Research in Science Teaching, 49*(10), 1211–1239.

Alvarado, C., Cañada, F., Garritz, A., & Mellado, V. (2015). Canonical pedagogical content knowledge by CoRes for teaching acid–base chemistry at high school. *Chemistry Education Research and Practice, 16*(3), 603–618.

An, S., Kulm, G., & Wu, Z. (2004). The pedagogical content knowledge of middle school, mathematics teachers in China and the U.S. *Journal of Mathematics Teacher Education, 7*(2), 145–172.

Avraamidou, L., & Zembal-Saul, C. (2005). Giving priority to evidence in science teaching: A first-year elementary teacher's specialized practices and knowledge. *Journal of Research in Science Teaching, 42*(9), 965–986.

Aydin, S., & Boz, Y. (2013). The nature of integration among PCK components: A case study of two experienced chemistry teachers. *Chemistry Education Research and Practice, 14*(4), 615–624.

Aydin, S., Demirdöğen, B., Akin, F. N., Uzuntiryaki-Kondakçı, E., & Tarkin, A. (2015). The nature and development of interaction among components of pedagogical content knowledge in practicum. *Teaching and Teacher Education, 46*, 37–50.

Aydin, S., Demirdöğen, B., Tarkin, A., Kutucu, E. S., Ekiz, B., Akin, F. N., et al. (2013). Providing a set of research-based practices to support preservice teachers' long-term professional development as learners of science teaching. *Science Education, 97*(6), 903–935.

Aydin, S., Friedrichsen, P. J., Boz, Y., & Hanuscin, D. L. (2014). Examination of the topic-specific nature of pedagogical content knowledge in teaching electrochemical cells and nuclear reactions. *Chemistry Education Research and Practice, 15*(4), 658–674.

Barendsen, E., & Henze, I. (2017). Relating teacher PCK and teacher practice using classroom observation. *Research in Science Education*.

Barnett, E., & Friedrichsen, P. J. (2015). Educative mentoring: How a mentor supported a preservice biology teacher's pedagogical content knowledge development. *Journal of Science Teacher Education, 26*(7), 647–668.

Baxter, J. A., & Lederman, N. G. (1999). Assessment and measurement of pedagogical content knowledge. In J. Gess-Newsome & N. G. Lederman (Eds.), *Examining pedagogical content knowledge: The construct and its implications for science education* (pp. 147–161). Dordrecht: Kluwer Academic.

Bektas, O., Ekiz, B., Tuysuz, M., Kutucu, E. S., Tarkin, A., & Uzuntiryaki-Kondakçı, E. (2013). Pre-service chemistry teachers' pedagogical content knowledge of the nature of science in the particle nature of matter. *Chemistry Education Research and Practice, 14*(2), 201–213.

Bennett, J., Lubben, F., & Hogarth, S. (2007). Bringing science to life: A synthesis of the research evidence on the effects of context-based and STS approaches to science teaching. *Science Education, 91*(3), 347–370.

Bergqvist, A., & Chang Rundgren, S.-N. (2017). The influence of textbooks on teachers' knowledge of chemical bonding representations relative to students' difficulties understanding. *Research in Science & Technological Education, 35*(2), 215–237.

Bergqvist, A., Drechsler, M., & Chang Rundgren, S.-N. (2016). Upper secondary teachers' knowledge for teaching chemical bonding models. *International Journal of Science Education, 38*(2), 298–318.

Berry, A., Depaepe, F., & van Driel, J. H. (2016). Pedagogical content knowledge in teacher education. In J. Loughran & M. L. Hamilton (Eds.), *International handbook of teacher education* (Vol. 1, pp. 347–386). Singapore: Springer Singapore.

Berry, A., Friedrichsen, P. J., & Loughran, J. (2015). *Re-examining pedagogical content knowledge in science education*. New York: Routledge.

Bindernagel, J. A., & Eilks, I. (2009). Evaluating roadmaps to portray and develop chemistry teachers' PCK about curricular structures concerning sub-microscopic models. *Chemistry Education Research and Practice, 10*(2), 77–85.

Blömeke, S., Suhl, U., & Kaiser, G. (2011). Teacher education effectiveness: Quality and equity of future primary teachers' mathematics and mathematics pedagogical content knowledge. *Journal of Teacher Education, 62*(2), 154–171.

Boesdorfer, S., & Lorsbach, A. (2014). PCK in action: Examining one chemistry teacher's practice through the lens of her orientation toward science teaching. *International Journal of Science Education, 36*(13), 2111–2132.

Bravo, P., & Cofré, H. (2016). Developing biology teachers' pedagogical content knowledge through learning study: The case of teaching human evolution. *International Journal of Science Education, 38*(16), 2500–2527.

Brown, P., Friedrichsen, P. J., & Abell, S. K. (2013). The development of prospective secondary biology teachers PCK. *Journal of Science Teacher Education, 24*(1), 133–155.

Burton, E. P. (2013). Student work products as a teaching tool for nature of science pedagogical knowledge: A professional development project with in-service secondary science teachers. *Teaching and Teacher Education, 29,* 156–166.

Cetin-Dindar, A., Boz, Y., Yildiran Sonmez, D., & Demirci Celep, N. (2018). Development of pre-service chemistry teachers' technological pedagogical content knowledge. *Chemistry Education Research and Practice*.

Chan, K. K. H., & Yung, B. H. W. (2015). On-site pedagogical content knowledge development. *International Journal of Science Education, 37*(8), 1246–1278.

Chan, K. K. H., & Yung, B. H. W. (2017). Developing pedagogical content knowledge for teaching a new topic: More than teaching experience and subject matter knowledge. *Research in Science Education*, 1–33.

Charalambous, C. Y. (2016). Investigating the knowledge needed for teaching mathematics. *Journal of Teacher Education, 67*(3), 220–237.

Chen, B., & Wei, B. (2015). Examining chemistry teachers' use of curriculum materials: in view of teachers' pedagogical content knowledge. *Chemistry Education Research and Practice, 16*(2), 260–272.

Cochran, K. F., DeRuiter, J. A., & King, R. A. (1993). Pedagogical content knowing: An integrative model for teacher preparation. *Journal of Teacher Education, 44*(4), 263–272.

Cochran, K. F., & Jones, L. L. (1998). The subject matter knowledge of preservice science teachers. In B. J. Fraser & K. G. Tobin (Eds.), *International handbook of science education* (Vol. 2, pp. 707–718). Dordrecht: Kluwer.

Cohen, R., & Yarden, A. (2009). Experienced junior-high-school teachers' PCK in light of a curriculum change: "The cell is to be studied longitudinally". *Research in Science Education, 39*(1), 131–155.

Cook, S. D. N., & Brown, J. S. (1999). Bridging epistemologies: The generative dance between organizational knowledge and organizational knowing. *Organization Science, 10*(4), 381–400.

Daehler, K. R., & Shinohara, M. (2001). A complete circuit is a complete circle: Exploring the potential of case materials and methods to develop teachers' content knowledge and pedagogical content knowledge of science. *Research in Science Education, 31*(2), 267–288.

Davidowitz, B., & Potgieter, M. (2016). Use of the Rasch measurement model to explore the relationship between content knowledge and topic-specific pedagogical content knowledge for organic chemistry. *International Journal of Science Education, 38*(9), 1483–1503.

Davis, E., & Krajcik, J. (2005). Designing educative curriculum materials to promote teacher learning. *Educational Researcher, 34*(3), 3–14.

Demirdöğen, B. (2016). Interaction between science teaching orientation and pedagogical content knowledge components. *Journal of Science Teacher Education, 27*(5), 495–532.

Demirdöğen, B., Hanuscin, D. L., Uzuntiryaki-Kondakçı, E., & Köseoğlu, F. (2016). Development and nature of preservice chemistry teachers' pedagogical content knowledge for nature of science. *Research in Science Education, 46*(6), 831–855.

Demirdöğen, B., & Uzuntiryaki-Kondakçı, E. (2016). Closing the gap between beliefs and practice: Change of pre-service chemistry teachers' orientations during a PCK-based NOS course. *Chemistry Education Research and Practice, 17*(4), 818–841.

Denzin, N. K. (1989). *The research act: A theoretical introduction to sociological methods* (3rd ed.). Englewood Cliffs, NJ: Prentice Hall.

Depaepe, F., Verschaffel, L., & Kelchtermans, G. (2013). Pedagogical content knowledge: A systematic review of the way in which the concept has pervaded mathematics educational research. *Teaching and Teacher Education, 34*, 12–25.

Diezmann, C. M., & Watters, J. J. (2015). The knowledge base of subject matter experts in teaching: A case study of a professional scientist as a beginning teacher. *International Journal of Science and Mathematics Education, 13*(6), 1517–1537.

Donnelly, D. F., & Hume, A. (2015). Using collaborative technology to enhance pre-service teachers' pedagogical content knowledge in Science. *Research in Science & Technological Education, 33*(1), 61–87.

Driver, R., Asoko, H., Leach, J., Scott, P., & Mortimer, E. (1994). Constructing scientific knowledge in the classroom. *Educational Researcher, 23*(7), 5–12.

Faikhamta, C., & Clarke, A. (2012). A self-study of a Thai teacher educator developing a better understanding of PCK for teaching about teaching science. *Research in Science Education, 43*(3), 955–979.

Falk, A. (2012). Teachers learning from professional development in elementary science: Reciprocal relations between formative assessment and pedagogical content knowledge. *Science Education, 96*(2), 265–290.

Fernández-Balboa, J.-M., & Stiehl, J. (1995). The generic nature of pedagogical content knowledge among college professors. *Teaching and Teacher Education, 11*(3), 293–306.

Findlay, M., & Bryce, T. G. K. (2012). From teaching physics to teaching children: Beginning teachers learning from pupils. *International Journal of Science Education, 34*(17), 2727–2750.

Fischer, H. E., Borowski, A., & Tepner, O. (2012). Professional knowledge of science teachers. In B. J. Fraser, K. Tobin, & C. J. McRobbie (Eds.), *Second international handbook of science education* (pp. 435–448). Dordrecht: Springer Netherlands.

Förtsch, C., Werner, S., von Kotzebue, L., & Neuhaus, B. J. (2016). Effects of biology teachers' professional knowledge and cognitive activation on students' achievement. *International Journal of Science Education, 38*(17), 2642–2666.

Fraser, S. P. (2016). Pedagogical content knowledge (PCK): Exploring its usefulness for science lecturers in higher education. *Research in Science Education, 46*(1), 141–161.

Friedrichsen, P. J., Abell, S. K., Pareja, E. M., Brown, P. L., Lankford, D. M., & Volkmann, M. J. (2009). Does teaching experience matter? Examining biology teachers' prior knowledge for teaching in an alternative certification program. *Journal of Research in Science Teaching, 46*(4), 357–383.

Friedrichsen, P. J., van Driel, J. H., & Abell, S. K. (2011). Taking a closer look at science teaching orientations. *Science Education, 95*(2), 358–376.

Geddis, A. N., Onslow, B., Beynon, C., & Oesch, J. (1993). Transforming content knowledge: Learning to teach about isotopes. *Science Education, 77*(6), 575–591.

Gess-Newsome, J. (1999). Pedagogical content knowledge: An introduction and orientation. In J. Gess-Newsome & N. G. Lederman (Eds.), *Examining pedagogical content knowledge: The construct and its implications for science education* (pp. 3–20). Dordrecht: Kluwer Academic.

Gess-Newsome, J. (2015). A model of teacher professional knowledge and skill including PCK: Results of the thinking from the PCK Summit. In A. Berry, P. J. Friedrichsen, & J. Loughran (Eds.), *Re-examining pedagogical content knowledge in science education* (pp. 28–42). New York: Routledge.

Gess-Newsome, J., Taylor, J. A., Carlson, J., Gardner, A. L., Wilson, C. D., & Stuhlsatz, M. A. M. (2017). Teacher pedagogical content knowledge, practice, and student achievement. *International Journal of Science Education*, 1–20.

Grossman, P. L. (1990). *The making of a teacher: Teacher knowledge and teacher education*. New York: Teachers College Press.

Großschedl, J., Harms, U., Kleickmann, T., & Glowinski, I. (2015). Preservice biology teachers' professional knowledge: Structure and learning opportunities. *Journal of Science Teacher Education, 26*(3), 291–318.

Großschedl, J., Mahler, D., Kleickmann, T., & Harms, U. (2014). Content-related knowledge of biology teachers from secondary schools: Structure and learning opportunities. *International Journal of Science Education, 36*(14), 2335–2366.

Guerriero, S. (2017). *Pedagogical knowledge and the changing nature of the teaching profession*. Paris: OECD Publishing.

Hallman-Thrasher, A., Connor, J., & Sturgill, D. (2017). Strong discipline knowledge cuts both ways for novice mathematics and science teachers. *International Journal of Science and Mathematics Education*.

Hanuscin, D. L. (2013). Critical incidents in the development of pedagogical content knowledge for teaching the nature of science: A prospective elementary teacher's journey. *Journal of Science Teacher Education, 24*(6), 933–956.

Hanuscin, D. L., Lee, M. H., & Akerson, V. L. (2011). Elementary teachers' pedagogical content knowledge for teaching the nature of science. *Science Education, 95*(1), 145–167.

Hashweh, M. Z. (2005). Teacher pedagogical constructions: A reconfiguration of pedagogical content knowledge. *Teachers and Teaching: Theory and Practice, 11*(3), 273–292.

Heller, J. I., Daehler, K. R., Wong, N., Shinohara, M., & Miratrix, L. W. (2012). Differential effects of three professional development models on teacher knowledge and student achievement in elementary science. *Journal of Research in Science Teaching, 49*(3), 333–362.

Henze, I., van Driel, J. H., & Verloop, N. (2008). Development of experienced science teachers' pedagogical content knowledge of models of the solar system and the universe. *International Journal of Science Education, 30*(10), 1321–1342.

Hill, H. C., Ball, D. L., & Schilling, S. G. (2008). Unpacking pedagogical content knowledge: Conceptualizing and measuring teachers' topic-specific knowledge of students. *Journal for Research in Mathematics Education, 39*(4), 372–400.

Jackson, P. W. (1986). *Life in classrooms*. New York: Holt, Rinehart & Winston.

Jin, H., Shin, H., Johnson, M. E., Kim, J., & Anderson, C. W. (2015). Developing learning progression-based teacher knowledge measures. *Journal of Research in Science Teaching, 52*(9), 1269–1295.

Jüttner, M., & Neuhaus, B. J. (2012). Development of items for a pedagogical content knowledge test based on empirical analysis of pupils' errors. *International Journal of Science Education, 34*(7), 1125–1143.

Kaiser, G., Blömeke, S., König, J., Busse, A., Döhrmann, M., & Hoth, J. (2017). Professional competencies of (prospective) mathematics teachers—Cognitive versus situated approaches. *Educational Studies in Mathematics, 94*(2), 161–182.

Kanter, D. E., & Konstantopoulos, S. (2010). The impact of a project-based science curriculum on minority student achievement, attitudes, and careers: The effects of teacher content and pedagogical content knowledge and inquiry-based practices. *Science Education, 94*(5), 855–887.

Käpylä, M., Heikkinen, J. P., & Asunta, T. (2009). Influence of content knowledge on pedagogical content knowledge: The case of teaching photosynthesis and plant growth. *International Journal of Science Education, 31*(10), 1395–1415.

Kaya, O. N. (2009). The nature of relationships among the components of pedagogical content knowledge of preservice science teachers: 'Ozone layer depletion' as an example. *International Journal of Science Education, 31*(7), 961–988.

Keller, M. M., Neumann, K., & Fischer, H. E. (2017). The impact of physics teachers' pedagogical content knowledge and motivation on students' achievement and interest. *Journal of Research in Science Teaching, 54*(5), 586–614.

Kellner, E., Gullberg, A., Attorps, I., Thorén, I., & Tärneberg, R. (2011). Prospective teachers' initial conceptions about pupils' difficulties in science and mathematics: A potential resource in teacher education. *International Journal of Science and Mathematics Education, 9*(4), 843–866.

Kersting, N. B. (2008). Using video clips of mathematics classroom instruction as item prompts to measure teachers' knowledge of teaching mathematics. *Educational and Psychological Measurement, 68*(5), 845–861.

Kersting, N. B., Sutton, T., Kalinec-Craig, C., Stoehr, K. J., Heshmati, S., Lozano, G., et al. (2016). Further exploration of the classroom video analysis (CVA) instrument as a measure of usable knowledge for teaching mathematics: Taking a knowledge system perspective. *ZDM Mathematics Education, 48*(1), 97–109.

Khourey-Bowers, C., & Fenk, C. (2009). Influence of constructivist professional development on chemistry content knowledge and scientific model development. *Journal of Science Teacher Education, 20*(5), 437–457.

Kind, V. (2009). Pedagogical content knowledge in science education: Perspectives and potential for progress. *Studies in Science Education, 45*(2), 169–204.

Kind, V. (2017). Development of evidence-based, student-learning-oriented rubrics for pre-service science teachers' pedagogical content knowledge. *International Journal of Science Education*, 1–33.

Kirschner, S., Borowski, A., Fischer, H. E., Gess-Newsome, J., & von Aufschnaiter, C. (2016). Developing and evaluating a paper-and-pencil test to assess components of physics teachers' pedagogical content knowledge. *International Journal of Science Education, 38*(8), 1343–1372.

Knievel, I., Lindmeier, A. M., & Heinze, A. (2015). Beyond knowledge: Measuring primary teachers' subject-specific competences in and for teaching Mathematics with items based on video vignettes. *International Journal of Science and Mathematics Education, 13*(2), 309–329.

König, J., Blömeke, S., Klein, P., Suhl, U., Busse, A., & Kaiser, G. (2014). Is teachers' general pedagogical knowledge a premise for noticing and interpreting classroom situations? A video-based assessment approach. *Teaching and Teacher Education, 38,* 76–88.

Krepf, M., Plöger, W., Scholl, D., & Seifert, A. (2017). Pedagogical content knowledge of experts and novices—What knowledge do they activate when analyzing science lessons? *Journal of Research in Science Teaching*, 1–23.

Lin, J.-W. (2016). Do skilled elementary teachers hold scientific conceptions and can they accurately predict the type and source of students' preconceptions of electric circuits? *International Journal of Science and Mathematics Education, 14*(2), 287–307.

Lin, J.-W. (2017). A comparison of experienced and preservice elementary school teachers' content knowledge and pedagogical content knowledge about electric circuits. *EURASIA Journal of Mathematics, Science and Technology Education, 13*(3).

Lin, J.-W., & Chiu, M. H. (2010). The mismatch between students' mental models of acids/bases and their sources and their teacher's anticipations thereof. *International Journal of Science Education, 32*(12), 1617–1646.

Loughran, J., Milroy, P., Berry, A., Gunstone, R., & Mulhall, P. (2001). Documenting science teachers' pedagogical content knowledge through PaP-eRs. *Research in Science Education, 31*(2), 289–307.

Loughran, J., Mulhall, P., & Berry, A. (2004). In search of pedagogical content knowledge in science: Developing ways of articulating and documenting professional practice. *Journal of Research in Science Teaching, 41*(4), 370–391.

Lucero, M. M., Petrosino, A. J., & Delgado, C. (2017). Exploring the relationship between secondary science teachers' subject matter knowledge and knowledge of student conceptions while teaching evolution by natural selection. *Journal of Research in Science Teaching, 54*(2), 219–246.

Luft, J. A. (2009). Beginning secondary science teachers in different induction programmes: The first year of teaching. *International Journal of Science Education, 31*(17), 2355–2384.

Luft, J. A., Firestone, J. B., Wong, S. S., Ortega, I., Adams, K., & Bang, E. (2011). Beginning secondary science teacher induction: A two-year mixed methods study. *Journal of Research in Science Teaching, 48*(10), 1199–1224.

Magnusson, S., Krajcik, J., & Borko, H. (1999). Nature, sources, and development of pedagogical content knowledge for science teaching. In J. Gess-Newsome & N. G. Lederman (Eds.), *Examining pedagogical content knowledge: The construct and its implications for science education* (pp. 95–132). Dordrecht: Kluwer Academic.

Mahler, D., Großschedl, J., & Harms, U. (2017). Using doubly latent multilevel analysis to elucidate relationships between science teachers' professional knowledge and students' performance. *International Journal of Science Education, 39*(2), 213–237.

Marks, R. (1990). Pedagogical content knowledge: From a mathematical case to a modified conception. *Journal of Teacher Education, 41*(3), 3–11.

Marshall, J. C., Smart, J., & Alston, D. M. (2016). Development and validation of Teacher Intentionality of Practice Scale (TIPS): A measure to evaluate and scaffold teacher effectiveness. *Teaching and Teacher Education, 59,* 159–168.

Matthew, J. K., Tae Seob, S., & Punya, M. (2012). How do we measure TPACK? Let me count the ways. In N. R. Robert, R. R. Christopher, & L. N. Margaret (Eds.), *Educational technology, teacher knowledge, and classroom impact: A research handbook on frameworks and approaches* (pp. 16–31). Hershey, PA, USA: IGI Global.

Mavhunga, E. (2016). Transfer of the pedagogical transformation competence across chemistry topics. *Chemistry Education Research and Practice, 17*(4), 1081–1097.

Mavhunga, E., & Rollnick, M. (2013). Improving PCK of chemical equilibrium in pre-service teachers. *African Journal of Research in Mathematics, Science and Technology Education, 17*(1–2), 113–125.

Mavhunga, E., & Rollnick, M. (2016). Teacher- or learner-centred? Science teacher beliefs related to topic specific pedagogical content knowledge: A South African case study. *Research in Science Education, 46*(6), 831–855.

McNeill, K. L., González-Howard, M., Katsh-Singer, R., & Loper, S. (2016). Pedagogical content knowledge of argumentation: Using classroom contexts to assess high-quality PCK rather than pseudoargumentation. *Journal of Research in Science Teaching, 53*(2), 261–290.

Mellado, V., Bermejo, M. L., Blanco, L. J., & Ruiz, C. (2007). The classroom practice of a prospective secondary biology teacher and his conceptions of the nature of science and of teaching and learning science. *International Journal of Science and Mathematics Education, 6*(1), 37–62.

Melo-Niño, L. V., Cañada, F., & Mellado, V. (2017a). Exploring the emotions in pedagogical content knowledge about the electric field. *International Journal of Science Education, 39*(8), 1025–1044.

Melo-Niño, L. V., Cañada, F., & Mellado, V. (2017b). Initial characterization of colombian high school physics teachers' pedagogical content knowledge on electric fields. *Research in Science Education, 47*(1), 25–48.

Meschede, N., Fiebranz, A., Möller, K., & Steffensky, M. (2017). Teachers' professional vision, pedagogical content knowledge and beliefs: On its relation and differences between pre-service and in-service teachers. *Teaching and Teacher Education, 66,* 158–170.

Miller, M. (2007). Pedagogical content knowledge. In G. Bodner & M. Orgill (Eds.), *Theoretical frameworks for research in chemistry/science education* (pp. 86–106). Upper Saddle River, NJ: Pearson Prentice Hall.

Monet, J. A., & Etkina, E. (2008). Fostering self-reflection and meaningful learning: Earth science professional development for middle school science teachers. *Journal of Science Teacher Education, 19*(5), 455–475.

Moodley, K., & Gaigher, E. (2017). Teaching electric circuits: Teachers' perceptions and learners' misconceptions. *Research in Science Education.*

Mthethwa-Kunene, E., Onwu, G. O., & de Villiers, R. (2015). Exploring biology teachers' pedagogical content knowledge in the teaching of genetics in Swaziland science classrooms. *International Journal of Science Education, 37*(7), 1140–1165.

National Council for Accreditation of Teacher Education. (2008). *Professional standards for the accreditation of teacher preparation institutions*. Washington, DC: National Council for Accreditation of Teacher Education.
National Research Council. (1996). *The national science education standards*. Washington, DC: National Academy Press.
Nilsson, P. (2014). When teaching makes a difference: Developing science teachers' pedagogical content knowledge through learning study. *International Journal of Science Education, 36*(11), 1794–1814.
Nilsson, P., & Elm, A. (2017). Capturing and developing early childhood teachers' science pedagogical content knowledge through CoRes. *Journal of Science Teacher Education*, 1–19.
Nilsson, P., & Loughran, J. (2012). Exploring the development of pre-service science elementary teachers' pedagogical content knowledge. *Journal of Science Teacher Education, 23*(7), 699–721.
Nilsson, P., & van Driel, J. H. (2010). Teaching together and learning together—Primary science student teachers' and their mentors' joint teaching and learning in the primary classroom. *Teaching and Teacher Education, 26*(6), 1309–1318.
Nilsson, P., & Vikström, A. (2015). Making PCK explicit—Capturing science teachers' pedagogical content knowledge (PCK) in the science classroom. *International Journal of Science Education, 37*(17), 2836–2857.
Oh, P. S., & Kim, K. S. (2013). Pedagogical transformations of science content knowledge in Korean elementary classrooms. *International Journal of Science Education, 35*(9), 1590–1624.
Park, S., & Chen, Y.-C. (2012). Mapping out the integration of the components of pedagogical content knowledge (PCK): Examples from high school biology classrooms. *Journal of Research in Science Teaching, 49*(7), 922–941.
Park, S., Jang, J.-Y., Chen, Y.-C., & Jung, J. (2011). Is pedagogical content knowledge (PCK) necessary for reformed science teaching?: Evidence from an empirical study. *Research in Science Education, 41*(2), 245–260.
Park, S., & Oliver, J. S. (2008a). National Board Certification (NBC) as a catalyst for teachers' learning about teaching: The effects of the NBC process on candidate teachers' PCK development. *Journal of Research in Science Teaching, 45*(7), 812–834.
Park, S., & Oliver, J. S. (2008b). Revisiting the conceptualisation of pedagogical content knowledge (PCK): PCK as a conceptual tool to understand teachers as professionals. *Research in Science Education, 38*(3), 261–284.
Park, S., Suh, J., & Seo, K. (2017). Development and validation of measures of secondary science teachers' PCK for teaching photosynthesis. *Research in Science Education*, 1–25.
Paulick, I., Großschedl, J., Harms, U., & Möller, J. (2016). Preservice teachers' professional knowledge and its relation to academic self-concept. *Journal of Teacher Education, 67*(3), 173–182.
Piliouras, P., Plakitsi, K., Seroglou, F., & Papantoniou, G. (2017). Teaching explicitly and reflecting on elements of nature of science: A discourse-focused professional development program with four fifth-grade teachers. *Research in Science Education*, 1–26.
Qhobela, M., & Kolitsoe Moru, E. (2014). Examining secondary school physics teachers' beliefs about teaching and classroom practices in Lesotho as a foundation for professional development. *International Journal of Science and Mathematics Education, 12*(6), 1367–1392.
Rollnick, M. (2017). Learning about semi conductors for teaching—The role played by content knowledge in Pedagogical Content Knowledge (PCK) development. *Research in Science Education*, 1–36.
Rollnick, M., Bennett, J., Rhemtula, M., Dharsey, N., & Ndlovu, T. (2008). The place of subject matter knowledge in pedagogical content knowledge: A case study of South African teachers teaching the amount of substance and chemical equilibrium. *International Journal of Science Education, 30*(10), 1365–1387.
Rosenkränzer, F., Hörsch, C., Schuler, S., & Riess, W. (2017). Student teachers' pedagogical content knowledge for teaching systems thinking: Effects of different interventions. *International Journal of Science Education*, 1–20.

Roth, K. J., Garnier, H. E., Chen, C., Lemmens, M., Schwille, K., & Wickler, N. I. Z. (2011). Videobased lesson analysis: Effective science PD for teacher and student learning. *Journal of Research in Science Teaching, 48*(2), 117–148.

Sadler, P. M., Sonnert, G., Coyle, H. P., Cook-Smith, N., & Miller, J. L. (2013). The Influence of teachers' knowledge on student learning in middle school physical science classrooms. *American Educational Research Journal, 50*(5), 1020–1049.

Salloum, S. L., & BouJaoude, S. (2008). Careful! It is H_2O? Teachers' conceptions of chemicals. *International Journal of Science Education, 30*(1), 33–64.

Scharfenberg, F.-J., & Bogner, F. X. (2016). A new role change approach in pre-service teacher education for developing pedagogical content knowledge in the context of a student outreach lab. *Research in Science Education, 46*(5), 743–766.

Schmelzing, S., van Driel, J. H., Jüttner, M., Brandenbusch, S., Sandmann, A., & Neuhaus, B. J. (2013). Development, evaluation and validation of a paper-and-pencil test for measuring two components of biology teachers' pedagogical content knowledge concerning the "cardiovascular system" *International Journal of Science and Mathematics Education, 11*(6), 1369–1390.

Schneider, R. M., & Plasman, K. (2011). Science teacher learning progressions. *Review of Educational Research, 81*(4), 530–565.

Settlage, J. (2013). On acknowledging PCK's shortcomings. *Journal of Science Teacher Education, 24*(1), 1–12.

Seung, E. (2013). The process of physics teaching assistants' pedagogical content knowledge development. *International Journal of Science and Mathematics Education, 11*(6), 1303–1326.

Shavelson, R. J., Ruiz-Primo, M. A., & Wiley, E. W. (2005). Windows into the mind. *Higher Education, 49*(4), 413–430.

Shulman, L. S. (1986). Those who understand: Knowledge growth in teaching. *Educational Researcher, 15*(2), 4–14.

Shulman, L. S. (1987). Knowledge and teaching: Foundations of the new reform. *Harvard Educational Review, 57*(1), 1–22.

Shulman, L. S. (2015). PCK: Its genesis and exodus. In A. Berry, P. J. Friedrichsen, & J. Loughran (Eds.), *Re-examining pedagogical content knowledge in science education* (pp. 3–13). New York: Routledge.

Sickel, A. J., & Friedrichsen, P. J. (2017). Using multiple lenses to examine the development of beginning biology teachers' pedagogical content knowledge for teaching natural selection simulations. *Research in Science Education*, 1–42.

Smit, R., Rietz, F., & Kreis, A. (2017). What are the effects of science lesson planning in peers?—Analysis of attitudes and knowledge based on an actor–partner interdependence model. *Research in Science Education*, 1–18.

Smit, R., Weitzel, H., Blank, R., Rietz, F., Tardent, J., & Robin, N. (2017). Interplay of secondary pre-service teacher content knowledge (CK), pedagogical content knowledge (PCK) and attitudes regarding scientific inquiry teaching within teacher training. *Research in Science & Technological Education*, 1–23.

Smith, S. P., & Banilower, E. R. (2015). Assessing PCK: A new application of the uncertainty principle. In A. Berry, P. J. Friedrichsen, & J. Loughran (Eds.), *Re-examining pedagogical content knowledge in science education* (pp. 88–103). New York: Routledge.

Sorge, S., Kröger, J., Petersen, S., & Neumann, K. (2017). Structure and development of pre-service physics teachers' professional knowledge. *International Journal of Science Education*, 1–28.

Stasinakis, P. K., & Athanasiou, K. (2016). Investigating Greek biology teachers' attitudes towards evolution teaching with respect to their pedagogical content knowledge: Suggestions for their professional development. *EURASIA Journal of Mathematics, Science & Technology Education, 13,* 1605–1617.

Stender, A., Brückmann, M., & Neumann, K. (2017). Transformation of topic-specific professional knowledge into personal pedagogical content knowledge through lesson planning. *International Journal of Science Education*, 1–25.

Suh, J., & Park, S. (2017). Exploring the relationship between pedagogical content knowledge (PCK) and sustainability of an innovative science teaching approach. *Teaching and Teacher Education, 64,* 246–259.

Supprakob, S., Faikhamta, C., & Suwanruji, P. (2016). Using the lens of pedagogical content knowledge for teaching the nature of science to portray novice chemistry teachers' transforming NOS in early years of teaching profession. *Chemistry Education Research and Practice, 17*(4), 1067–1080.

Tamir, P. (1988). Subject matter and related pedagogical knowledge in teacher education. *Teaching & Teacher Education, 4,* 99–110.

Tay, S. L., & Yeo, J. (2017). Analysis of a physics teacher's pedagogical 'micro-actions' that support 17-year-olds' learning of free body diagrams via a modelling approach. *International Journal of Science Education*, 1–30.

Tepner, O., Borowski, A., Dollny, S., Fischer, H. E., Jüttner, M., Kirschner, S., et al. (2012). Modell zur Entwicklung von Testitems zur Erfassung des Professionswissens von Lehrkräften in den Naturwissenschaften [Item development model for assessing professional knowledge of science teachers]. *Zeitschrift für Didaktik der Naturwissenschaften, 18,* 7–28.

Uzuntiryaki-Kondakçı, E., Demirdöğen, B., Akin, F. N., Tarkin, A., & Aydin, S. (2017). Exploring the complexity of teaching: The interaction between teacher self-regulation and pedagogical content knowledge. *Chemistry Education Research and Practice, 18*(1), 250–270.

van Dijk, E. M. (2009). Teachers' views on understanding evolutionary theory: A PCK-study in the framework of the ERTE-model. *Teaching and Teacher Education, 25*(2), 259–267.

van Dijk, E. M. (2014). Understanding the heterogeneous nature of science: A comprehensive notion of PCK for scientific literacy. *Science Education, 98*(3), 397–411.

van Dijk, E. M., & Kattmann, U. (2007). A research model for the study of science teachers' PCK and improving teacher education. *Teaching and Teacher Education, 23*(6), 885–897.

van Driel, J. H., & Abell, S. K. (2010). Science teacher education. In P. Peterson, E. Baker, & B. McGaw (Eds.), *International encyclopedia of education* (3rd ed., pp. 712–718). Oxford: Elsevier.

van Driel, J. H., & Berry, A. (2010). The teacher education knowledge base: Pedagogical content knowledge. In P. Peterson, E. Baker, & B. McGaw (Eds.), *International encyclopedia of education* (3rd ed., pp. 656–661). Oxford: Elsevier.

van Driel, J. H., & Berry, A. (2012). Teacher professional development focusing on pedagogical content knowledge. *Educational Researcher, 41*(1), 26–28.

van Driel, J. H., & Berry, A. (2017). Developing pre-service teachers' pedagogical content knowledge. In D. J. Clandinin & J. Husu (Eds.), *The SAGE handbook of research on teacher education* (Vol. 2, pp. 561–576). Thousand Oaks: California SAGE Publications Ltd.

van Driel, J. H., Berry, A., & Meirink, J. A. (2014). Research on science teacher knowledge. In N. G. Lederman & S. K. Abell (Eds.), *Handbook of research on science education* (Vol. 2, pp. 848–870). New York, NY: Routledge.

van Driel, J. H., Verloop, N., & de Vos, W. (1998). Developing science teachers' pedagogical content knowledge. *Journal of Research in Science Teaching, 35*(6), 673–695.

Veal, W. R., & MaKinster, J. (1999). Pedagogical content knowledge taxonomies. *Electronic Journal of Science Education, 3*(4).

Walan, S., Nilsson, P., & Ewen, B. M. (2017). Why inquiry? Primary teachers' objectives in choosing inquiry- and context-based instructional strategies to stimulate students' science learning. *Research in Science Education*, 1–20.

Wang, Y.-L., Tsai, C.-C., & Wei, S.-H. (2015). The sources of science teaching self-efficacy among elementary school teachers: A mediational model approach. *International Journal of Science Education, 37*(14), 2264–2283.

Willermark, S. (2017). Technological pedagogical and content knowledge: A review of empirical studies published from 2011 to 2016. *Journal of Educational Computing Research*, 1–28.

Wilson, S. M., Shulman, L. S., & Richert, E. R. (1987). '150 different ways' of knowing: Representations of knowledge in Teaching. In J. Calderhead (Ed.), *Exploring teachers' thinking*. London: Cassell.

Wongsopawiro, D. S., Zwart, R. C., & van Driel, J. H. (2017). Identifying pathways of teachers' PCK development. *Teachers and Teaching, 23*(2), 191–210.
Zembylas, M. (2007). Emotional ecology: The intersection of emotional knowledge and pedagogical content knowledge in teaching. *Teaching and Teacher Education, 23*(4), 355–367.
Zhou, S., Wang, Y., & Zhang, C. (2016). Pre-service science teachers' PCK: Inconsistency of pre-service teachers' predictions and student learning difficulties in Newton's Third Law. *EURASIA Journal of Mathematics, Science and Technology Education, 12*(3), 373–385.

Chapter 2
The Refined Consensus Model of Pedagogical Content Knowledge in Science Education

Janet Carlson, Kirsten R. Daehler, Alicia C. Alonzo, Erik Barendsen, Amanda Berry, Andreas Borowski, Jared Carpendale, Kennedy Kam Ho Chan, Rebecca Cooper, Patricia Friedrichsen, Julie Gess-Newsome, Ineke Henze-Rietveld, Anne Hume, Sophie Kirschner, Sven Liepertz, John Loughran, Elizabeth Mavhunga, Knut Neumann, Pernilla Nilsson, Soonhye Park, Marissa Rollnick, Aaron Sickel, Rebecca M. Schneider, Jee Kjung Suh, Jan van Driel, and Christopher D. Wilson

Abstract This chapter chronicles the developmental journey of a model for teacher pedagogical content knowledge (PCK) in science education, now identified as the Refined Consensus Model (RCM) of PCK, that represents the contributions and

---

Illustrated by
Jennifer Mendenhall

The original version of this chapter was revised: Authorship were changed. The correction to this chapter is available at https://doi.org/10.1007/978-981-13-5898-2_15

J. Carlson (✉)
Center to Support Excellence in Teaching, Stanford University, Palo Alto, USA
e-mail: janet.carlson@stanford.edu

K. R. Daehler
WestEd, San Francisco, USA
e-mail: kdaehle@wested.org

A. C. Alonzo
Michigan State University, East Lansing, USA

E. Barendsen
Radboud University, Nijmegen, Netherlands

Open University of the Netherlands, Heerlen, Netherlands

A. Berry
Monash University, Melbourne, Australia

A. Borowski
University of Potsdam, Potsdam, Germany

J. Carpendale
University of Waikato, Hamilton, New Zealand

A. Hume et al. (eds.), *Repositioning Pedagogical Content Knowledge in Teachers' Knowledge for Teaching Science*, https://doi.org/10.1007/978-981-13-5898-2_2

collective thinking of two dozen international researchers in science teacher education. This journey starts by recounting the process that led to an update and significant revisions to the model of teacher professional knowledge and skills including PCK (informally known as the 2012 Consensus Model (CM)). Then, we unpack and describe the different components of the model in both diagrammatic form and in explanatory text. The RCM describes the complex layers of knowledge and experiences that shape and inform teachers' practice and mediate student outcomes. A key feature of this model is the identification of three distinct realms of PCK—collective PCK, personal PCK, and enacted PCK. These realms are used to situate the specialised professional knowledge held by different science educators in different settings ranging from the collected knowledge understood by many to the unique subset of knowledge an individual teacher draws upon. The model also recognises that the broader professional knowledge bases are foundational to teacher PCK while the learning context a teacher is working in can greatly influence the teaching and learning that takes place.

K. Kam Ho Chan
The University of Hong Kong, Hong Kong, Hong Kong

R. Cooper
Faculty of Education, Monash University, AustraliaMelbourne

P. Friedrichsen
University of Missouri, Columbia, USA

J. Gess-Newsome
Oregon State University-Cascades, Bend, USA

I. Henze-Rietveld
Delft University of Technology, Delft, Netherlands

A. Hume
University of Waikato, Hamilton, New Zealand

S. Kirschner
Justus Liebig University, Giessen, Germany

S. Liepertz
Kiel, Germany

J. Loughran
Faculty of Education, Monash University, AustraliaMelbourne

E. Mavhunga
Science Education Division, School of Education, University of the Witwatersrand, Johannesburg, South Africa

Introduction

In this chapter, we introduce a model for teacher pedagogical content knowledge (PCK) that represents the contributions and collective thinking of two dozen international researchers in science teacher education. We begin by chronicling the process that led to an update and significant revisions to the model of teacher professional knowledge and skills including PCK that came to be known informally as the "2012 Consensus Model (CM)" (Berry, Friedrichsen, & Loughran (Eds.), 2015). Figure 2.1 shows the graphic of the model developed during the PCK Summit in 2012 and later published in the edited volume above (see Chap. 3 by Gess-Newsome, p. 31). A second contribution was the definition of PCK as "the knowledge of, reasoning behind, and planning for teaching a particular topic in a particular way for a particular purpose to particular students for enhanced student outcomes" (Gess-Newsome, 2015, p. 36). As researchers worked with that model, their understanding of its affordances and limitations became clearer. In December 2016, a second (2nd) Summit was convened that included participants in science teacher education research from the first (1st) Summit[1], as well as additional science PCK researchers from around the world. The purpose of the 2nd PCK Summit was to share instruments and tools for measuring PCK, the resulting data, and assess the alignment of people's work with the 2012 CM. In the process of doing this, we discovered what was and was not working with the extant model.

K. Neumann
Leibniz Institute for Science and Mathematics Education, Kiel University, Kiel, Germany

P. Nilsson
Halmstad University, Halmstad, Sweden

S. Park
Department of STEM Education, College of Education, North Carolina State University, Raleigh, NC, USA

A. Sickel
Melbourne, Australia

M. Rollnick
University of Witwatersrand, Johannesburg, South Africa

R. M. Schneider
University of Toledo, Toledo, OH, USA

J. K. Suh
Curriculum and Instruction, The University of Alabama, Tuscaloosa, AL, USA

J. van Driel
Melbourne Graduate School of Education, The University of Melbourne, Melbourne, Australia

C. D. Wilson
Biological Sciences Curriculum Study, Colorado Springs, CO, USA

[1]Details of the 1st Summit can be found here: http://pcksummit.bscs.org/.

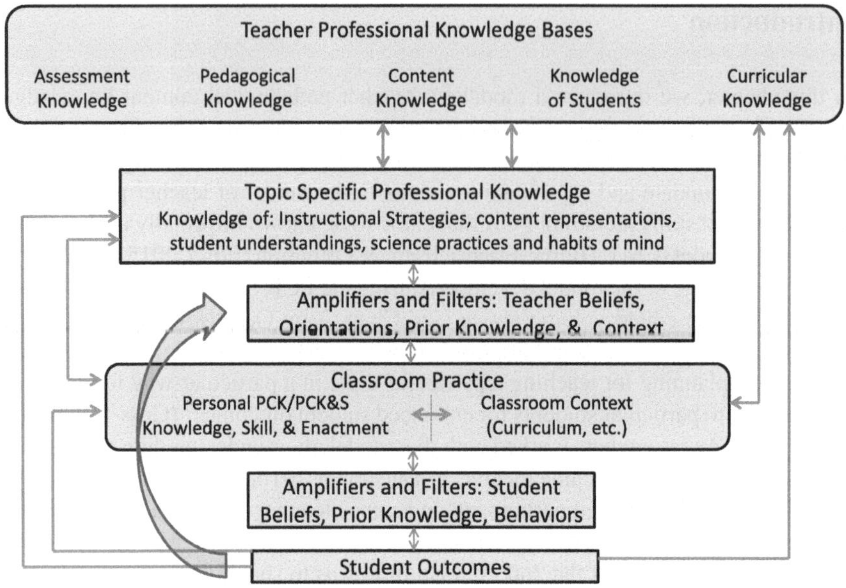

Fig. 2.1 A model of teacher professional knowledge and skills including PCK created in 2012 at the 1st PCK Summit, Colorado Springs, USA (referred to as the 2012 Consensus Model (CM)) (Gess-Newsome, 2015, page 31)

The science PCK researchers present at the 2nd PCK Summit noted that one of the key limitations of the 2012 CM was the minimal detail about PCK in the model itself. More specifically, the 2012 CM is in fact an illustration of teacher professional knowledge and skills, which includes situating PCK and the influences of PCK on classroom practice and student outcomes in that bigger picture (see Fig. 2.1). In particular, the "classroom practice" box positions a teacher's personal PCK in the context of other professional knowledge bases and acknowledges how those bases and other contextual information influence classroom practice. More explicitly, the 2012 CM articulated two elaborations that differed from prior models: (1) the addition of teaching practice (skills) to the definition of PCK, indicating that PCK is dynamic and encompasses more than static knowledge, and (2) the idea that personal PCK and skills (PCK&S) could and should be articulated separately from the more canonical topic-specific professional knowledge (TSPK).

One trade-off of situating PCK in a larger picture of professional knowledge and skills, however, was that the PCK component of the model was underspecified, meaning the model did not do enough to unpack and represent the variables, layers, and complexities of PCK. While the 2012 CM differentiated between TSPK and personal PCK&S, the distinction in terms of PCK was not sufficiently clear and researchers struggled with the specifics, such as where to put knowledge about instructional strategies. Was that PK or TSPK or personal PCK&S? When and under which conditions? As the 2nd PCK Summit attendees came to terms with this realisation, they started to articulate what needed to be specified so that a consensus model of PCK was more useful for guiding a wide variety of research studies that might develop a deeper understanding of teachers' PCK in science and the implications for teacher education, curriculum, and policy.

By building from the previous 2012 CM, as well as other frameworks such as Shulman (1986) and the Magnussen model (Magnussen, Krajcik, & Borko, 1999), we developed the Refined Consensus Model (RCM) of PCK in 2017. One aim was to provide researchers with a means to situate studies of student science learning in relationship to PCK by focusing on teachers and classrooms. Another aim was to provide science teacher educators a means to situate theories about the development of teacher PCK through formal education, in-service professional learning, and first-hand teaching experiences. Consequently, the RCM takes a practitioner perspective with *enacted PCK* at the centre. In addition, the conversations at the 2nd PCK Summit began to shift in purpose from driving consensus of a definition for PCK to a visualisation of where to situate science teacher education research studies and how to shape science teacher education and professional learning programmes. This shift created the need to refine and clearly articulate nuances among different types of PCK and the relationship among these types of PCK, as well as other types of knowledge, so that a broad range of studies could be considered. Therefore, it was time to reconceptualise the model and how to represent the group's updated thinking about PCK in the context of science teacher education.

Process of Model Development

Towards the end of the 2nd PCK Summit, a small working group sketched the group's ideas for a revision of the CM on a whiteboard, as shown in Fig. 2.2. In this version of the model, we sought to add greater specificity by addressing the role of grain sizes of science PCK (e.g. discipline, topic, concept); showing how PCK develops dynamically for individuals through feedback loops between experiences in the classroom and professional contributions; elaborating on the concept of *PCK in practice* as teachers' application of PCK during the pedagogical reasoning cycles of instruction (planning, enacting, reflecting); and by contextualising this practice of PCK within the larger context of a teacher's personal PCK and the greater collective of science PCK knowledge held beyond an individual. While this updated model highlights various aspects of PCK, the group did not see it as a replacement of other models such as the Magnussen model or the 2012 CM. Rather, the resulting RCM described in this chapter builds on these frameworks and incorporates new ideas to elaborate and clarify the 2012 model for science teacher education research.

By the end of the 2nd Summit, participants recognised that the newly emerging PCK model would need further articulation of the components and an improved visual representation. As such, Kirsten Daehler and Janet Carlson (functioning as a small subcommittee and principal authors of this chapter) agreed to help the group progress

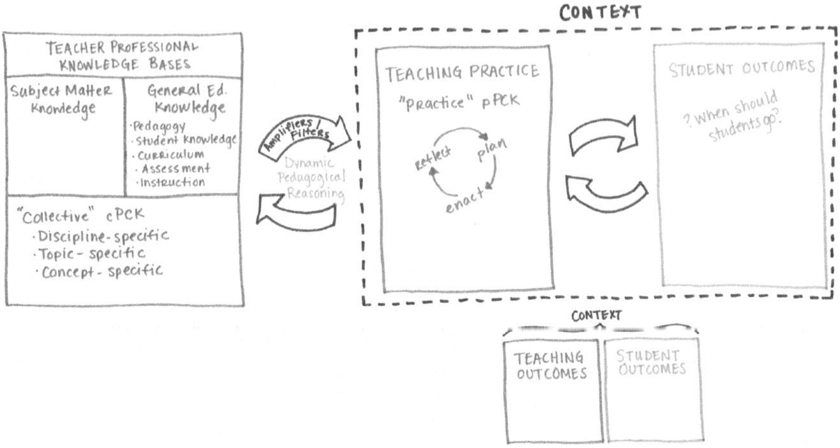

Fig. 2.2 Draft representation of the Revised Consensus Model of PCK generated during the 2nd Summit

by working with a graphic artist and drafting descriptive text, with the understanding that the intended outcome was an updated consensus model incorporating ideas that emerged from the 2nd PCK Summit. In subsequent weeks, this subcommittee worked in tandem with a graphic artist, who interviewed them about what was important to represent in the updated PCK model and then drafted various diagrams (e.g., box and arrows, concentric circles) to determine the best representation to convey the ideas that emerged in the 2nd Summit. This process involved multiple iterations and revisions, from which we chose concentric circles representation, as it seemed the best model to convey the complexity of relationships among PCK elements, especially the way in which an individual teacher's PCK in science was dependent on other larger bodies of knowledge along with specialised knowledge gained from classroom experiences.[2] By placing teachers and students at the centre of the concentric circles, the representation elevates the lens of the practitioner and reflects the field's attention to the act of science teaching within a particular classroom (Berry et al., 2016). It also reinforces the value of research on teachers' pedagogical reasoning and the dynamic nature of enacted PCK or PCK *in* action.

Several months later, the draft model and descriptive text were shared electronically with all participants from the 2nd PCK Summit, who provided comments and editorial suggestions. A few weeks later, Carlson and Daehler presented the draft model and its evolution in a session at the National Association of Research in Science Teaching (NARST) Conference in March 2017 with participants from both the 1st and 2nd Summits attending. To allow for further discussion about what was,

[2]While the whiteboard sketch in Fig. 2.2 is visually different from the resulting RCM shown with concentric circles in Fig. 2.3, after iterative feedback and discussion second Summit participants generally embrace the revised consensus model and agree that it represents discussions we had about the nature of PCK at the 2nd Summit.

and was not, working about this draft of the model, more than a dozen science PCK researchers met for a follow-up discussion during NARST. In addition, the subcommittee reached out to other PCK researchers who attended the 1st Summit, to invite their feedback. Ideas from these combined discussions and feedback sessions were then taken into account as the subcommittee again revised the images and refined the text, with more than a half dozen additional iterations. The revised text and images were again shared electronically with the broader science education community of participants from both PCK Summits and later revised to best incorporate participants' comments. Carlson presented this revision at the 2017 European Association of Science Education (ESERA) Conference in Dublin, Ireland. After this conference, there was an additional comment period in which others interested in PCK could weigh in on the text and visual representation. This chapter represents our synthesis of the feedback received in writing, in discussion, and as questions after each presentation. While it remains a less-than-perfect model (as all models are), the updated RCM of PCK reflects our current and best representation and description of the group's collective ideas about teachers' PCK in science education at this moment in time, as described in the following sections of this chapter.

Overview of the Refined Consensus Model (RCM) of PCK

The RCM of PCK is centred around the practice of science teaching. The model describes the complex layers of knowledge and experiences that shape and inform teachers' science practice throughout their professional journey and, in turn, mediate student outcomes. A key feature of this model is the identification of three distinct realms of PCK—*collective PCK (cPCK), personal PCK (pPCK),* and *enacted PCK (ePCK)*—which describe the specialised professional knowledge held by multiple educators in a field, to the personalised professional knowledge held by an individual teacher in science, and the unique subset of knowledge that a teacher draws on to engage in pedagogical reasoning during the planning of, teaching of, and reflecting on a lesson. Inherent in the development of these layers of PCK are the contributions of teachers, students, peers, and others. At the same time, a teacher's knowledge and skills are utilised throughout practice in ways that mediate student outcomes, as shown in Fig. 2.3. The model recognises that the broader *professional knowledge bases* (e.g. content knowledge, pedagogical knowledge) are foundational to teacher PCK in science and the *learning context* (e.g. federal and school policies, community values, and student attributes) influences the teaching and learning that takes place.

Throughout a science teacher's professional journey, a two-way *knowledge exchange* (⇄) takes place between the various concentric circles in the model. The knowledge and skills a teacher possesses from each realm are filtered and amplified in ways that shape a teacher's *personal PCK (pPCK)* in science over time. Teacher attitudes and beliefs about students, the nature of science content knowledge, or the role of the teacher are examples of beliefs and attitudes that can amplify and/or filter how a teacher develops pPCK for science teaching.

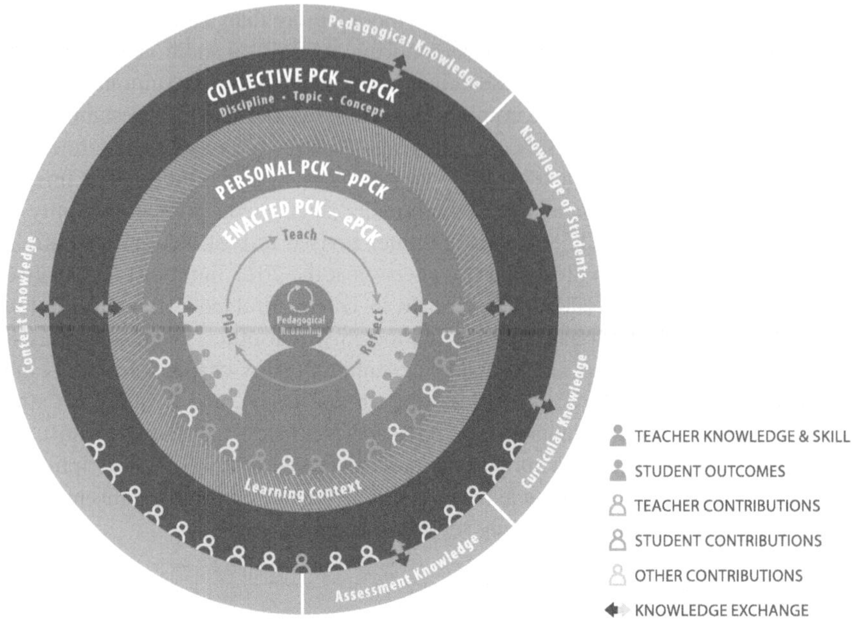

Fig. 2.3 Representation of the Refined Consensus Model (RCM) of PCK resulting from conversations at the 2nd PCK Summit, along with feedback sessions at NARST and ESERA 2017 and via electronic exchanges

Knowledge exchanges also take place when a teacher makes instructional choices related to teaching particular content to particular students in a particular context, again moderated by the teacher's own amplifiers and filters that inform the specific professional knowledge utilised in the practice of teaching known as *enacted PCK (ePCK)*. Similarly, experiences gained from the practice of science teaching provide feedback that further develops and shapes a teacher's pPCK in science. An individual teacher, through conversation and sharing, can then contribute to the *collective PCK (cPCK)* in science constructed by a group of teachers or add more broadly to the field's collective canonical knowledge by participating in a science education research project or an organised professional learning community, which ultimately may inform the broader professional knowledge bases (e.g. knowledge of students, curricular knowledge). This flow of knowledge and skills, in and out and through the concentric circles, is another key component of the RCM.

Components of the Refined Consensus Model (RCM) of PCK

The following is a description of the RCM, its components, and their interactions, beginning with the interior circle that focuses on the action of science teaching and working outward through the concentric circles. We begin with the teacher and students in that teacher's class to reflect the conversations we had at the 2nd PCK Summit about how science PCK develops within each teacher and is influenced by many factors including other humans and the environment.

Enacted PCK (ePCK)

When engaging in the practice of science teaching—whether planning instruction, carrying out that plan, or reflecting on instruction and student outcomes—a teacher utilises *enacted PCK (ePCK)*, as shown in Fig. 2.4. This ePCK is the specific knowledge and skills utilised by an individual teacher (👤) in a particular setting, with a particular student (👤) or group of students (👥), with a goal for those students to learn a particular concept, collection of concepts, or a particular aspect of the discipline. It is important to note that enactment in this model not only applies to the knowledge of and reasoning behind the act of teaching when interacting directly with students (reflection *in* action), but also to the acts of planning instruction and reflecting on instruction and student outcomes (reflection *on* action).

The centre circle of the model acknowledges that the pedagogical cycle of teaching is dynamic and the pedagogical reasoning that takes place during all aspects of teaching is unique to each teacher and every teaching moment. A teacher's ePCK for any given teaching moment reflects the context of the school, the classroom, and each individual student interacting with the teacher and the teacher's particular understanding of the science subject matter as well as his/her pedagogical knowledge and skills.

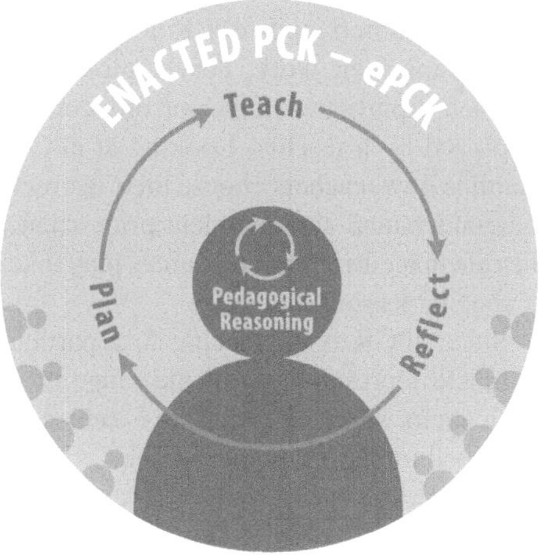

Fig. 2.4 Enacted PCK (ePCK): the interior circle of the model representing the specific knowledge and skills utilised by a teacher in a particular setting to achieve particular student outcomes

This conception of ePCK mirrors key ideas articulated during the 1st PCK Summit, particularly the description of "personal PCK" in science teaching as "the *knowledge* of, reasoning behind, and *planning* for teaching a particular topic in a particular *way* for a particular *purpose* to particular *students* for enhanced *student outcomes*". Furthermore, "skills" was included in the 1st Summit consensus definition acknowledging that "Personal PCK&S is the *act of teaching* a particular *topic* in a particular *way* for a particular *purpose* to particular *students* for enhanced *student outcomes*." (Gess-Newsome, 2015, p. 36).

The role of student outcomes in teachers' PCK in science is important from both a practice and research perspective. Student outcomes are one way of determining the effectiveness of a teacher's instruction; therefore, one way to assess the usefulness of a construct, such as PCK, is to determine if that construct predicts student learning outcomes. The student icons () in the inside circle represent both the students the teacher is interacting with and the *outcomes* of those interactions.

A key feature of ePCK is the way a teacher's actions draw on his/her knowledge to meet the unique needs of students in the classroom during a given instructional period. These actions involve the teacher pulling from his/her range of knowledge and practice bases to achieve a particular instructional goal. To provide meaningful instruction in a way that all students have access to learn what is being taught, a teacher needs to have the time to plan, teach, and reflect on all these components on a regular basis. Ideally, this aspect of pedagogical reasoning is done by working with others for planning, co-teaching, and/or reflecting on the evidence of learning. As teachers meet the needs of the learners in their classroom, they are employing pedagogical reasoning in the act of teaching that draws from a larger shared understanding. This means ePCK is visible in the teacher's expression of knowledge, choice of instructional strategies and representations, articulation of rationale for specific pedagogical moves, and the integration of multiple factors in pedagogical reasoning (e.g., knowledge of students, curricular saliency, assessment knowledge).

As researchers study ePCK in science teaching, there are many aspects to consider for framing their work. For example, one could study how teachers plan, reflect, or teach a particular topic; or one could study how all the parts of the cycle are employed by a teacher. Looking at ePCK with another lens, one might decide to examine how teachers choose their instructional strategies for particular topics, what they understand about student prior science conceptions or learning difficulties, or the role that curriculum resources play in teachers' planning as well as their teaching of specific lessons and units.

Since ePCK is so specific to a particular science teaching episode, it is not a complete representation of the range of a given teacher's PCK. The 2nd Summit participants situated ePCK as a subset of a teacher's personal PCK (pPCK), as described in the following section.

Personal PCK (pPCK)

A teacher's *pPCK* is the cumulative and dynamic pedagogical content knowledge and skills of an individual teacher that reflects the teacher's own teaching and learning experiences, along with the contributions of others ( ), such as teaching colleagues, educational researchers, scientists, and other content specialists in the form of professional exchanges, journal articles, social media, coursework, and professional learning experiences, as well as contributions from all students ( ) the teacher has ever taught, as shown in Fig. 2.5.

This pPCK serves as a reservoir of knowledge and skills that the teacher can draw upon during the practice of teaching. As aspects of the larger realm of pPCK are accessed and utilised, it becomes ePCK. In other words, ePCK is a subset of pPCK. More specifically, when responding to a particular instructional situation, teachers cannot possibly use all of the information and skills that reside in their knowledge banks, nor will teachers employ their entire repertoire of pedagogical moves in any given teaching moment. Rather, a teacher will use what makes sense for particular learners in a particular context, based on their prior experiences, along with collegial advice, educational preparation, and input from students. This interplay between teachers' pPCK in science, their reasoning, their ePCK, and how teachers respond to students, as well as what students learn, is at the centre of many PCK research projects in science teacher education.

In the RCM, the *knowledge exchange* ( ) between ePCK and pPCK operates in both directions—the insight a teacher takes away from each interaction with students further informs the teacher's pPCK, and the ePCK a teacher brings into practice for a specific science learning moment depends on the teacher's specific knowledge and

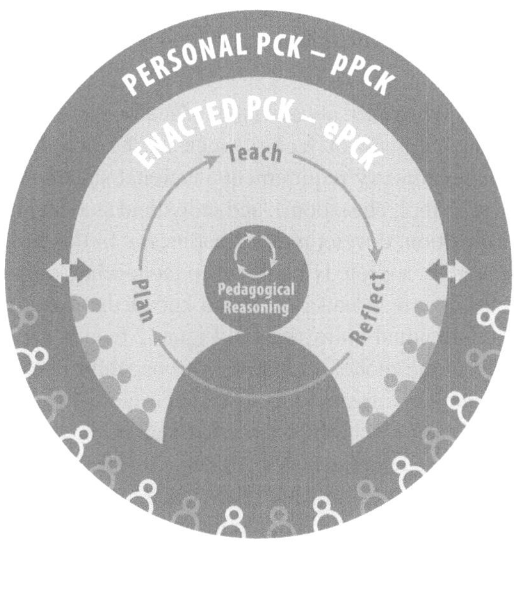

Fig. 2.5 Personal PCK (pPCK): a teacher's personal knowledge and unique expertise about teaching a given subject area, resulting from the cumulative experiences with and contributions from students, peers, and others

skill, which is amplified and filtered through pedagogical reasoning. As such, pPCK is both informed by and informs ePCK. Similarly, pPCK in teaching science is both informed by and can inform the learning context in which a teacher operates.

A teacher's pPCK in teaching science is developed, shaped, and refined over time through formal education, teaching experiences, and professional sharing. The result is a specialised knowledge and set of skills for teaching particular science topics for particular students in particular learning contexts. Given that every teacher has varied educational experiences and classroom interactions, it follows that pPCK is unique to each individual. While some teachers might have significant overlaps among their pPCK profiles, especially if they have participated in similar teacher preparation programmes or co-taught the same group of students, the pPCK for each teacher would necessarily be different. Even teachers with similar experiences will have differing attitudes and beliefs that serve to amplify and/or filter what each teacher brings to and takes away from an experience, thus contributing to teachers' varied pPCK.

During the 1st PCK Summit, the participants distinguished PCK as a personalised knowledge and skill set held by individual teachers, while other knowledge bases were considered external and public (see Gess-Newsome, 2015, page 36 for elaboration). As the participants attending the 2nd Summit began to specify PCK in more detail, we began to see a continuum from privately held PCK to publicly shared PCK with ePCK being the most private and pPCK having some aspects that are relatively private or individual and other aspects that are more likely to be shared knowledge with colleagues. As is evident in the sections above, the RCM takes ideas from the 2012 CM version of personal PCK&S and incorporates them in both pPCK and ePCK as part of the work to better specify key ideas from that earlier model.

Learning Context

Science learning always takes place in a context—a space and time that is defined by a multitude of factors, such as the broader educational climate (e.g. federal policy, ministry requirements, national standards), a specific learning environment (e.g. school, classroom), and individual student attributes (e.g. language proficiency, disposition, developmental readiness). In this model, the *learning context* is represented by a circle lying between the teacher knowledge and practice at the personal level and the broader PCK and knowledge bases that represent what others beyond the individual know and do. Figure 2.6 adds the learning context circle as the next ring in the model's sequence of concentric circles to symbolically situate science teaching and learning in space and time—a context that serves to both amplify and filter each teacher's knowledge and skills and to mediate teachers' actions. This circle also telegraphs how essential it is for teachers to have a deep knowledge of the learning context in which they teach, including knowledge of contexts that are both distal and more proximal to their students.

Fig. 2.6 Learning context: a multitude of factors that define and mediate learning, including everything from the broader educational climate to individual student attributes

Another factor that influences teachers' actions and affects student outcomes is the actual *classroom environment* in which student learning takes place, whether in a traditional school with a classroom full of students, as part of an afterschool programme with an individual learner, at a museum with a small group of students, or any other formal or informal learning environment. In this case, the term "classroom" is used metaphorically and is not limited to science learning that takes place in a physical classroom space inside a school. It can refer to the myriad of other factors that set the context for learning, such as the nature of teacher–student interactions, intended learning outcomes, curricular materials used, and group norms or dynamics among learners.

From the lens of PCK, *student attributes* are perhaps the most important aspect of the learning context, including factors such as students' age, grade level, prior experiences, dispositions, developmental readiness, language proficiency, and cultural beliefs. A skilled teacher recognises the value in knowing the attributes of each student and draws on this knowledge to facilitate learning. Knowledge of students therefore has a great influence on a teacher's pedagogical reasoning and their cycle of planning, teaching, and reflecting. The processes of attending to student attributes in both reasoning and instruction can richly inform a teacher's decisions in response to individual students, such as during a discussion or in comments on written assignments.

While the *learning context* serves as an amplifier and a filter that may inform an individual teacher's PCK, it is also what separates a teacher's private or pPCK from the more public or collective PCK (cPCK), as described in the following section.

Collective PCK (cPCK)

This form of PCK is an amalgam of multiple science educators' contributions ( ), including the teacher's own contributions ( ) and those from the combined professional knowledge bases and varied teaching experiences within a given subject matter as understood and documented by multiple people. The resulting cPCK is a specialised knowledge base for science teaching that has been articulated and is shared among a group of professionals, which is related to teaching that particular subject matter knowledge to particular students in a particular learning context. Most importantly, this knowledge can be shared and articulated in ways that encourage conversations among researchers, teachers, and other education professionals. This cPCK is the realm of PCK that has been in the literature since Lee Shulman's presidential address to the American Education Research Association (AERA) in 1986.

In other words, cPCK encompasses the knowledge that more than one person possesses, meaning knowledge that is not private, but rather knowledge that is public and held collectively. While this public knowledge is often what is documented in academic circles (e.g. in the form of published research articles and conference presentations), participants at the 2nd Summit acknowledged that teacher PCK also includes knowledge articulated and used by any group of educators and/or researchers. For example, the knowledge expressed in a Content Representation (CoRe) put together by a group describes some of their collective knowledge. While cPCK often builds from what is known from research (i.e., canonical PCK), cPCK represents a continuum of knowledge held by a group that extends beyond what is in the literature and recognises that knowledge about science teaching is also developed within school districts, school sites, departments, grade-level teacher teams, and professional learning communities (i.e., local collective PCK). Alternately, it is also worth considering the more situated knowledge that may go unrecognised in the local way of doing things. For example, when a team of teachers develops a shared teaching sequence this tacit knowledge also represents cPCK and may be one of the main drivers for professional learning in that school setting.

This specialised knowledge in science teaching can range in grain size from *discipline-specific* to *topic-specific* to *concept-specific* PCK, as shown in Fig. 2.7. While various research teams may have different definitions of these terms, what is important to note is that content knowledge used within cPCK can be specified along a continuum of broad to narrow ideas. To illustrate how ideas might fall on a continuum of PCK, consider these examples:

- Knowledge of effective ways to introduce young children to scientific argumentation (discipline-specific PCK)
- Knowledge of typical prior conceptions held by secondary students about photosynthesis (topic-specific PCK)
- Knowledge of strategies to help middle-grade students understand the concept that matter is neither created or destroyed (concept-specific PCK).

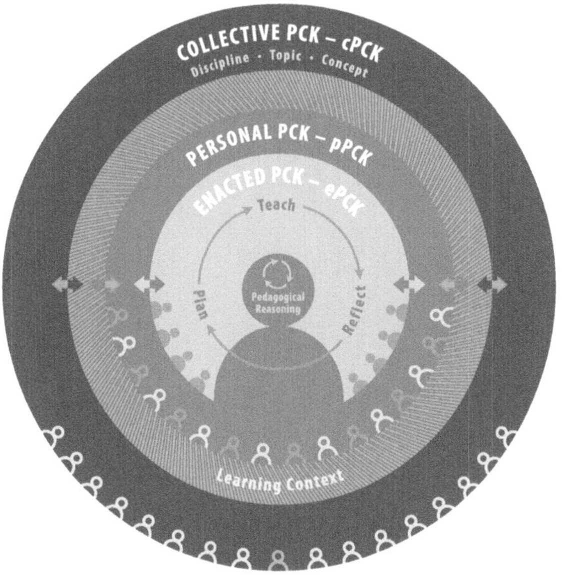

Fig. 2.7 Collective PCK (cPCK): the knowledge held by a group of people and considered generalisable to some degree, which is why this layer is situated after the learning context layer

It is important to note that while the grain size of specialised content is shown in only the cPCK realm, this notion is relevant to all other aspects of PCK, including a teacher's pPCK and ePCK. For reasons of readability and space limitations, the discipline–topic–concept notation is omitted from the inner circles of the diagram.

As historically defined, teacher PCK in teaching science stems from the intersection of content knowledge and other professional knowledge bases (e.g. pedagogical knowledge and knowledge of students). In the next section, we describe how those knowledge bases relate to the realms of PCK described in the model so far.

Professional Knowledge Bases

The outermost layer of the model represents different aspects of a teacher's broader professional knowledge bases including science content knowledge, pedagogical knowledge, knowledge of students, curricular knowledge, and assessment knowledge, as shown in Fig. 2.8.

Content knowledge describes the academic content of a given discipline (e.g. earth science, biology, chemistry, or physics). This expertise includes having discipline-specific knowledge and skills, such as an understanding of the nature of science and how to write scientific explanations, along with an understanding of a given domain within the discipline, and the relationship among domains, along with related topics and concepts within a domain. Often, but not always, teachers obtain their content knowledge in classes or settings separate from learning about how to teach the content. It is also the case, especially in science, that teachers may be asked

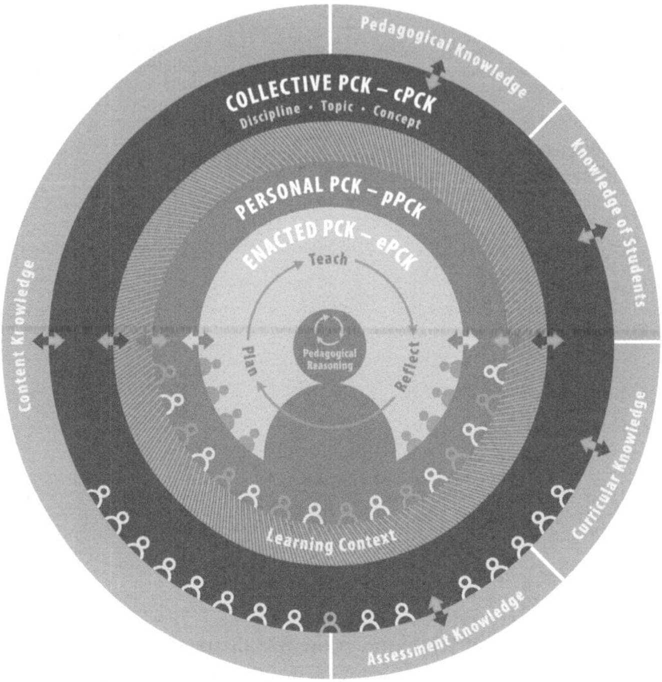

Fig. 2.8 Professional knowledge bases: these knowledge bases (e.g. pedagogical knowledge, knowledge of students) are essential foundations for teachers to become experts. Without these knowledge bases, teachers' PCK is quite limited

to teach a subject they have not formally studied, thus requiring them to learn this content in a self-directed or informal way.

In addition to content knowledge, there are a number of knowledge bases that define what it means to teach. Teachers need to understand the developmental readiness of their students, the nature of curriculum, the nature of different types of assessment, and a range of pedagogical skills and strategies that enable them to reach each student effectively. This educational knowledge is generic and not connected to a specific discipline. For example, curricular knowledge might include knowing how to sequence lessons to develop student conceptual understanding, while general pedagogical knowledge may include knowing how to set up the classroom and use strategies to engage students in collaborative learning or to support the needs of second-language learners. These knowledge bases are often developed through more formal routes, such as teacher preparation programmes, and then strengthened through teaching experiences and professional learning activities.

Implications of Updating a Model

Given that it has only been five years since the development of the 2012 CM and two years from the publication of the book on the topic, some might argue it is too soon to update a consensus model because people are just beginning to use it in their work. Yet, a dedicated and diverse group of researchers in science teacher education found enough limitations with the 2012 model that an update was the prudent response. While we do not presume the RCM to be a perfect model or even a true consensus model, we think it takes the thinking about PCK further and could be a useful tool for researchers who study PCK, including those who conduct research in other domains.

This model visually connects more static knowledge to the dynamic enactment of PCK, which provides opportunities for situating research studies of PCK in different realms in the model and helps to identify elements to study. It is important to note that this model does not specify the mechanisms and pathways by which teachers strengthen their PCK for teaching science, change their teaching, or connect various knowledge bases. Neither does the model assert a specific relationship between teachers' actions and student learning. By identifying three realms of PCK and the relationship to other knowledge bases, PCK researchers can place their studies in the most appropriate realm and begin to define mechanisms and pathways.

This model also addresses the conflict inherent in the question "Is PCK general or individual?" by proposing relationships among those "categories". From these proposed relationships, one can generate hypotheses about relationships that can be studied. The model also offers a way to think about how to support teacher development over a career trajectory from preservice to expert leadership by considering the role of experience, students, and colleagues in the development of PCK for teaching science at the individual level.

There is work to be done connecting this model to other models that define other aspects of PCK that are not explicitly articulated in this model. For example, one might argue that the Magnusson et al. (1999), and variations of it, could be a very useful way to further unpack what teachers are doing and thinking about during the development and use of pPCK and ePCK.

Researchers and practitioners from around the world contributed to this model. As such, it is intended to withstand scrutiny in different countries, be relevant across different policy environments, be useful for different research paradigms, and inform a wide range of teacher preparation and professional learning programmes both in science and potentially other domains. Now, it is time to test the model.

References

Berry, A., Depaepe, F., & van Driel, J. (2016). Pedagogical content knowledge in teacher education. In J. Loughran & M. L. Hamilton (Eds.), *International handbook of teacher education* (Vol. 1, pp. 347–386). Singapore: Springer Singapore.

Berry, A., Friedrichsen, P., & Loughran, J. (Eds.). (2015). *Re-examining pedagogical content knowledge in science education.* New York, London: Routledge.

Gess-Newsome, J. (2015). A model of teacher professional knowledge and skill including PCK: Results of the thinking from the PCK summit. In A. Berry, P. Friedrichsen, & J. Loughran (Eds.), *Re-examining pedagogical content knowledge in science education.* New York, London: Routledge.

Magnusson, S., Krajcik, J., Borko, H. (1999). Nature, sources, and development of pedagogical content knowledge for science teaching. In: J. Gess-Newsome & N. G. Lederman (Eds.) *Examining pedagogical content knowledge. Science & technology education library* (Vol. 6). Dordrecht: Springer.

Shulman, L. (1986). Those who understand: Knowledge growth in teaching. *Educational Researcher, 15*(2), 4–14.

Chapter 3
Vignettes Illustrating Practitioners' and Researchers' Applications of the Refined Consensus Model of Pedagogical Content Knowledge

Janet Carlson, Rebecca Cooper, Kirsten R. Daehler, Patricia J. Friedrichsen, Joan I. Heller, Sophie Kirschner, Nicole Lusiani Elliott, Karen Marangio, and Nicole Wong

Abstract The goal of this chapter is to provide examples of how the Refined Consensus Model (RCM) of pedagogical content knowledge (PCK) in science teacher education research moves from an abstract visualisation to a tool used by researchers and educators alike. The chapter offers a collection of short vignettes from a variety of settings and perspectives, each of which brings the RCM to life, either by illustrating ways the model is being used in science teacher education or

J. Carlson (✉) · N. L. Elliott
Center to Support Excellence in Teaching, Stanford University, Palo Alto, USA
e-mail: janet.carlson@stanford.edu

N. L. Elliott
e-mail: nlusiani@stanford.edu

R. Cooper · K. Marangio
Faculty of Education, Monash University, Melbourne, Australia
e-mail: rebecca.cooper@monash.edu

K. Marangio
e-mail: Karen.Marangio@monash.edu

K. R. Daehler
WestEd, San Francisco, USA
e-mail: kdaehle@wested.org

P. J. Friedrichsen
University of Missouri, Columbia, USA
e-mail: friedrichsenp@missouri.edu

J. I. Heller · N. Wong
Heller Research Associates, Oakland, USA
e-mail: jheller@gordonheller.com

N. Wong
e-mail: nwong@gordonheller.com

S. Kirschner
Justus Liebig University, Giessen, Germany
e-mail: Sophie.Kirschner@zfl.uni-giessen.de

A. Hume et al. (eds.), *Repositioning Pedagogical Content Knowledge in Teachers' Knowledge for Teaching Science*, https://doi.org/10.1007/978-981-13-5898-2_3

by situating studies and PCK instruments within the layers of the model. The first and second vignettes provide a pre-service perspective, yet from distinctly different learning contexts and science teacher educator experiences in Australia and the USA. The third vignette shares insights into the use of the RCM during a teacher professional learning course presented to teachers in Russia, while the fourth and fifth vignettes offer a researcher lens with examples from large-scale science PCK studies conducted in Germany and the USA. Each vignette follows a similar structure and includes these three parts:

- Context: A description of the setting (e.g., pre-service, professional learning, research), participants, and goals of the research or the teacher education programme
- Connections to the model: A description of how the RCM informed(s) the work or the application
- Reflections: A discussion of the implications of the model for the work and/or how the model was received and/or how it may shape future work.

As with any model, the true test of its power lies in its utility. Only by employing the RCM in a host of settings—to situate research questions, plan and conduct studies, develop explanations about findings, design education programmes, explore the mechanisms of PCK development, and reflect on teaching and learning—will we come to understand the benefits and limitations of the model. We look forward to learning from each other as we put this RCM of PCK to the test.

Describing the Possibilities Presented by the Refined Consensus Model for Pre-service Science Teacher Education

Rebecca Cooper and Karen Marangio

Context

At Monash University in Australia, our General Science Education method units for pre-service teachers span a full year (two 12-week semesters) and are always team-taught. Each unit consists of nine weeks of coursework made up of a one-hour lecture and a two-hour workshop per week, along with a three-week placement in schools each semester. We often run two classes of this unit as we usually have around 50 pre-service teachers each year. For over a decade, we (the authors) have taught the General Science Education method units and we have constantly looked to reimagine these units to incorporate new materials and approaches.

R. Cooper · K. Marangio
Clayton, Australia

In recent years, we have paid considerable attention to pre-service teachers developing ideas about pedagogy. We favour the work of Morine-Dershimer and Kent (1999) and introduce their model of pedagogical knowledge (PK) at the very beginning of the unit as it is an accessible way for pre-service teachers to explore the ideas of learning, teaching, and pedagogy for science. Morine-Dershimer and Kent's (1999) model of PK promotes the interplay between the internal dimensions of being a science teacher (such as personal beliefs and perceptions of science and science education) and external dimensions (such as teaching practice, building relationships with students), with reflection as a central facet. Later in the course, we introduce pedagogical content knowledge (PCK) after our student teachers have had the opportunity to develop a unit of work around a specific topic and have implemented their unit in a classroom. This strategy provides pre-service teachers with an actual teaching experience thus better equipping them to work with the important, but sometimes overwhelming, idea of PCK.

Our intention is to provide a challenge to what pre-service teachers view as science from the very start of our unit and to help them to appreciate the influence of these views on their teaching practice in science classrooms. Towards this end, we are always looking to support our pre-service teachers' growth as both learners and teachers, with both science and teaching at the heart of what is being explored. For instance, we see the nature of science (NOS) as influencing several of the facets of Morine-Dershimer and Kent's (1999) model. We make this influence explicit to our pre-service teachers by articulating our pedagogy whenever possible and welcoming our students to question our teaching practice. We also emphasise the links between their notions of the NOS and the choices they/we make as a teacher (especially through the beliefs facet). Additionally, we encourage them to talk and write about ideas that surface related to the NOS and the development of their understanding of PK.

As the year progresses, towards the end of our second unit, we shift to having our pre-service teachers think about PCK. We do this shift after they have completed two three-week placements in schools. We find that asking them to consider PCK, which we view as, "... the knowledge of, reasoning behind, and enactment of the teaching of particular topics in a particular way with particular students for particular reasons for enhanced student outcomes" (Carlson, Stokes, Helms, Gess-Newsome, & Gardner, 2015) is initially overwhelming and somewhat uncomfortable. We find they cannot fully embrace the idea of PCK until they have observed and engaged with some planning, teaching, assessment, and, most importantly, reflection. Thus, we introduce PCK the week after our pre-service teachers return to university from their second school placement. We use PCK as a way of reflecting on what they have observed and experienced in terms of quality science learning and teaching. We do this introduction through the use of Content Representations (CoRes) and Pedagogical and Professional-experience Repertoires (PaP-eRs) (Loughran, Berry, & Mulhall, 2012), where our pre-service teachers engage in creating a group CoRe and completing a "writing on the reading" task involving two PaP-eRs, one written by an experienced teacher and one written by a pre-service teacher. The pre-service teachers' final assessment is to write their own PaP-eR based on a lesson they taught while on the placement.

Connections to the Model

What the RCM offers us, and our students, is the ability to show a pathway of growth and development of PCK over the course of a year and across the learning that takes place in our units. By starting with the knowledge bases in the form of PK and later moving to PCK through CoRes and PaP-eRs, the RCM offers a single model that draws these ideas together and provides some clarity around the links between the theory we offer in the coursework, the assessment opportunities, the professional experience opportunities, connections with practicing teachers, and the need for personal and public reflection. By making these links explicit, the links the model provides help our students to make sense of all of these ideas and to consider what it means for each individual pre-service teacher as they develop their professional knowledge and skills throughout the course and into/during their career.

In our teacher education programme, we want to stimulate pre-service teachers' thinking about ongoing professional learning and have them consider the notion of expertise and what it takes to develop PCK throughout their careers. We also want to get them away from the idea of their teaching career being the "one year of experience many times over" and switch them to thinking more about what an expert teacher does, but further, how they can develop their own expertise and what are the significant components of this expertise. The RCM provides a framework for thinking about the components of teacher knowledge and skill; monitoring their own growth and development; supporting an individual in becoming a part of the collective; and remaining focused on student learning. The model encourages teachers to move between reflecting on more focused aspects within each component to exploring the connected and complex relationships between these components in order to grasp the bigger picture of what it truly means to be an expert science teacher.

Reimagining our units by making use of the RCM will allow us to use a single model throughout the year. This change will offer consistency in the language we use with our pre-service teachers, but also the consistency of language with teachers in schools who mentor our pre-service teachers.

Reflections

In a previous year, just after we introduced PCK in one of our workshops, a pre-service teacher raised her hand and said: "This all makes so much sense now! PCK really pulls it altogether… why didn't you tell us this sooner". While there were some nods of agreement in the room, the astonished faces of other pre-service teachers suggested this revelation was not the case for everyone. We see that the RCM model can provide us with a way to introduce ideas step by step in a way that supports and guides our pre-service teachers' induction into teaching as a profession. An extra benefit is that it also brings in and positions PCK for those pre-service teachers who benefit from seeing the bigger picture earlier in their professional pathway to help make sense of what they are learning and how it all comes together.

One of the teachers we work with hosts nearly half of our pre-service teachers during their school placement. This science teacher is experienced and incredibly thoughtful. He commented to us during a reflection session at the end of the year, "There was a black hole between the institutions that were giving us the students and us; the school…I had no idea, beyond my own experiences at uni[versity] what was happening at uni[versity], but now I know where you're coming from, I know where the students are coming from and I know what is being taught, it's just a plus, plus, plus and smashes down the gates". We feel that the RCM could further support this shared understanding between universities and schools about how we all contribute to the development and growth of future science teachers. Of particular note is the way the model values collective PCK (cPCK) as something that is public, shared among a group of professionals, and encourages conversations among a range of education professionals—in our case, schools and universities.

In summary, we see the RCM as offering many possible ways to enhance our General Science Education method units, but also to consolidate and sustain our relationship with a school to improve the learning of science pre-service teachers.

Illustrating Application of the Refined Consensus Model of PCK: A Biology Methods Course

Patricia J. Friedrichsen

Context

Until a few years ago, at the University of Missouri, pre-service teachers seeking certification in biology, chemistry, earth science, and physics enrolled in the same set of three sequential middle/secondary science methods courses. These courses emphasised science-specific PCK, rather than topic-specific. In 2015, due to pressure to reduce course credits in our undergraduate programmes, we reduced the sequence to two courses. Informed by my work as a PCK researcher, I sought out discipline-specific experiences for the science education majors in place of the third methods course. These experiences vary from discipline to discipline, depending on the resources in each of the affiliated departments. In the physics department, physics education students enrol in a course originally designed as a recruiting course

P. J. Friedrichsen
Columbia, USA

to encourage physics majors to explore high school physics teaching careers. The course focuses on exploring topics taught in the ninth-grade physics first course offered at many high schools in the state. In the chemistry department, our education majors are hired as teaching assistants for freshmen-level chemistry laboratory sections. My previous appointment in the biology department and the larger enrolment in biology education mean we are able to offer a biology methods course specifically for pre-service teachers.

The overarching goal of this biology methods course is informed by the Gess-Newsome (2015) model of teacher professional knowledge and skill, also referred to as the Consensus Model, in that teachers can develop pedagogical content knowledge and skills (PCK & S) through the study of topic-specific professional knowledge. More specifically, the course aims to develop pre-service teachers' topic-specific PCK associated with the four life science disciplinary core ideas (DCIs) presented in the national *Next Generation Science Standards'* (*NGSS*) (NGSS Lead States, 2013). In conjunction with this course, the pre-service teachers enrol in a field practicum in a local middle school or high school biology classroom. As an instructor of this biology methods course, I strive to develop a professional learning community (PLC) among the pre-service teachers and myself. To do this, I draw upon Kruse, Louis, and Bryk's (1995) five essential features of PLCs—reflective dialogue, de-privatisation of practice, a collective focus on student learning, collaboration, and shared norms and values. Given that the majority of our graduates will be participating in PLCs at their new schools, I find this approach makes sense during their pre-service education and I want to prepare them for this highly collaborative work.

Connections to the Model

The design of the biology methods course aligns well with the Refined Consensus Model (RCM) of PCK. This model, with enacted PCK (ePCK) in the centre, was intended to "elevate the lens of the practitioner and reflects the field's attention to the act of teaching within a particular classroom" (see Chap. 2 in this book). As the pre-service teachers have limited opportunities to teach lessons at their practicum, and hence, develop their ePCK, there are many other aspects of PCK that are attended to during the science methods courses. In this section, I describe the alignment of the biology methods course with the RCM by starting with the outer rings of the model and working inward (see Fig. 3.1).

The realm of *collective PCK* (*cPCK*), as a component of the RCM, aligns with and informs my PLC-oriented approach to pre-service teacher education. In this model, cPCK is not only represented in academic publications but can also be "knowledge articulated and used by a group of educators and/or researchers" (see Chap. 2 in this book). When our biology methods class studies each of the four *NGSS* life science DCIs, we develop cPCK at both the discipline level and the topic level. At the disciplinary level, we explore the *NGSS* crosscutting concepts (CCCs) and

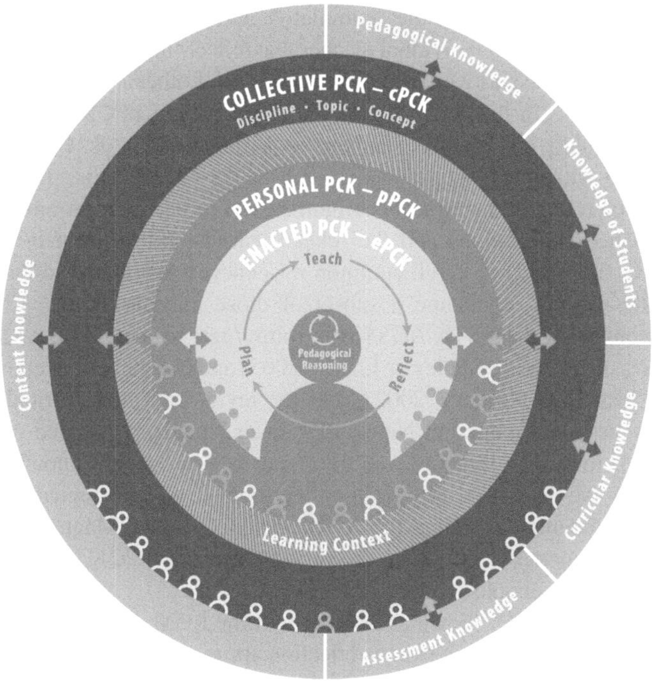

Fig. 3.1 Representation of the 2017 Refined Consensus Model (RCM) of pedagogical content knowledge (PCK) used to evaluate alignment with the pre-service biology methods course

science and engineering practices (SEPs). As the class develops our cPCK, we draw upon content knowledge and other professional knowledge bases (e.g., assessment knowledge, curricular knowledge.) To enrich our cPCK, I invite experienced biology teachers to share their curriculum related to a particular DCI. As a class, we also attend the state-level conference of the National Science Teachers Association, another reification of cPCK.

Learning context is the next inward layer of the RCM that applies to the biology methods course. While the learning context varies throughout the course, it is always heavily influenced by national reform documents, such as *A Framework for K-12 Science Education* (National Research Council [NRC], 2012) and *NGSS,* as our state has adopted a closely aligned version of *NGSS*. As our pre-service teachers do not yet have their own classrooms, the biology methods classroom is their primary learning context. When completing assignments, such as those described below, the learning context is the pre-service teachers' envisioned future classroom. The field practicum serves as another learning context; however, it offers limited learning opportunities for the pre-service teachers to teach. Each of these learning contexts, working synergistically, facilitates the dynamic interaction between the group's cPCK and each teacher's personal PCK (pPCK).

Personal PCK is defined as the "cumulative and dynamic pedagogical content knowledge and skills of an individual teacher that reflects the teacher's own teaching and learning experience, as well as the contributions of others, … including coursework" (see Chap. 2 in this book, p. X). To develop their pPCK, our pre-service teachers individually complete assignments in a digital notebook, OneNote, which they can share with their instructor. For each of the *NGSS* DCIs, the pre-service teachers complete the following set of assignments: (1) read, interpret, and summarise the performance expectations and the related learning progression in the frameworks (NRC, 2012); (2) research common misconceptions related to the DCI and plan specific ways to challenge a subset of those misconceptions; (3) critique several lessons from the NSTA *NGSS* Hub (http://ngss.nsta.org); (4) review topic-specific assessments (e.g., Keeley & Eberle, 2005) and conceptual inventories (e.g., Anderson, Fisher, & Norman, 2002); and (5) review selected curriculum units, both locally designed units and National Science Foundation-funded curriculum. The nature of these assignments was informed by professional knowledge bases represented by the outermost grey circle of the RCM (e.g., pedagogical knowledge, knowledge of students), as well as the PCK components of the Magnusson, Krajcik, and Borko PCK model (1999). The pre-service teachers' pPCK is reified in their individual digital notebooks.

Enacted PCK (ePCK), shown at the centre of the RCM, is represented in terms of cycles of planning, teaching, and reflection. In the biology methods course, I have participants plan a lesson, teach it to their peers, and reflect on the lesson. When teaching the lessons, the pre-service teacher takes on the role of a professional developer or lead teacher who is sharing a lesson, instructional strategy, or resource with our class PLC, and who then facilitates a follow-up discussion with his/her peers. As one example of this strategy, during the fall of 2017, the pre-service teachers participated in a citizen science project in which they observed plant and animal behaviour during the solar eclipse. As a follow-up, each pre-service teacher researched biology-related citizen science projects and selected one that he or she might use with a future group of 6th–12th grade students. Pre-service teachers then took turns presenting their selected project to our group, as if they were persuading teachers in a hypothetical future PLC to have their students participate in the project. As a biology methods class PLC, we discussed the added value of each citizen science project, along with cost, safety concerns, and implementation logistics. A primary goal of this activity and others completed during the biology methods course is to help the pre-service teachers become more comfortable sharing lessons and ideas with their peer teachers in their future PLCs.

Reflections

The RCM builds upon past PCK models, particularly the Gess-Newsome (2015) model and the Magnusson et al. (1999). Likewise, the design of the biology methods course was influenced by this same set of PCK models. The Gess-Newsome model influenced my thinking about developing teacher's personal PCK through studying the professional knowledge bases. The Magnusson model influenced the design of the digital notebook assignments focusing on common misconceptions, instructional strategies, topic-specific assessments, and curricula. These aspects of the two earlier models are represented in the RCM of PCK for science teacher education research. While the Gess-Newsome model combined PCK & Skill, the RCM makes a distinction between ePCK and pPCK, which I find useful. In the biology methods course, more time is spent on developing pre-service teachers' pPCK through the digital notebook assignments. Through leading professional development sessions, the pre-service teachers begin to develop ePCK. Another contribution of the RCM is the inclusion of the knowledge developed within teachers' professional learning communities as part of cPCK. The concept of cPCK has shaped our classroom norms for the biology methods class. As I reflect on the design of this course, the following research questions arise: in their future classrooms, in what ways will the current pre-service teachers draw from the pPCK and cPCK they developed from this course? And conversely, in what ways do their ePCK inform their existing pPCK? As beginning teachers, in what ways do they feel empowered to contribute to their PLCs and the development of cPCK in their PLC groups?

Using the Refined Consensus Model of PCK to Structure a Professional Learning Experience

Janet Carlson and Nicole Lusiani Elliott

Context

This vignette describes the use of the Refined Consensus Model (RCM) of PCK for teaching science during an introductory professional learning institute with experienced teachers from different academic disciplines. The model was used to situate and focus the goals of the learning experience, titled "Improving Instruction to Promote Excellence for All". Our premise for using the visual representation of the model to situate different parts of the learning was to provide a means by which the teachers could bridge theory and practice by seeing how the different activities we did connected back to building their own PCK. Our plan included five days of onsite instruction during the summer with follow-up work that had not yet been defined.

The learning goals of the summer institute were for the teachers to:

- Develop and deepen their understanding of pedagogical content knowledge (PCK) and apply PCK as a framework for thinking about effective instruction

J. Carlson · N. L. Elliott
Stanford, USA

- Deepen their understanding of students by applying the principles of how people learn and considering the influence of rings of culture
- Deepen understanding of how to plan coherent lesson arcs to have more coherent and articulated units of instruction
- Develop a general understanding of what core practices of teaching are and why they are important
- Develop a deeper understanding of two core practices—conducting an academic discussion and using formative assessment
- Build greater community that promotes a collegial atmosphere for planning curriculum and strengthening instruction.

We were working with 25 middle and high school teachers of all disciplines based in Moscow, Russia, who would be teaching gifted students to meet the standards of both the Russian Federation and the International Baccalaureate curriculum. The mission of the school was to "inspire bright students to achieve their intellectual and creative potential, develop strong moral principles and become leaders with a deep sense of responsibility towards their nation and the world".

Connections to the Model

We first introduced teachers to the RCM by asking them to look at the diagram of the model and discuss with a partner what they thought it meant. We also asked them to describe what they thought PCK was specifically. After listening to their ideas, which were quite accurate, we unpacked the layers of the RCM diagram. Then, we drew circles around the terms "content knowledge", "curricular knowledge", "knowledge of students", and "assessment knowledge" (as shown in Fig. 3.2) and asked these two questions:

- How do you already use the four circled categories in your instructional planning?
- How might understanding each category improve your practice?

Teachers wrote their ideas in journals and then had a discussion with each other. This exercise was designed to serve as an advance organiser for the week's activities. For the work related to content knowledge, the teachers work in discipline-specific groups to discuss these questions:

- What are the ***central questions*** surrounding the topics in your discipline?
- What are the ***concepts*** you want students to understand deeply?
- What ***skills*** do you want students to engage in?
- Why is it important to know and understand ***concepts*** and ***skills*** in addition to facts and basic knowledge?

They used ideas from their discussions to make a list of 3–5 central questions for their discipline, identify one major concept for each question, and identify 3–5 important skills for their discipline. To share their ideas, teachers created posters

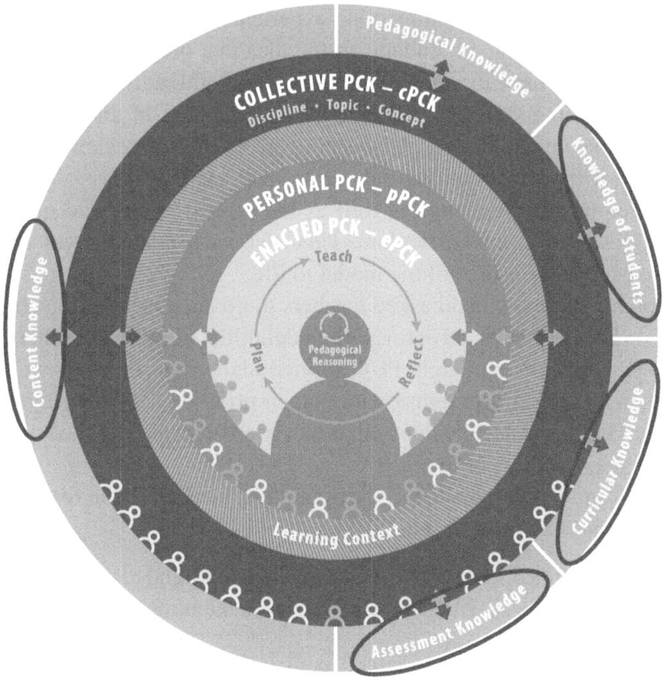

Fig. 3.2 Image of PCK model with knowledge of subject matter, curriculum, students, and assessment circled

to share their ideas, which were hung in the classroom for the week and used as reference points as we developed the concept of personal PCK (pPCK). Our goal was to support the teachers in recognising and developing their knowledge and skills for supporting students to learn those major concepts effectively.

One essential component to helping students learn effectively is knowing the students well and understanding their context as well as the learning context in the classroom. To help develop teachers' knowledge of students, we used an equity perspective and asked them to consider these questions:

- What are the reading, writing, and speaking skill levels and styles of your students?
- What do your students generally understand, or misunderstand, about the subject matter in your discipline?
- Who are your students?
  - What are your students' interests?
  - How might interests relate to the subject matter?
  - How might interests inform instructional activities?
  - What can you do with instruction that values who they are, how they learn, and supports their academic growth?

In this part of the professional learning experience, we worked with the teachers to connect their knowledge of students to the idea that pedagogical decisions need to be strategically chosen to validate and affirm cultural and school behaviours in addition to supporting students' expanding understanding of disciplinary content. We linked this to content knowledge by asking the teachers to review their curricular choices to see if their instructional materials were representative of how students see themselves in their learning, as well as the typical voices and examples that are (over)represented in the curriculum. This exercise provided a way to bring the "knowledge exchange" aspect of the model to life.

At this point, we paused and asked teachers to go back to what they had written about PCK on Day 1 and expand or revise their definition of PCK. They then discussed with a colleague how they thought PCK applied to their teaching. As we supported the teachers in making connections between their professional knowledge bases and their personal PCK, we saw evidence that they were considering the nature of *how* to teach in concert with *what* to teach.

Next, we examined knowledge of curriculum by reviewing the features of high-quality curriculum that include the following:

- An overt and planned learning sequence designed around a central question
- A coherent flow of ideas with factual information intentionally connected to larger concepts and contexts
- Carefully supported opportunities for students to develop the practices of the discipline
- An emphasis on how to develop evidence-based interpretations, explanations, and arguments
- An explicit attention to how people learn.

Teachers reviewed the list of central questions made during the discussion on PCK and subject matter knowledge and choose one that represented an essential idea in their discipline. This task was then connected to their knowledge of pedagogy by focusing on academic discussion. Academic discussion is a conversation about the content students are learning in which the students are making sense of what they are learning. Using this particular teaching practice and a unit of their choosing, we asked the teachers to put their PCK into action by choosing a particular topic for their particular students where the students could engage in a rich academic discussion that would include

- Making sense of key ideas
- Using evidence to interpret
- Developing an explanation based on evidence
- Making an argument using evidence
- Using academic terms that are appropriate to the subject matter.

Because the teachers planned an academic discussion related to a particular topic within a specific unit, this exercise provided an opportunity for teachers to link their knowledge of students, curriculum flow, and specific content together, representing the planning part of the pedagogical reasoning cycle in the centre of the RCM, i.e. the ePCK realm.

We then created an opportunity for teachers to teach and reflect on the use of academic discussion by asking them to rehearse the academic discussion they designed with their peers. This rehearsal is an opportunity to practice the enactment of the discussion they had just planned by trying it out with a small group of fellow teachers. During the rehearsal, they were able to discuss teaching, work through potential challenges of the lesson, anticipate what students might do in the discussion, and consider how to best support students in their success.

Finally, we spent some time developing teachers' knowledge of assessment by focusing on formative assessment strategies. Teachers explored the roles of both formal and informal procedures used during the learning process to determine how a teacher needs to change instruction to maximise learning for students. We emphasised how PCK links together the information gathered during formative assessment and knowing what to do with that information. We explored how teachers can adjust instruction in the moment as well as adjusting longer-term plans to improve student learning.

Reflections

We used the visual of the RCM in August 2017 as a test of its sensibility with and for practitioners. This group of teachers who participated in the professional learning were characterised by being reasonably well read in the area of learning theory, though no one was able to define PCK at the beginning of the summer workshop. Since they were new to the general concept of PCK, we did not get into detailed discussions of the differences among cPCK, pPCK, and ePCK. We did this intentionally to prevent creating an unnecessary overload of terminology that was not in the teachers' first language. We situated the majority of the PCK connections for the teachers in the knowledge exchanges between their professional knowledge bases and their personal PCK.

This group was also characterised by being intellectual risk-takers—they were starting a new school and highly engaged in that work and hungry to consider the characteristics of effective instruction. In this somewhat unique and special group of teachers, the model was an effective means of helping them see relationships among the many knowledge and skill bases that comprise the art of teaching. Daily feedback from the teachers confirmed that the model helped them situate the range of activities we did to better understand the connections among curriculum, students, subject matter, and instruction. Based on this experience, we expect to integrate the use of the RCM into future work with intellectually curious teachers.

Considering a PCK Instrument Through the Lens of the Refined Consensus Model of PCK

Sophie Kirschner

Context

The Refined Consensus Model (RCM) of PCK aims to be one model that is applicable to very diverse approaches used to capture and assess physics teachers' PCK. In this vignette, I explore how an existing paper-and-pencil test that is used to assess physics teachers' PCK aligns with the RCM. The development, evaluation, and validation of this instrument were the central focus of my dissertation and related publications (see Kirschner, Borowski, Fischer, Gess-Newsome, and Aufschnaiter, 2016), which was part of a bigger project with the broader goal of understanding in-service science teachers' professional knowledge [known as ProwiN; see Tepner et al. (2012)]. In the first phase of the study, we developed subject-specific content knowledge (CK) and PCK paper-and-pencil tests for biology, chemistry, and physics teachers and two subject-independent PK tests. In the second phase, researchers investigated the relationship between PCK (assessed using the paper-and-pencil tests from phase I), teachers enacting a specific lesson with their class, and student outcomes (see Cauet, Liepertz, Borowski, & Fischer, 2015, for physics). The physics tests were developed from 2009 to 2010, the main study was conducted in 2010, and data were analysed through 2015. Today, single items from the PCK test are still being used, for example, to assess the knowledge of pre-service science teachers at Justus Liebig University Giessen.

The physics PCK test was used with a large number of physics teachers (N = 186) and test persons for validation (N = 107). The items were designed to fit a model concerning the knowledge areas (declarative, procedural, and conditional knowledge), the physics topics (mainly mechanics), and the facets (knowledge about student understanding and knowledge about instructional strategies and representations). A sample item, coding scheme, and teacher responses are shown in Fig. 3.3. A discussion of the model and a translation of the entire PCK test can be found in Kirschner et al. (2016).

Two raters coded all open-ended items using a coding manual. For every item, a coding scheme was developed in order to identify correct answers (2 points), partially correct answers (1 point), or incorrect answers (0 points). Partial credit was given for knowledge of part of the solution. Possible correct answers were derived from the literature and from teachers' answers, along with contributions from the researchers' experiences and reasoning. The coding scheme was specified for every single item when an answer was good enough to be rated as "correct" or "partially correct."

S. Kirschner
Giessen, Germany

Sample Item
Students may have misconceptions having to do with the physics concepts of *speed* and *velocity*. Write down as many misconceptions as possible.

Coding scheme
Expectations:
- Misconceptions related to the relationship between distance and time
  - o v = s/t always can be used for calculation
  - o The formula is v = s*t
  - o The relationship between v, s, and t is vague [no real misconception]
  - o Average speed and mean speed are the same
- Misconceptions related to force
  - o A body in motion can cause something / has force; it has more force when it moves faster
  - o Without force there is no motion
  - o A uniform movement requires a force
  - o Bodies become slower by themselves
  - o High speed is the result of a large force (neglecting the time aspect)
- Misconceptions related to the direction
  - o Velocity and speed are the same
  - o Velocity has no direction
  - o In a circular motion the velocity is constant
  - o Two bodies have the same direction of motion when the have the same goal
  - o The direction of force/acceleration and velocity are the same
  - o Negative velocity is not meaningful

Example of an incorrect answer:
- Students have trouble understanding velocity as a derivative with respect to time of distance.

Rating:
- No correct answer: 0 points
- At least one correct answer: 1 point

Correct answers out of two or three categories: 2 points

Translations of teachers answers [and coding]
Teacher A:
1) A long way is automatically connected with high speed.
2) A short distance cannot be connected with "small/little speed" – you cover a short distance fast.
This is important because considerations from other relationships are often connected incorrect (for example mapping distance-force)
[0 points with the coding scheme even if it seems to be a description of a misconception]

Teacher B:
- "speed" needs permanent force (students do not recognize that for a constant motion force is only needed to compensate friction)
- Energy of a moving body increases linear with its speed (students do not recognize that e. g. a cars stopping distance does not increases linear with its speed)
[1 point for "A uniform movement requires a force"]

Teacher C:
- Scalar
- Decreases without push
- „Natural" speed/velocity of a body is static/rest
- Direction is easy to change (kink in a path is possible)
- Speed needs force
The chance to develop adequate concepts lies in the differentiation to the naïve concept of speed/velocity respectively in the insight that corrections are necessary.
[2 points for "Velocity has no direction" and "Without force there is no motion"]

Fig. 3.3 Example item, coding scheme (Kirschner et al., 2016), and translated teachers answers

Connections to the Model

While the ProwiN study was planned (and completed) prior to the development of the RCM, there are clear connections between our research and this new model. More specifically, I would suggest that the PCK test instrument is a measure of collective PCK (cPCK), with an emphasis on teachers' *knowledge of instructional strategies* and *knowledge of student understandings*. The items on the PCK test instrument cover a mix of grain sizes including topic level to concept level. In addition, the instrument asks physics teachers about their planning, teaching, and reflections on teaching episodes. So, you may be wondering: Why does it make sense to locate the PCK test instrument within the realm of cPCK? Why is it not a measure of personal PCK (pPCK)? Or enacted PCK (ePCK)? Does it make sense to separate out the evaluation of teacher's cPCK from pPCK and ePCK? I will try to answer these questions related to our study.

Why not ePCK? When completing our written PCK test instrument, the exercise triggers teachers to search their own personal knowledge base; however, we cannot be sure that the knowledge they access to respond to the PCK test items is the same knowledge they would actually use when planning, teaching, and reflecting on their science teaching in an authentic setting with their own students. We tried to analyse this link with validation studies and by exploring the connection in the second phase of the ProwiN project. We found when introducing the term "enacted PCK", a validation problem emerges—prior to the RCM, researchers may have expected that their PCK instruments captured knowledge indicative of the knowledge used by science teachers when they were teaching, when in fact researchers may have been measuring some other components of PCK that was not in fact enacted PCK (ePCK). With the RCM, it is advantageous that we now have more precise language to help address this issue about the type of PCK that is being captured and evaluated (e.g. cPCK, pPCK, ePCK).

Why not pPCK? In our study, even if we had wanted to capture the knowledge base of individual science teachers, not of a group, when coding their responses on the PCK test, we would still have to check the knowledge of every analysed teacher against a form of group knowledge (cPCK), for example, in our case using the coding scheme to decide if a teacher's response demonstrates expert PCK or not. In this way of thinking, it is only possible to capture pPCK if we portray knowledge without judging.

Why cPCK? Individual knowledge becomes collective knowledge in the moment we read a teacher answer and decide if it is in line with expert PCK or not (or is "correct" or "incorrect"). By adding "answers" into our coding scheme from individual teachers that seemed to be reasonable, but were not previously part of (the researcher's) collective PCK, we were trying to incorporate new aspects of pPCK in the test instrument. In the RCM, this modification to the test could be thought of as an example of "knowledge transfer" from pPCK to cPCK (as shown by the arrows from pPCK to cPCK in the model). In order to analyse cPCK, one "only" has to analyse the knowledge of a group of professionals—which in a group is a normative decision—and then compare the relative amount of knowledge held by an individual teacher to this group's knowledge.

Reflections

After analysing where to locate our PCK test instrument in the RCM, I wanted to reflect on a thought about the validity of the model as it relates to PCK test instruments. I see the model as comprehensive—it has to be in order to cover such a complex construct. I argued why I think our paper-and-pencil PCK test instrument only covers cPCK, but even when evaluating only cPCK the potential application of the model is still huge when considering what the cPCK realm can encompass. This breadth and depth of the model mean if we want to assess PCK validly, then we would have to pick some aspects of the model and examine cPCK systematically. More specifically, this choice is especially relevant to the grain size (e.g. analysing cPCK concept by concept) and the phases of teaching (e.g. planning, teaching, and reflecting). Each of these aspects would have to be analysed separately and with many different tests because of a limited number of items per test (assuming we used paper-and-pencil tests). The RCM gives us a more precise language to describe different studies in science teacher education research. As a next step in research, we will have to use this expanded language to compare studies and to come to a comprehensive picture of science PCK assessed by different instruments.

The Applications and Implications of the Refined Consensus Model of PCK to Research About Making Sense of SCIENCE Professional Learning

Kirsten R. Daehler, Joan I. Heller and Nicole Wong

Context

Over the past decade, researchers from WestEd and Heller Research Associates have partnered to investigate the development of teachers' pedagogical content knowledge (PCK) for teaching science as a result of their participation in Making Sense of SCIENCE professional learning, an approach that is intentionally designed to

K. R. Daehler
San Francisco, USA

J. I. Heller · N. Wong
Oakland, USA
e-mail: nwong@gordonheller.com

strengthen teacher PCK. In early studies, data were collected via pre-/post-interviews from a small number of elementary school teachers. Later, written paper-and-pencil instruments were developed to accommodate large sample sizes in causal studies with both elementary and middle school teachers. Most recently, through the use of classroom video and teacher interviews, we investigated ways teachers utilised or enacted their professional knowledge when engaged in content-specific teaching. Related research questions included:

- Did the professional learning produce changes in teacher PCK?
- Is there evidence that the impact of the professional learning on teachers' PCK accounts in part for the impact on student outcomes?
- What changes in teacher PCK may have contributed to increases in student science achievement?

Connections to the Model

While this research was carried out prior to the development of the Refined Consensus Model (RCM) of PCK, it is helpful to situate our work in this newer model. Using this lens, we see our research contributions as focusing squarely on teacher practitioners and the inner layers of the model—personal PCK (pPCK) and enacted PCK (ePCK). More specifically, some components of the structured interviews and written paper-and-pencil instruments provide insight into elements of teachers' pPCK, for example, when we ask them to describe what they anticipate being difficult for students when learning about a specific topic (e.g. electric circuits at the fourth grade). With both types of instruments, we also ask teachers to describe their learning goals for students related to a given topic and to explain how they might go about teaching students a given concept (e.g., a force is a push or pull interaction). Given that the written and interview prompts inquire about teaching students at a specific grade level, the resulting data also tap into each teacher's personal knowledge, which may be informed by their prior teaching experiences as well as their own learning experiences.

By sampling teachers before and after the science professional learning intervention and identifying changes in the depth and sophistication of their responses, we acknowledge the dynamic nature of teachers' pPCK as it grows and changes over time. We further hypothesise that participation in the Making Sense of SCIENCE professional learning—which utilises teaching cases of practice, analysis of student work samples, and structured conversations with colleagues—is the mechanism that shapes teachers' pPCK, informs their ePCK and classroom practices, and ultimately enhances student achievement.

In order to gather data about teachers' ePCK, which we view as a key link in the cascade of influences from teacher professional learning to student outcomes, we have used measures such as written paper-and-pencil instruments, structured interviews, and classroom videos in conjunction with interviews conducted before and after instruction. Both the written instrument and structured interviews situate teachers in a specific classroom context (e.g., teaching electric circuits to fourth graders) and then ask them to engage in several authentic tasks—analysing a sample of student work and planning next steps in instruction to address that student's understanding. Teachers' responses are analysed for how they applied their ePCK, as evidenced by their identification of common student misconceptions and gaps in understanding and by their explicit selection and sequencing of multiple, accurate, and grade-appropriate strategies to support student understanding (e.g., engaging students in observable phenomena, utilising drawings and other representations, explaining underlying mechanisms to explain "why") of a specific topic (e.g., complete circuits). While these PCK instruments do not provide direct observations of teachers' ePCK as applied to teaching their own students, we view this approach as a proxy measure for ePCK in situ, that is, similar to the way in which a flight simulator is an indicator of a pilot's performance in the air because as the teacher (or pilot) responds to the information provided they must draw on their prior knowledge and skill (pPCK) to take action.

In contrast, data collected from classroom videos in conjunction with teacher interviews conducted before and after instruction provide a more situated window into teachers' ePCK as this approach looks more directly at "the act of teaching a particular topic in a particular way for a particular purpose to particular students for enhanced student outcomes" (Gess-Newsome, 2015, p.36). This video interview approach attends to teachers' planning, enacting, and reflecting on their own teaching with their own students, which help us learn about teachers' underlying pedagogical reasoning related to the observed lesson. In our studies, the videos were rated in relation to five dimensions of classroom practice (e.g., domain-specific representations are used to support sense-making, teacher elicits and attends to student thinking). Together, these dimensions provide information about teachers' ePCK and the extent to which they are able to draw on their pPCK to actively engage students in sense-making around core science concepts, which we believe lead to the observed enhanced student outcomes (e.g., student achievement).

Reflections

As we consider our prior science teacher education research through the lens of the RCM, it pushes us to think more deeply about what specific aspects of PCK our research addresses. The RCM also points out ways in which our instruments could be revised to expand on the specific aspects of PCK being measured. For example, while our written instrument focuses on ePCK and the act of *planning* to teach a particular concept to a particular student with the explicit purpose of enhancing that student's learning outcomes, we could add prompts to elicit information about a teacher's *pedagogical reasoning*.

From the pragmatic side of research into teachers' PCK in science, the model has led us to wonder if there are less costly ways than using classroom observations with teacher interviews to examine ePCK. More specifically, we wonder if there is a high correlation between how teachers perform on written measures of pPCK (with regard to planning instruction for a hypothetical student or class) and their ePCK as demonstrated when authentically planning lessons for students in their actual class. As we discuss the RCM and our own PCK research, we can see significant challenges associated with measuring ePCK as it relates to the act of teaching, in contrast with the seemingly more manageable task of assessing teacher's ePCK related to act of planning to teach and reflecting on their teaching. Another notable challenge for researchers assessing teachers' ePCK and pPCK is taking into account the myriad of amplifiers and filters—attitudes, beliefs, knowledge, and experiences from their own classrooms—that shape each teacher's responses. In summary, the new language and constructs introduced by the RCM present interesting challenges that are exciting to us and suggest a host of new research directions.

References

Anderson, D. L., Fisher, K. M., & Norman, G. J. (2002). Development and evaluation of the conceptual inventory of natural selection. *Journal of Research in Science Teaching, 39*(10), 952–978.

Carlson, J., Stokes, L., Helms, J., Gess-Newsome, J., & Gardner, A. (2015). The PCK summit: A process and structure for challenging current ideas, provoking future work, and considering new directions. In A. Berry, P. Friedrichsen, & J. Loughran (Eds.), *Re-examining pedagogical content knowledge in science education* (pp. 14–27). New York, London: Routledge.

Cauet, E., Liepertz, S., Borowski, A., & Fischer, H. E. (2015). Does it matter what we measure? Domain-specific professional knowledge of physics teachers. *Schweizerische Zeitschrift für Bildungswissenschaften, 37*(3), 463–480.

Gess-Newsome, J. (2015). A model of teacher professional knowledge and skill including PCK. In A. Berry, P. Friedrichsen, & J. Loughran (Eds.), *Re-examining pedagogical content knowledge in science education* (pp. 28–42). New York, NY: Routledge.

Keeley, P., & Eberle, F. (2005). *Uncovering student ideas in science: Another 25 formative assessment probes* (Vol. 3). NSTA press.

Kirschner, S., Borowski, A., Fischer, H. E., Gess-Newsome, J., & Aufschnaiter, C. V. (2016). Developing and evaluating a paper-and-pencil test to assess components of physics teachers' pedagogical content knowledge. *International Journal of Science Education,* 1343–1372.

Kruse, S., Louis, K. S., & Bryk, A. (1995). An emerging framework for analyzing school-based professional community. In K. S. Louis & S. Kruse (Eds.), *Professionalism and community: Perspectives on reforming urban schools.* Corwin: Long Oaks, CA.

Loughran, J. J., Berry, A., & Mulhall, P. (2012). *Understanding and developing science teachers' pedagogical content knowledge* (2nd ed.). Rotterdam, the Netherlands: Sense Publishers.

Magnusson, S., Krajcik, J., & Borko, H. (1999). Nature, sources and development of pedagogical content knowledge for science teaching. In J. Gess-Newsome & N. G. Lederman (Eds.), *Examining pedagogical content knowledge* (pp. 95–132). Dordrecht, The Netherlands: Kluwer Academic Publishers.

Morine-Dershimer, G., & Kent, T. (1999). The complex nature and sources of teachers' pedagogical knowledge. In J. Gess-Newsome & N. G. Lederman (Eds.), *Examining pedagogical content knowledge* (pp. 21–50). Dordrecht, The Netherlands: Kluwer.

National Research Council [NRC]. (2012). *A framework for K-12 science education: Practices, crosscutting concepts, and core ideas.* In H. Schweingruber, T. Keller & H. Quinn (Eds.). National Academies Press.

NGSS Lead States. (2013). *Next generation science standards: For states, by states*. Washington, DC: The National Academies Press.

Tepner, O., Borowski, A., Dollny, S., Fischer, H. E., Jüttner, M., Kirschner, S.... Wirth, J. (2012). Modell zur Entwicklung von Testitems zur Erfassung des Professionswissens von Lehrkräften in den Naturwissenschaften [Item development model for assessing professional knowledge of science teachers]. *Zeitschrift für Didaktik der Naturwissenschaften*, 18, 7–28.

Part II
Reimagining the Existing

Rebecca Cooper

Overview

The introduction of the Refined Consensus Model (RCM) of pedagogical content knowledge in Chap. 2 of this volume has provided researchers in science teacher education with an opportunity to reconsider their existing research, and to reflect on how this new model may allow for new connections to be made and understandings of PCK deepened. Chapters in this part have been contributed by individual authors and by research teams who have taken retrospective views of their existing work and considered how this work can be reimagined in light of the RCM of PCK.

We start with Chap. 4 by Park, where she discusses how two extant PCK models (the pentagon model of PCK and the indispensable and idiosyncratic PCK model) can be situated within the RCM of PCK. Park's rethinking of her significant work using these models, as it relates to the RCM, highlights the importance of professional communities and shared expertise to development of PCK. She sees collaborative interactions among science teachers as essential for the development of PCK as they make teachers' knowledge public and more visible to colleagues. Such findings tie in closely with the realms of PCK (i.e., collective, personal, and enacted PCK) depicted in the RCM and the knowledge exchanges and transformations occurring in both directions from realm to realm as teachers interact about their teaching practices in more public arenas.

Mavhunga's work in Chap. 5 seeks to demonstrate how her ongoing research about topic-specific PCK can be reimagined through the RCM. She now revisits her examination of PCK at topic level and calls on PCK researchers to be more specific about grain size when exploring PCK and to be less ambiguous in their references to PCK. In her view, the RCM's identification and inclusion of grain sizes ranging from *discipline-specific* to *topic-specific* to *concept-specific* PCK through all realms and aspects of PCK acknowledge this call.

Chapter 6 by Sorge, Stender, and Neumann sought to learn more about the exchanges of knowledge occurring between the broader professional knowledge bases and the three realms of PCK depicted in the RCM as science teachers develop their professional competence. In reworking data from two of their previous studies using the RCM to guide their analyses and interpretations, they found that the RCM confirmed and validated the worth of their earlier work when viewed as explorations of relationships between collective PCK (cPCK) and personal and enacted PCK (pPCK and ePCK).

In Chap. 7, Schneider identifies the possibilities the RCM presents for untangling the complexity of PCK by describing the three realms of PCK, from the professional knowledge of a community of science teachers, to that of an individual science teacher, to the ideas used to inform and the actions taken in an instance of teaching. Schneider then reconsiders her learning study research in relation to the RCM, as an approach to develop and illustrate science teachers' pedagogical knowledge from a teacher educator's stance. She finds that it is helpful to situate the work within the three realms of PCK as described in the RCM.

Finally, in Chap. 8, Park and Suh focus on how the RCM provides a conceptual framework for informing a methodological approach to PCK research called "PCK mapping." They find that the RCM is a useful tool for examining the pedagogical reasoning and action that drives the knowledge exchange between pPCK and ePCK of science teaching, highlighting the support that the RCM offers for making the abstract and complex structure of PCK more visible, explicit, and accessible.

Chapter 4
Reconciliation Between the Refined Consensus Model of PCK and Extant PCK Models for Advancing PCK Research in Science

Soonhye Park

Abstract In this chapter, I discuss how two pedagogical content knowledge (PCK) models known as the pentagon model of PCK and the indispensable and idiosyncratic PCK model can be situated within the Refined Consensus Model (RCM) of PCK as I reflect on examples of my earlier research in science teacher education. To guide my previous research, I used the pentagon model of PCK to *capture* and *portray* PCK and the indispensable and idiosyncratic PCK model to *measure* and *assess* PCK. I also illustrate how research methods drawn from these two existing models, including approaches such as PCK mapping, in-depth analysis of PCK, PCK surveys, and PCK rubrics, align with the RCM and what insights the RCM provides for improving these methods and advancing PCK research. The body of this chapter is structured around four distinctive features of the RCM, compared to the earlier Consensus Model (CM), that emerged through a critical comparison of the new model with the two extant PCK models, i.e. the RCM's (1) emphasis on learning context for capturing PCK, (2) explicit visual representation of the link between PCK and the enactment of PCK, (3) distinction between personal PCK and collective PCK, and (4) shift in focus towards PCK development. Major methodological suggestions emerging from this critique for future research into science teacher education using the RCM include data collection encompassing the entire pedagogical cycle and greater attention to contextual factors, student learning, and pedagogical reasoning.

Introduction

As an outcome of the second pedagogical content knowledge (PCK) summit in 2016 and follow-up discussions, the participants developed the refined consensus model (RCM) of PCK, building on the 2012 consensus model of teacher professional knowledge and skills (Gess-Newsome, 2015) and incorporating new ideas that emerged during the summit. Whereas the former consensus model aimed to

S. Park (✉)
Department of STEM Education, College of Education, North Carolina State University, Raleigh, NC, USA
e-mail: spark26@ncsu.edu

A. Hume et al. (eds.), *Repositioning Pedagogical Content Knowledge in Teachers' Knowledge for Teaching Science*, https://doi.org/10.1007/978-981-13-5898-2_4

reach agreement in defining PCK, the RCM intends to help researchers identify areas to study for advancing PCK research through situating their studies within the context of teaching practices (see Chap. 2). In this chapter, I describe how my research on PCK for science teaching can be reimagined within the context of the RCM and the insights this updated model provides for future PCK research, especially those relating to research methodology. In particular, I look for areas of compatibility and differences between the RCM and two extant PCK models that have previously guided my research on PCK, i.e. the pentagon model of PCK (Park & Oliver, 2008a) and the indispensable and idiosyncratic PCK model (Park, Suh, & Seo, 2017). Then, drawing on areas where the PCK models diverge from one another, I discuss how the research methods used to portray and assess PCK, derived from the extant models, can be situated within and applied to the RCM.

The body of this chapter is structured according to four distinctive features of the RCM that emerged through critical comparison with the two extant PCK models: (1) emphasis on learning context in capturing PCK, (2) explicit visual representation of the link between PCK and the enactment of PCK, (3) distinction of personal PCK from collective PCK, and (4) shifted focus towards PCK development. While discussing each of the features highlighted by the RCM, important insights into the conceptualising of PCK for science teaching and methodological approaches to PCK research are also provided.

Emphasis on Learning Context in Capturing PCK

One critical difference between the RCM and the original Consensus Model (CM) (Gess-Newsome, 2015) is more explicit and greater emphasis on learning context as an amplifier and a filter of teacher PCK in science, as well as, a mediator for teacher actions. In the RCM, the learning context encompasses a wide range of factors influencing teacher PCK from the broader education sector to individual student attributes (see Chap. 2). The prominence of the learning context in the RCM suggests important implications for methodological approaches, especially for those that I have previously used to capture PCK. For example, the PCK map approach was developed to capture interactions among PCK constituent components by a series of quantifications and visualisations of PCK episodes (i.e. instances of a teacher's PCK in use) that were identified in science classroom observations and teacher interviews (Park & Chen, 2012).

The PCK map approach employs the pentagon model of PCK as both a conceptual and analytic framework that defines PCK as an integration of five constituent components, which are (1) orientations towards teaching science, (2) knowledge of students' understanding in science, (3) knowledge of science curriculum, (4) knowledge of instructional strategies and representations, and (5) knowledge of assessment of science learning (Park & Oliver, 2008a). To accentuate the interrelatedness among them, the components are presented in a pentagon shape in the model, as shown in

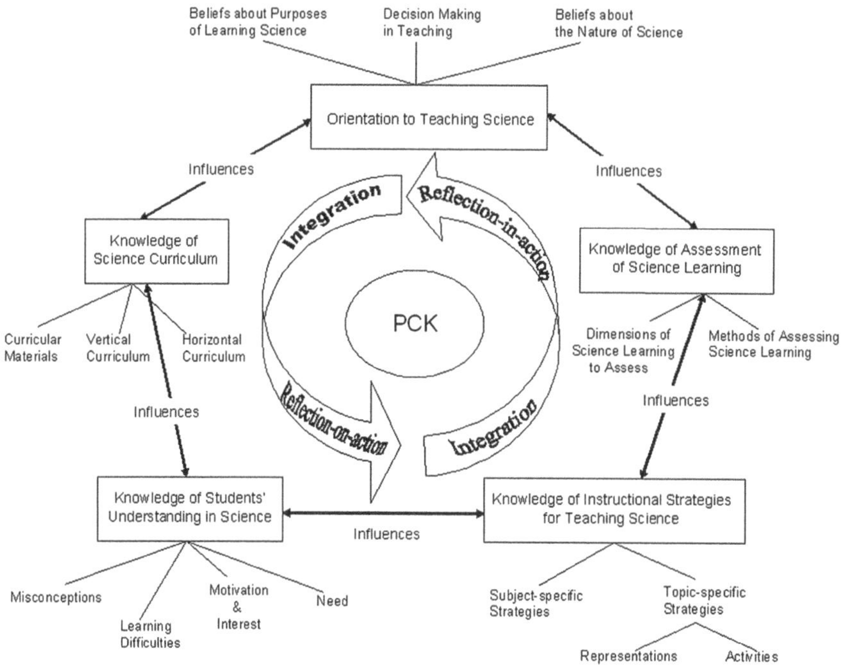

Fig. 4.1 Pentagon model of pedagogical content knowledge for teaching science (Park & Oliver, 2008b)

Fig. 4.1, instead of a linear manner. The components are integrated through complementary and ongoing readjustments by both reflection-in-action and reflection-on-action.

Given the interplay and integration of PCK components, the PCK map approach draws greater attention to individual science teachers' cognitive processes than to the teachers' interactions with contextual factors. Specifically, as shown in Fig. 4.2, during the first step of the PCK map approach, segments that include PCK components are identified from observation and interview data and then synthesised into a PCK episode through the in-depth analysis of explicit PCK. This analysis procedure involves both de-contextualisation and re-contextualisation of data (Tesch, 1990). In other words, extracting data segments that retain PCK components separates them from their original contexts (i.e. de-contextualisation), and these segmented data are then reassembled into a PCK episode through re-contextualisation (Coffey & Atkinson, 1996). As part of the re-contextualisation process, the PCK episode is assembled in terms of what was happening in the context in which the episode occurred and the PCK components involved in the episode. Interestingly, reflecting on this analysis procedure in the light of the RCM, I have come to realise that contextual factors included in the description of a PCK Episode were often limited to a specific classroom environment and did not include sufficient descriptions of broader contextual

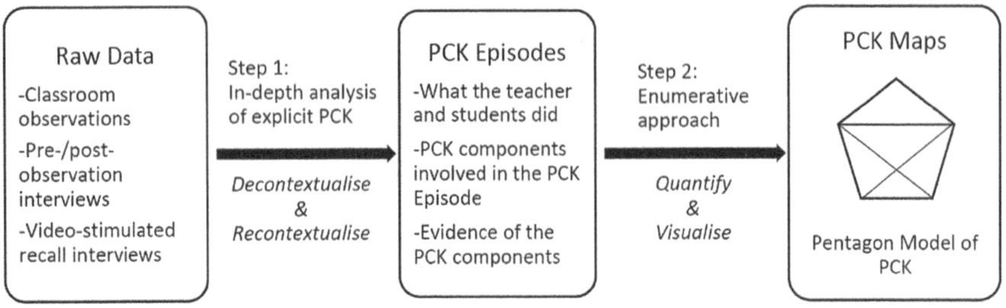

Fig. 4.2 Analysis procedures of the PCK map approach

factors like educational policies, school climate, administrative requirements, and peer interactions.

Once a PCK episode is described in the PCK map approach, interactions among the identified PCK components are represented using the pentagon model of PCK to create the outcome known as a PCK map (Park & Chen, 2012; Suh & Park, 2017). Since the PCK map approach was first developed to answer particular research questions related to interactions among the PCK components, this step centres on visualising only the components' relationships. Consequently, contextual factors associated with the PCK episode are not presented. Recognising the importance of the learning context in the RCM, I acknowledge that the PCK map, as the final product of this analytic method, may unintentionally reinforce a static view of PCK. The impression forms that teachers can develop and apply PCK independently from the learning context (Berry, Depaepe, & van Driel, 2016) even though the mapping process takes into consideration contextual factors. In this regard, modifications to how a PCK map is depicted are suggested to enable those contextual factors directly influencing the PCK components to be visually represented in the PCK map.

Explicit Visual Representation of the Link Between Teachers' PCK and the Enactment of PCK

A central feature of the RCM is that it attempts to clearly illustrate PCK in practice (i.e., enacted PCK (ePCK)) as teachers' application of PCK during a pedagogical cycle (see Chap. 2). This effort to show the linkage directly responds to criticisms levelled at PCK research for its concentration on investigating what teachers know, without relating it to what teachers actually do and what students learn (Settlage, 2013). Recently, Shulman (2015) reminded us that PCK was not coined as a cognitive construct that resides in individual teachers' heads, but as a dynamic construct that describes the complex processes that teachers apply during the act of teaching particular subjects for particular purposes to particular students within particular settings. In this regard, the RCM signifies the idea that PCK comprises what a teacher does and the pedagogical reasoning that guides the teacher's actions, as well as what the teacher knows (Baxter & Lederman, 1999).

The conceptualisation of ePCK as a distinct form of PCK in the RCM is congruent with the conception underpinning the pentagon model of PCK; that is, PCK consists of two dimensions—understanding and enactment (Park, 2005). It can be argued that ePCK from the RCM conceptually corresponds to the enactment dimension of PCK in the pentagon model and pPCK (personal PCK) with the understanding dimension of PCK. However, there are significant differences in the way that the respective models define the relationships between ePCK and pPCK and the understanding dimension and the enactment dimension. In the RCM, ePCK is defined as *a subset* of pPCK that is manifested when a teacher utilises pPCK, not only when interacting directly with students through reflection-in-action, but also when planning instruction and reflecting on instruction and student outcomes through reflection-on-action. In this sense, PCK encompasses both knowledge and skills (see Chap. 2; Gess-Newsome, 2015). On the other hand, in the pentagon model, the understanding and enactment dimensions are described as *two complementary aspects* of a teacher's PCK that are demonstrated, rather than one being a part of the other. In other words, the two dimensions indicate different facets of PCK that constitute a teacher's whole PCK construct.

This conceptual variation necessitates different approaches to measuring PCK. The pentagon model of PCK suggests that measuring a teacher's PCK for science teaching requires measuring both cognitive and enacting dimensions of PCK. Consistent with this view, the PCK measures that my research team developed consisted of two types, each for measuring one of the two dimensions (Park & Suh, 2015). Specifically, the PCK survey directly measures "what teachers know" using teachers' responses to a paper–pencil-type survey, whereas the PCK rubric indirectly measures PCK through inferences from teachers' enacted PCK, focusing on "what teachers do" and their underlying pedagogical reasoning. Assuming that one measure only partially estimates an individual's PCK, individual teachers' PCK scores ought to be determined by the sum of their scores on both measures (Park et al., 2017).

On the other hand, the RCM suggests the possibility of estimating a teacher's science pPCK by measuring ePCK, given that ePCK is an expression of pPCK. How then can we measure ePCK? What data sources are necessary to measure ePCK? How should those data be analysed to estimate an individual's ePCK or pPCK in reliable and valid manners? Although those questions call for further empirical investigation, I believe that the RCM provides substantial direction for such research efforts. First, ePCK is manifested, evident, and utilised throughout the three phases of the pedagogical cycle: plan, enact, and reflect (see Chap. 2). This understanding of ePCK implies that data sources and collection methods should be planned to elicit PCK from each of the three stages. For example, teaching observations can capture elements of ePCK during the enact phase, while lesson plans, interviews on planning instruction, and reflective interviews can be useful to identify other elements of ePCK during the planning and reflection phases. Thus, to capture ePCK holistically, it is suggested researchers collect data from multiple sources at different points in time that span a full pedagogical cycle.

Second, a two-way knowledge exchange between pPCK and ePCK should be considered as a critical component of ePCK. Particularly, the pedagogical reasoning behind a teacher's use and synthesis of pPCK into a form of ePCK ought to be considered in gauging the teacher's PCK. Interviews anchored to videotaped lessons or reflection journal entries are examples of data sources from which to draw understanding of a teacher's pedagogical reasoning through both reflection-in-action and reflection-on-action (Schön, 1983). Finally, measuring PCK necessitates a normative stance in which some forms of PCK are more highly regarded than others (Park & Suh, 2015). The RCM recommends that impact on students' science learning should be counted towards norms for determining the quality of PCK. Building on the conceptualisation of personal PCK and skills in the original CM (Gess-Newsome, 2015), ePCK is a representation of personal PCK (pPCK) in the act of teaching a particular science topic in a particular way for a particular *purpose* to particular students for enhanced *student outcomes* (see Chap. 2). This conceptualisation indicates that comparative judgments made about teacher PCK in science need to include their students' science learning outcomes in relation to the teaching purpose. In this regard, I propose a modification to the PCK rubric that includes student learning outcomes as a component by which a teacher's enacted PCK in science is determined. To this end, how to measure student learning outcomes, in relation to teacher PCK, in reliable and valid ways must be a central question driving my future research involving the measuring of PCK. Besides measuring PCK, research studies that aim to portray PCK should also expand their scope of study, embracing the influence that a teacher's PCK in science exerts on his or her students' science learning.

Distinction of Personal PCK from Collective PCK

Another unique feature of the RCM to emerge is the differentiation between personal PCK (pPCK) and collective PCK (cPCK). Collective PCK in science represents a compendium of knowledge held by a group that extends beyond the knowledge of an individual science teacher and embodies more than what is known from research about teaching particular science subject matter to particular students in a particular learning context (see Chap. 2). Stated differently, cPCK is a collection of pPCK shared by a group of teaching professionals related to teaching a specific discipline, a specific topic, or a specific concept. For pPCK to become cPCK, the RCM requires sharing, articulation, and communication of that personal knowledge amongst a group of professionals. Although there is no explicit mention of validation of this shared knowledge through standardised processes, the RCM differentiates cPCK from pPCK and ePCK by virtue of the vetting of this knowledge by peers and other professionals through formal and informal channels. In this regard, cPCK in science comes to signify the perspective of PCK as knowledge *for* teachers (i.e. what teachers need to know) rather than knowledge *of* teachers (i.e. what teachers know) (Fenstermacher, 1994). With that said, cPCK is likely to be assessed in a comparative and normative way (Berry et al., 2016).

The concept of cPCK reflects the idea of indispensable PCK as illustrated in the indispensable and idiosyncratic model of PCK (Park et al., 2017). As shown in Fig. 4.3, the model distinctly incorporates both the personal idiosyncratic knowledge *of* individual science teachers, along with the indispensable knowledge *for* teachers that is necessary to execute effective instruction. Specifically, the indispensable PCK refers to the aspects of PCK for effective teaching of subject matter that are considered necessary across different teachers and a variety of educational contexts (Park & Suh, 2015). Thus, the indispensable PCK is measurable in a normative way, distinguishing between teachers with sophisticated and shallow PCK for teaching science. To clarify norms for determining the quality of the indispensable PCK for teaching science, the indispensable and idiosyncratic PCK model posits two major criteria: (1) appropriate representation of canonical science and (2) purposeful application of empirically supported consensus knowledge about effective instruction grounded in research on learners and learning (Donovan & Bransford, 2005). Both the PCK survey and PCK rubric tools target the measurement of indispensable PCK (Park et al., 2017).

I consider idiosyncratic PCK in science as an essential part of teacher knowledge that exemplifies teachers' professionalism, demonstrating their autonomy and ability to be responsive to diverse students within a specific social, cultural, and educational context, through adapting and tailoring instructional materials and strategies (Barnett

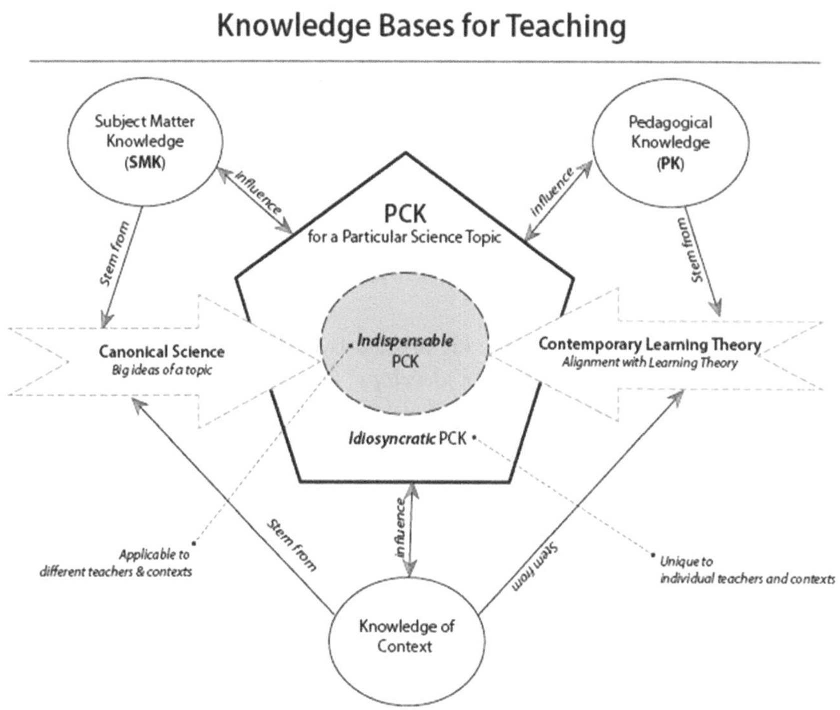

Fig. 4.3 Conceptual model of indispensable PCK and idiosyncratic PCK (Park et al., 2017)

& Hodson, 2000; Donnelly, 2001; Park & Oliver, 2008b). However, I also assume the presence of a collection of PCK in science that cuts across diverse contexts and which serves as foundational knowledge upon which teachers build the personal idiosyncratic PCK for science teaching that is unique to the learning context where their knowledge base is put into practice. This indispensable PCK is what cPCK intends to capture in the RCM.

However, I contend that in its current form, the RCM does not sufficiently unpack the meaning of cPCK in a way that will help conceptualise standards against which individual teachers' PCK in science is compared. Well-defined conceptualisations of PCK must be central to determining what is to be assessed, how it is to be assessed, and theorising important assumptions about the outcomes of PCK measures (Park & Suh, 2015). Hence, in order to better guide a line of research on measuring PCK, further clarification and articulation of cPCK that encapsulate the notions of knowledge *for* teachers of science and what constitutes indispensable PCK are imperative.

Nonetheless, I believe that the inception of cPCK by the RCM is a noteworthy step forward in advancing PCK research, in that it sheds light on the public aspect of PCK that transcends the knowledge bases of individual teachers of science. Professional knowledge must be public and represented in a form that enables it to be accumulated and shared with others (Hiebert, Gallimore, & Stigler, 2002). cPCK is, indeed, the core of PCK as the professional knowledge of teachers. Yet, professional knowledge also requires a system for verification and improvement (Hiebert et al., 2002). As I mentioned before, however, the knowledge exchanges in the RCM do not clearly demonstrate mechanisms through which shared pPCK can be publicly examined, verified, refuted, or modified. Clarification on those procedures will fill the gaps in our understanding about how pPCK becomes the profession's collective canonical knowledge of effective science teaching.

Shifted Focus Towards PCK Development

The last and most important feature highlighted by the RCM is the shift in focus from *what PCK is* towards *how PCK develops*. Accordingly, the model offers a different way to think about how teachers develop PCK for effective science teaching. Specifically, it draws special attention to the importance of colleagues, students, professional organisations, and contextual factors in developing PCK. By doing so, the RCM underscores that the development of PCK goes beyond individual science teachers' cognitive processes and requires their social interactions with colleagues, students, and others when negotiating the complexities of the learning context.

In this regard, the RCM suggests three new directions for science teacher education research on PCK, especially the development of PCK that has been a dominant line of PCK research (Depaepe, Verschaffel, & Kelchtermans, 2013). First, researchers need to further consider relevant external and contextual factors closely associated with pPCK and ePCK development to fully understand how teachers grow and change in their PCK in science over time. Previous studies about factors that influence the development of PCK primarily focused on those more proximal to teacher knowledge and practice, such as teaching experience, students, mentors, and professional development programmes (Park, 2005). Few researchers have examined broad and distal factors in relation to PCK such as federal policy, ministry requirements, national/state standards, school culture, and collegial dynamics that are also crucial components of the whole learning context. As the RCM indicates, a teacher develops and applies PCK through complex processes mediated by a multitude of factors. However, it is neither easy nor feasible to consider all factors in a single study. Close attention to these under-researched factors will contribute to building a complete picture of how multiple factors interact in shaping the development of teachers' PCK in science and provide significant implications for the design of interventions to improve PCK.

Second, considering knowledge exchanges between different components in the RCM, researchers need to attend to the causes of PCK growth and mechanisms of the cause–effect relationship. In particular, a clear understanding of the factors that can be best leveraged to create changes in PCK and the mechanisms through which they work will advance our understanding of how to design learning opportunities for teachers of science and how to assess teacher PCK in science. Last, the RCM represents PCK development for teaching science embedded in the larger milieu of the learning context, interaction with students and colleagues, and interplay among broader professional knowledge bases (see Chap. 2). This embedded nature of PCK development implies that every aspect of teachers' daily work in the rich, complex, and constantly changing environments where they are situated impacts their PCK to varying degrees. Consequently, individual teachers may not always experience steady incremental growth in PCK along their career trajectory (Schneider & Plasman, 2011). Having said that, I postulate that we should view PCK development for effective science teaching as continual change across a broad span of time, rather than as a series of discrete changes resulting from particular training experiences or critical classroom incidents. In this respect, longitudinal studies are needed to illuminate pathways that teachers move through as they develop more sophisticated PCK through experience in contextualised situations. Those studies will provide invaluable insights into how to design professional development programmes and experiences that support teachers' continued PCK growth through different stages of their career.

Closing Remarks

I have discussed implications for research methodology and future research on PCK in the field of science education drawn from the RCM, concentrating on four unique and significant features of the model, which are the emphasis on learning context, acknowledgement of ePCK (PCK in practice), recognition of cPCK, and the focus on PCK development. To recap, regardless of the focus of the study, any study that intends to examine PCK should consider that a multitude of contextual factors exerts profound influence on PCK and design data collection and analysis methods that sufficiently capture the influence of these contextual factors. Moreover, data sources and data collection methods to elicit PCK need to encompass the entire pedagogical cycle, because PCK is applied and used in every phase of the cycle from planning a lesson, to enactment of the planned lesson, to reflection on the enacted lesson.

The RCM implies several directions for methodological work, especially for the line of research focused on measuring and assessing teachers' PCK in science. ePCK should be the target in measuring PCK, with careful attention paid to how PCK manifests itself as teacher actions and practices in the classroom, rather than treating PCK solely as a cognitive construct. Given the complementary, two-way knowledge exchange between pPCK and ePCK, methodological approaches to measuring ePCK will need to address the pedagogical reasoning behind a teacher's actions. Data collection methods to tap into pedagogical reasoning may include interviews combined with observations, video-stimulated recall interviews about critical classroom incidents, or written reflections. Scoring rubrics or checklists are the most prevalent analytical tools for quantifying those qualitative data when measuring PCK (Park, Jang, Chen, & Jung, 2011), but researchers are encouraged to devote more effort to developing innovative, yet robust, analytic methods to discriminate between different levels of sophistication in pedagogical reasoning and ePCK. Most importantly, students' science learning outcomes should take precedence over other measures in determining the quality of an individual teacher's PCK for science teaching. However, given the diversity of students, standardising the assessment process for variables associated with student learning will be a challenging but necessary task that requires rigorous scholarly endeavour. Similarly, research focusing on describing and capturing PCK should also give ample consideration to PCK contextualised in practice and the associated contextual factors; pedagogical reasoning for instructional decisions and actions; and student learning outcomes in relation to PCK.

A significant insight to research on teachers' PCK for science teaching drawn from the RCM is the importance of professional communities and shared expertise to the development of PCK. This insight implies collaborative interactions among teachers become essential for the development of a teacher's PCK for science teaching because those interactions encourage teachers to make their knowledge public and understood by colleagues. However, little is known about how teachers' PCK evolves through interactions with other members of the profession and through feedback loops between classroom experience and professional contributions. In addition, there is a significant need for longitudinal studies with in-service teachers of science to better understand how to support teachers as lifelong learners who continuously advance to higher levels of PCK. Finally, more research on mechanisms that disentangle complex relations among the components within the RCM will propel the field forward in building a theoretical model of PCK rooted in firm empirical evidence that explains how to improve science learning for all students through promoting teachers' PCK and practice in science teaching (Park & Suh, 2015).

References

Barnett, J., & Hodson, D. (2000). Pedagogical context knowledge: Toward a fuller understanding of what good science teachers know. *Science Education, 85,* 426–453.
Baxter, J. A., & Lederman, N. G. (1999). Assessment and measurement of pedagogical content knowledge. In J. Gess-Newsome & N. G. Lederman (Eds.), *Examining pedagogical content knowledge* (pp. 147–161). Dordrecht, Netherlands: Kluwer Academic Publishers.
Berry, A., Depaepe, F., & van Driel, J. (2016). Pedagogical content knowledge in teacher education. In J. Loughran & M. L. Hamilton (Eds.), *International handbook of teacher education* (pp. 347–386). Singapore: Springer.
Coffey, A., & Atkinson, P. (1996). *Making sense of qualitative data: Complementary research strategies*. Thousand Oaks, CA: Sage.
Depaepe, F., Verschaffel, L., & Kelchtermans, G. (2013). Pedagogical content knowledge: A systematic review of the way in which the concept has pervaded mathematics educational research. *Teaching and Teacher Education, 34,* 12–25.
Donnelly, J. F. (2001). School science teaching as a profession: Past, present, and future. *School Science Review, 82*(300), 31–39.
Donovan, M. S., & Bransford, J. D. (Eds.). (2005). *How students learn: Science in the classroom*. Washington, DC: National Academy Press.
Fenstermacher, G. D. (1994). The knower and the known: The nature of knowledge in research on teaching. In L. Darling-Hammond (Ed.), *Review of research in education* (pp. 3–56). Washington, DC: American Educational Research Association.
Gess-Newsome, J. (2015). A model of teacher professional knowledge and skill including PCK. In A. Berry, P. Friedrichsen, & J. Loughran (Eds.), *Re-examining pedagogical content knowledge in science education* (pp. 28–42). New York: Routledge.
Hiebert, J., Gallimore, R., & Stigler, J. W. (2002). A knowledge base for the teaching profession: What would it look like and how can we get one? *Educational Researcher, 31*(5), 3–15.
Park, S. (2005). *A study of PCK of science teachers for gifted secondary students going through the national board certification process* (Unpublished doctoral dissertation). University of Georgia, Athens.
Park, S., & Chen, Y.-C. (2012). Mapping out the integration of the components of pedagogical content knowledge (PCK) for teaching photosynthesis and heredity. *Journal of Research in Science Teaching, 49*(7), 922–941.
Park, S., Jang, J., Chen, Y.-C., & Jung, J. (2011). Is pedagogical content knowledge (PCK) necessary for reformed science teaching?: Evidence from an empirical study. *Research in Science Education, 41,* 245–260.
Park, S., & Oliver, J. S. (2008a). Revisiting the conceptualisation of pedagogical content knowledge (PCK): PCK as a conceptual tool to understand teachers as professionals. *Research in Science Education, 38*(3), 261–284.
Park, S., & Oliver, J. S. (2008b). National board certification (NBC) as a catalyst for teachers' learning about teaching: The effects of the NBC process on candidate teachers' PCK development. *Journal of Research in Science Teaching, 45*(7), 812–834.
Park, S., & Suh, J. (2015). Trajectory from portraying toward assessing PCK: Drives, dilemmas, and directions for future research. In A. Berry, P. Friedrichsen, & J. Loughran (Eds.), *Re-examining pedagogical content knowledge in science education* (pp. 104–119). New York: Routledge.
Park, S., Suh, J., & Seo, K. (2017). Development and validation of measures of secondary science teachers' PCK for teaching photosynthesis. *Research in Science Education.*
Schneider, R. M., & Plasman, K. (2011). Science teacher learning progressions: A review of science teachers' pedagogical content knowledge development. *Review of Educational Research, 81*(4), 530–565.
Schön, D. A. (1983). The reflective practitioner: How professionals think in action. New York: Basic Books.

Settlage, J. (2013). On acknowledging PCK's shortcomings. *Journal of Science Teacher Education, 24,* 1–12.

Shulman, L. (2015). PCK: Its genesis and exodus. In A. K. Berry, P. Friedrichsen, & J. J. Loughran (Eds.), *Re-examining pedagogical content knowledge in science education* (pp. 3–13). New York: Routledge.

Suh, J., & Park, S. (2017). Exploring the relationship between pedagogical content knowledge (PCK) and sustainability of an innovative science teaching approach. *Teaching and Teacher Education, 64,* 246–259.

Tesch, R. (1990). *Qualitative research: Analysis types and software tools*. New York: Falmer Press.

Chapter 5
Exposing Pathways for Developing Teacher Pedagogical Content Knowledge at the Topic Level in Science

Elizabeth Mavhunga

Abstract This chapter seeks to illustrate how pedagogical content knowledge (PCK) at the topic level is positioned within the Refined Consensus Model (RCM) of PCK by retrospectively applying the RCM to an existing study. It demonstrates this positioning by tracking the development of chemistry pre-service teachers' PCK in electrochemistry across the newly conceptualised realms of collective, personal and enacted PCK. In this study, pre-service teachers are seen developing their personal PCK (pPCK) in the topic from a structured course based on collective PCK (cPCK) for the topic of electrochemistry and demonstrating part of their enacted PCK (ePCK) in the topic by developing a teaching programme on the topic. Data were collected using tools that were able to specifically measure the pre-service teachers' pPCK in electrochemistry (pre- and post-intervention tests) and from classroom tutorial assignments during the intervention. The analysis of test data to detect shifts in the quality of pPCK was enabled using a criterion-based rubric, while a qualitative in-depth content analysis was employed on the collected assignments. Findings indicated a differentiated improvement in the pre-service teachers' pPCK and in their ePCK for planning to teach the topic. Implications drawn for researchers in science teacher education include the importance of explicitly referencing the level or grain size of PCK under investigation in PCK studies and the realm in which the construct is located, as promoted in the RCM.

Introduction

What distinguishes science student teachers from others? A likely response to this question would refer to *learning* the *knowledge needed to teach* abstract and difficult science content for learner understanding despite the diversity in the classroom. Pedagogical content knowledge (PCK) is the teacher knowledge identified by Shulman (1986) as linked uniquely to the profession of teaching. PCK offers teachers the

E. Mavhunga (✉)
Science Education Division, School of Education, University of the Witwatersrand, Johannesburg, South Africa
e-mail: Elizabeth.Mavhunga@wits.ac.za

A. Hume et al. (eds.), *Repositioning Pedagogical Content Knowledge in Teachers' Knowledge for Teaching Science*, https://doi.org/10.1007/978-981-13-5898-2_5

knowledge to pedagogically restructure and package difficult and abstract content knowledge in formats accessible for learner understanding. The science education community has embraced PCK as the knowledge for teaching science (Abell, 2008; Darling-Hammond & Richardson, 2009), and the knowledge to be passed on to prospective science teachers. This transference of knowledge demands PCK to be addressed without ambiguity, as is the purpose of the collective chapters in this book. For example, it is important for pre-service teachers to be consciously aware of their learning in order to be aware of their own development. On the other end, it is important for science educators to distinguish the level of PCK (e.g., discipline- vs. topic-specific) in their teaching as these have a different composition and require emphasis on different knowledge types for their pedagogical value (Abell, 2008) to be realised. The Refined Consensus Model (RCM) of PCK, introduced in Chap. 2 of this book, brings back into focus the multidimensional nature of PCK and affords us a unison vocabulary for communicating the different levels of the construct and its applications.

The multidimensional nature of PCK emerges in the literature in different ways. Earlier studies have reported on the tacit nature of PCK and recommended for its assessment to occur in both planning and enacted classroom conditions as these contexts afford distinctly different insights about PCK (Aydeniz & Kirbulut, 2014; Park & Chen, 2012). A different set of studies are reported on the differentiated grain size of PCK evident when the teaching of a particular discipline and a specific topic is considered (Nezvalová, 2011; Veal & MaKinster, 1999). These authors suggested PCK exists at three levels, namely at discipline-, domain- and topic-specific levels. Other studies looked at PCK emerging from a perspective of a group of teachers (Loughran, Berry, & Mulhall, 2004) and recognised the emerging wisdom from the collective, while other studied PCK from individual teachers (Aydin, Friedrichsen, Bozc, & Hanuscinb, 2014). It could be reasonably argued that all the above-listed studies presented the multiple perspectives of the same theoretical construct, which has over the years posed a challenge on how to refer to each element without ambiguity (Aydin et al., 2014b). The RCM draws all these perspectives of dimensions into a single interlinked pictorial representation and illustrates the possible developmental path of the teacher's professional knowledge for teaching (PCK). It also shows how both practicing and prospective teachers may first draw on and grow their professional knowledge from a shared professional understanding about PCK, called collective PCK (cPCK). Such collective professional knowledge may be drawn from sources such as coursework, text or publications that all constitute a realm of understanding that is public and commonly shared as teacher knowledge for teaching. A pre-service teacher sitting in such courses would subsequently access and personalise the teaching as personal PCK (pPCK) to an extent made permissible by personal beliefs, context and other affective factors (referred to as filters and/or amplifiers in the RCM). pPCK consequently becomes the personal collection of ideas and understandings from which the pre-service teacher draws to inform the pedagogical reasoning involved in enacting out the planning, the actual teaching and lesson reflections—all referred to as enacted PCK (ePCK). Thus in the RCM, cPCK, pPCK

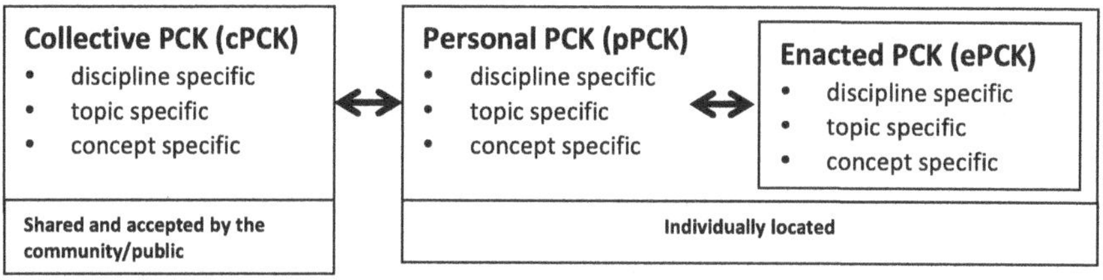

Fig. 5.1 A simplified illustration of the multidimensional nature of PCK in the RCM of PCK

and ePCK can be seen to represent three distinct but interrelated realms of PCK that map the path a teacher may follow in developing their professional knowledge.

The development of PCK through the above-mentioned three realms could be experienced at a discipline-specific, topic-specific or concept-specific level. These three levels reflect a continuum of grain sizes at which PCK could be considered at a particular time. The RCM of PCK acknowledges that this continuum occurs in each of the three PCK realms (cPCK, pPCK and ePCK) (see Chap. 2). The illustration below presents a simplified version of the RCM to make the continuum of PCK grain sizes within each of the three realms of PCK more explicit (Fig. 5.1).

This chapter contributes to the purpose of the book by demonstrating the positioning of PCK at a topic level with the grain size of PCK directed at a specific topic and available in each of the three PCK realms. Furthermore, the chapter serves to illustrate empirically how pre-service teachers in science can develop their pPCK in a specific science topic from a formalised structured course that is based on shared knowledge or cPCK for teaching that specific topic. The discussion on PCK at the topic level begins with the revisitation of its theoretical conceptualisation as provided below.

Explaining PCK in Science at a Topic-Specific Level

In the foundational literature on PCK, several science education researchers (Abell, 2007; Geddis & Wood, 1997; Van Driel, Verloop, & de Vos, 1998) pointed out PCK at a topic level when they first referred to the topic-specific nature of PCK in science. A number of sequel studies are built on this understanding and explored ways of examining the topic-specific nature of PCK (Aydin et al., 2014a) and the nature of the interactions among components that are visible when PCK is considered in specific topics in a classroom situation (Aydin, Demirdogen, Atkin, Uzuntiryaki-Kondakci, & Tarkin, 2015). Emerging from these studies came a growing understanding of the close proximity of the topic-specific nature of PCK to content knowledge of that topic. In this section, the conceptualisation of PCK in science at a topic level, as a theoretical construct that could be described, taught and measured, is revisited (Mavhunga & Rollnick, 2013) in its new representation as a grain size of PCK and referenced in the RCM as topic-specific PCK. The discussion forwarded here for the close links of PCK to specific content knowledge is drawn from earlier arguments in the literature. Shulman (1986) argued that for expert teachers, many of the pedagogical knowledge strategies they use are content-specific. This view was supported by Geddis, Onslow, Beynon, and Oesch (1993) who explicitly pointed out that 'An outstanding teacher is not just a teacher, but rather an "English teacher" or a "chemistry teacher"' (Geddis et al., 1993, p. 675). Both these authors referred to knowledge for teaching a discipline, which is reflected in the RCM as the highest level or grain size item in the continuum of PCK within each of the cPCK, pPCK and ePCK realms. Drawing from this argument, it implies that pre-service teachers should learn to teach a discipline and therefore the relevant core topics that constitute the discipline. The challenge, however, comes in the realisation that the knowledge generated to teach the topics of a discipline is different from topic to topic and not automatically transferable. Thus, the need to give attention to the knowledge for teaching a specific topic within a discipline arises. Such knowledge was referred to in our earlier study (Mavhunga & Rollnick, 2013) as topic-specific pedagogical content knowledge (TSPCK), which is now referred to, hence with, as topic-specific PCK in line with the terminology in the RCM.

In conceptualising topic-specific PCK as a theoretical construct, Shulman's statement (1987) that 'comprehended ideas must be transformed in some manner if they are to be taught' (p. 16) was used as a starting point. This view was supported by Geddis et al. (1993) who pointed out that teachers need to develop the awareness that teaching requires the transformation of their (topic) content knowledge in general. Geddis argued that, once this awareness was in place, the articulation of the kinds of knowledge needed to achieve such transformation becomes important, in that knowledge of a *multitude of particular things* about the content knowledge of a topic that are relevant to its teachability is required (Geddis et al., 1993, p. 676). The *particular things* or knowledge components that Geddis identified were (i) learner prior knowledge including misconceptions, (ii) curricular saliency, (iii) what makes a topic easy or difficult to understand, (iv) representations including analogies and (v) conceptual teaching strategies. In short, *learner prior knowledge* refers to common students' misconceptions and alternative conceptions, as well as correct knowledge about a particular content. *Curricular saliency* refers to the learning of the various topics, relative to the curriculum as a whole. Within a topic, it is about understanding what concepts are most core and which are peripheral, the needed prior knowledge and the sequencing in teaching concepts. The component *what makes a topic easy or difficult to understand* refers to the ability to identify gate-keeping concepts within a topic that are difficult to understand and not necessarily misconceptions, and it triggers dedicated awareness and possible interventions for teaching them. *Representations* refer to a range of representations including examples, illustrations, analogies, simulations and models that are relevant to a topic. Lastly, *conceptual teaching strategies* refer to effective instruction strategies for particular learner misconceptions, known areas of difficulty to learn, or known importance of concepts. The strategies involve the use of combinations of conceptual principles and rules of a topic as tools to confront potential confusion and misconception. However, the term does not refer to general pedagogical knowledge strategies.

The knowledge components, identified by Geddis, were referred to as content-specific components in a previous science PCK publication by Mavhunga and Rollnick (2013) because of their orientation to content knowledge in a topic. When a particular element of content knowledge in a topic (K) is thought about and reasoned through against these content-specific components as a collective, understanding for teaching is generated that is specific to that topic (K'), Fig. 5.2.

The quality of topic-specific PCK observed in planning and/or teaching is therefore linked to the extent to which demonstrated knowledge of the components and their interactions is used by teachers to generate coherent and rich explanations and responses. It is important to note that in the previous consensus model of PCK (Gess-Newsome, 2015), these content-specific components were incorporated into the layer of PCK referred to as the professional topic-specific knowledge. Thus, the idea of

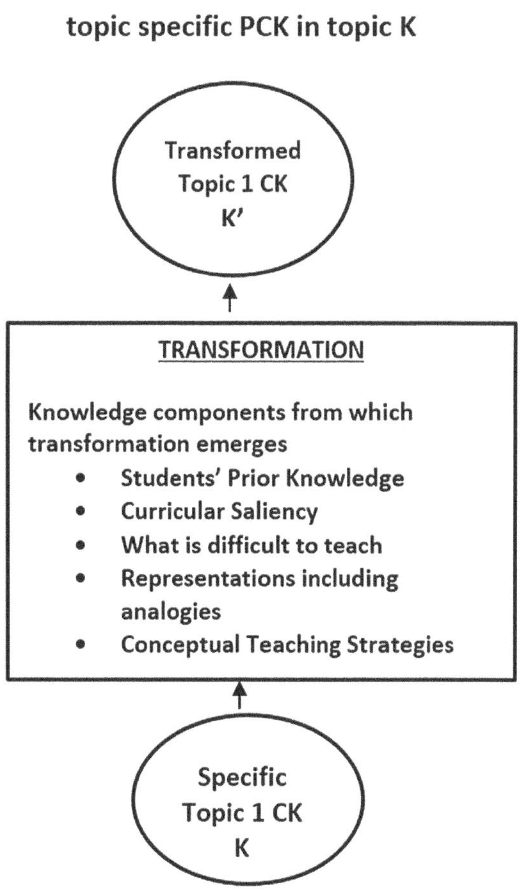

Fig. 5.2 Transformation of content knowledge in a topic from content-specific knowledge components of PCK

topic-specific PCK is not new, but it is more explicit and given greater clarity in the RCM of PCK for science teaching in relation to other grain sizes of PCK.

In summary, the location of topic-specific PCK within the RCM is explained and the composition of the construct revisited for purposes of distinguishing it from other levels in the continuum of PCK (i.e. discipline- and concept-specific PCK), as these are yet to be fully conceptualised—a challenge for future studies. An example of a study where pre-service teachers can be said to develop their pPCK and ePCK at a level of topic-specific PCK in electrochemistry is revisited to demonstrate how the development of their professional knowledge from a structured formal course based on cPCK in electrochemistry occurred. The example study illustrates the differentiated extent to which pre-service teachers develop their pPCK for the topic as influenced by personal beliefs and other factors. The research question explored in the example study can be reframed as: what factors promoted/hindered the development of pre-service teachers' pPCK at a level of topic-specific PCK in electrochemistry from a structured course based on cPCK in electrochemistry?

Method

Participants

The study was conducted at a large university in the Province of Gauteng in South Africa. It was conducted in a chemistry methodology component of a course that combines chemistry and physics as equal course components, called physical science. Participants in the course were 16 final (4th)-year bachelor of education pre-service teachers who were registered to major in physical science. The course had a main objective to improve the quality of PCK in several core topics of chemistry and physics. The topic used in this particular study was electrochemistry. The pre-service teachers were all from the disadvantaged black community who had come through three of the four academic years of the teacher qualification degree (B.Ed.) for teaching in secondary schools. The participants had received teaching on electrochemistry content knowledge in a separate but parallel content course before being exposed to the intervention described below.

Topic-Specific PCK Intervention

The intervention offered in the physical science course had been implemented and well documented in several other studies (Huang, Lubin, & Ge, 2011) prior to this study, thus offering a standardised sequence of events. In summary, the intervention introduces the idea of pedagogical transformation as the key activity in the process of pedagogical reasoning that science teachers do (Shulman, 1987). Topic-specific PCK is then introduced as the teacher knowledge that enables pedagogical transformation of the science content knowledge of topics. The five components of topic-specific PCK (Mavhunga & Rollnick, 2013) are viewed as working together in a coherent interaction to enable formulation of effective explanations. Pre-service teachers are provided with an opportunity to learn to reason through the science content of the topic from the perspective of each component and its interaction with others. The discussion unfolds over a period of 6 weeks with 3-hour periods in a week.

Data Collection

Several types of data were collected with the intent to illuminate the development of the pre-service teachers' pPCK at a topic level in electrochemistry. For the study, a specially designed PCK tool in electrochemistry was administered before and after the intervention to the whole class as pre- and post-tests. Tutorial work collected during the intervention served as additional data resources. The PCK tool is designed to measure the quality of pre-service teachers' pPCK at a topic level, against the collective understanding of PCK for teaching electrochemistry, which is within the cPCK realm. The tool is structured into five categories based on the five content-specific knowledge components of topic-specific PCK, respectively. Each category has 2–4 test items, which are teacher tasks seeking teacher responses. An extract of a test item on the component of learner prior knowledge is provided in Fig. 5.3.

Analysis

Data analysis was done in two parts. The first analysis considered the whole class as a sample and focused on establishing any shifts in the quality of their pPCK at a topic-specific level in electrochemistry as a direct result of the intervention by comparing the responses in the pre- and post-tests. This analysis served as a baseline from which to analyse the main research question on factors that promoted or hindered the development of pre-service teachers' personal PCK in electrochemistry. For this purpose, a criterion-based rubric with four categories of increasing quality of topic-specific PCK in electrochemistry (Mavhunga & Rollnick, 2013) was used to grade the written responses from the completed PCK tools (see Fig. 5.4 for an extract). The key strength of the rubric lies in the nature of the criteria in each progressive category of quality in the rubric. The criteria called for evidence of increasing personal knowledge of each topic-specific PCK component as well as its increasing interaction with other components in a provided teacher response, thus measuring the pre-service teachers' pPCK at topic level. The full rubric is provided on request.

The second analysis was in relation to the main research question, seeking to illuminate the development of pPCK at a topic-specific level in electrochemistry by exposing factors that promoted or hindered such development. For this purpose, a

CATEGORY A: LEARNERS' PRIOR KNOWLEDGE

1. How do you respond verbally to a learner who writes on a script:

"The electrons flow through the salt bridge to keep the galvanic cell neutral"

Response A: No, this is not the case; the electrons do not flow through the salt bridge to keep the galvanic cell neutral but through the external circuit. Only ions flow through the salt bridge.

Response B: No, this is not the case; electrons need a medium like a wire (solid) which is a good conductor for them to flow. The salt bridge contains a solution and only ions can flow within the salt bridge.

Response C: No, this is not the case; electrons flow through the external wire whereas the ions flow through the salt bridge. The flow of the ions through the salt bridge will maintain the galvanic cell electrically neutral.

Response D: None of the above. I have another response, which is...

Choose your response and indicate the reason(s) for choice in the space below:

Fig. 5.3 An extract of one of the test items on learner prior knowledge

	(1)Limited	(2) Basic	(3) Developing	(4)Exemplary
Learner Prior Knowledge	• No identification/No acknowledgement/No consideration of student prior knowledge or misconceptions • No attempt to address the misconception.	• Identifies misconception or prior knowledge • Provides standardized definition as a means to counteract the misconception • No evidence of drawing on other topic-specific PCK components.	• Identifies misconception or prior knowledge • Provides standardized knowledge as definition • Expands and re-phrase explanation using one other component of topic-specific PCK interactively.	• Identifies misconception or prior knowledge • Provides standardized knowledge as definition • Expands and re-phrases explanation correctly • Confronts misconceptions/confirms accurate understanding drawing on two or more other component of topic-specific PCK interactively.

Fig. 5.4 An extract of the topic-specific PCK rubric showing criteria for scoring learner prior knowledge

subset sample of three pre-service teachers was assembled by picking out a participant from the varied performance categories that emerged from the first analysis on shifts in PCK for electrochemistry. Further details on this selective process are provided in the findings section below. An in-depth qualitative analysis of their responses to particular teacher tasks items in the tool that demanded extensive explanations from the participants was conducted. Such responses were readily found from test items located within the topic-specific PCK components of learner prior knowledge, curricular saliency and representations. The responses across the three participants were analysed for similarities, differences and utterances reflecting factors contributing or impeding gains in topic-specific PCK. Emerging themes, within the findings from each case participant, were triangulated by identifying converging lines of evidence from the collected tutorials during the intervention (Patton, 2002). The tutorials provided insight of developing knowledge in each of the topic-specific components as they were discussed during the intervention.

Findings

Table 5.1 presents pre(post)-PCK scores from the completed tools measuring the quality of pPCK at a topic-specific level in electrochemistry. The names of the topic-specific PCK components were abbreviated for ease of reference as LPK—learner prior knowledge; CS—curricular saliency; WID—what is difficult to understand; RP—representations and CTS—conceptual teaching strategies.

Table 5.1 Overview shifts in the pre(post)-pPCK scores in electrochemistry

Pre-service teacher	LPK	CS	WID	REP	CTS	Person average
Jessie	2(4)	2(4)	2(3)	1(4)	1(3)	2(4)
Oumah	2(4)	1(3)	1(3)	2(4)	2(4)	2(4)
Annah	2(4)	1(3)	1(3)	2(3)	1(3)	1(3)
Bongi	1(3)	1(3)	1(2)	2(3)	1(3)	1(3)
Ludo	1(3)	1(3)	1(3)	2(2)	1(3)	1(3)
Imani	1(3)	2(2)	1(3)	1(4)	2(3)	1(3)
Khosi	2(3)	1(3)	1(3)	2(4)	1(3)	1(3)
Musa	1(3)	2(3)	1(2)	2(3)	1(3)	1(3)
Thulani	1(2)	2(3)	1(3)	1(2)	1(3)	1(3)
Ncebo	1(2)	2(3)	1(3)	2(3)	1(2)	1(3)
Sipho	1(2)	1(3)	2(2)	1(2)	1(1)	1(2)
Masekgo	2(2)	2(3)	1(2)	1(2)	1(1)	1(2)
Danisile	1(2)	1(2)	1(2)	2(2)	2(2)	1(2)
Vumani	1(2)	2(2)	1(1)	1(2)	1(1)	1(2)
Thabo	1(1)	2(2)	1(2)	1(2)	1(1)	1(2)
Xoli	1(1)	1(2)	1(2)	1(2)	1(1)	1(2)
Average scores per component	1(3)	1(3)	1(3)	2(3)	1(2)	1(3)
Overall Average						1(3)

Notes: Pseudonyms have been used to protect the identity of the participants. Scores in brackets are from the post-pPCK test for the electrochemistry topic

When comparing the pre(post)-intervention scores in Table 5.1, it is noticed that a positive gain in the quality of pPCK, at a topic level in electrochemistry, was experienced by all the participants as a result of the intervention. The extent of the gain, however, was found to be a stratified pattern showing the majority (10 out of 16) of the pre-service teachers registering a gain by two categories, where two of these (Jessie and Oumah) jumped from a 'basic' to the 'exemplary' category as per the rubric and the rest jumped from the 'limited' to the 'developing' category. The minority (6 of the 16) of the pre-service teachers registered a single category gain (from the 'limited' to 'basic' category). Pre-service teachers who experienced the single category improvement 'limited' to a 'basic' were, however, still in the lower quality levels of the PCK rubric for this topic. All the pre-service teachers were seen to have made more gain in three topic-specific PCK components: learner prior knowledge, curricular saliency and what is difficult to understand, and a lesser gain in the other two topic-specific PCK components: representations and conceptual teaching strategies. These findings point to the ease and ready interactive use of three of the five components by the teachers in formulating their explanations or responses. The performance patterns described above are similar to those in previously reported findings (Davidowitz & Potgieter, 2016) that pointed to an overall improvement in the quality of PCK at topic level to the maximum quality category in the rubric of 'developing', as a result of the explicit intervention.

The identification of factors affecting the development of pPCK at a topic-specific level in electrochemistry, as mentioned earlier, was based on a sub-sample comprised of three pre-service teachers selected purposefully from the 'exemplary' (Jessie), the 'developing' (Annah) and the 'basic' (Sipho) post-intervention scoring groups. Factors that assisted the development of pPCK at topic level and those that hindered were considered to be elements present or lacking in the sub-sample's written responses. The findings revealed three salient factors as influencing the developmental path towards pPCK at a topic-specific level. These are: (i) a developing competence to visualise a conceptual topic structure for teaching and personal beliefs, (ii) developing teacher identification competence beyond the source of learner doubt or misconception into connections and disconnections in understanding, and (iii) the competence to interweave representations into a transparent topic structure seamlessly. These factors are presented and explained in detail below.

Visualising a Topic Structure for Teaching and Personal Beliefs

The production of electrical energy from chemical reactions and vice versa is considered in the field as one of the key concepts to be taught in electrochemistry (cPCK for the topic). The responses of all three participants in the pretest showed an accurate understanding of this knowledge. However, the rest of their suggestions were a mixture of what could be considered as subordinate concepts in electrochemistry.

Examples of such instances include Jessie's comment 'ions carry charge in solution', which is an element of a more central concept on charge neutrality in a cell. An example of a subordinate concept offered by Annah as a core concept is 'calculation of cell potential', which is an algorithmic element related to the more central concept of oxidation/reduction half-reactions at the electrodes.

However, the first improvement was noticed in the extracts shown in Fig. 5.5 drawn from the responses in the tutorial on curricular saliency collected during the intervention. The activity in the tutorial requested pre-service teachers to draw a concept map reflecting statements of meaning they were to derive from most core concepts in electrochemistry. These would be similar to the concept of 'big ideas' introduced by Loughran et al. (2004) in tools capturing topic-specific PCK called Content Representations (CoRes).

All three pre-service teachers unanimously retained the reference to the relationship between chemical and electrical energy seen in the pretest. Both Jessie and Annah further referred to redox reactions as a core idea, thus making all their suggestions fit the requirement for the identification of concepts whose meaning is considered core teaching in the topic. Sipho's second example remained broad citing *links to real examples.*

In the post-intervention topic-specific PCK test, Jessie retained the reference to the core concept of the chemical/electrical relationship. In addition, and different to her pretest, reference was made to the more embodying concept of cell neutrality as a core concept to be taught (see Fig. 5.6).

In her reasoning, Jessie made explicit suggestions about how the listed statements of understanding on core concepts in the topic linked to related subordinate ideas, as seen in her statement

> learners need to understand that ions carry charges through the solution. These ionic flows maintain cell neutrality.

Her reasons reflected her description of how each of the concepts in the sequence allowed discussion of specific subordinate concepts like half-reactions. Furthermore, they reflected how the sequence allowed a strategically timed discussion on salt bridges to show its need (fit), and its links to core concepts like cell neutrality. Similarly, Annah's response had shifted from mere listing of pieces of content knowledge of the topic, such as *calculations of potential cell*, to building or visioning the struc-

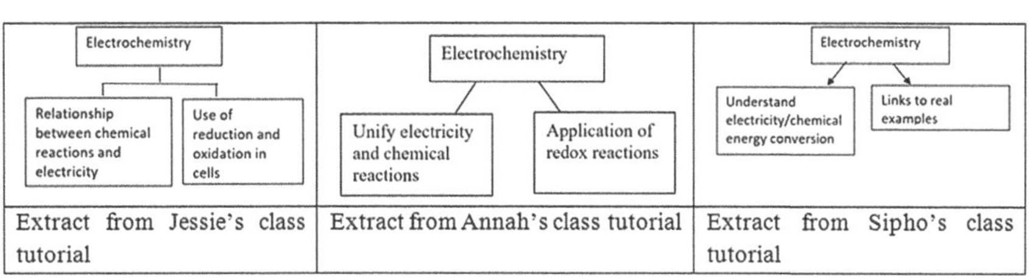

Fig. 5.5 Extracts of concept maps drawn from a tutorial

Suggested concepts and sequence	Reasons
1. Electricity can be used to produce a chemical reaction.	The concept of the relationship between chemical change and electrical energy is central to electrochemistry. This will inevitably lead to the notion of electron flow. To fully understand how half-cells interact, learners need to understand that ions carry charges trough the solution. These ionic flows maintain cell neutrality, emphasizing the need for a salt bridge and electrolytic ionic flow.
2. Electrical neutrality is preserved in a cell.	
3. Energy from chemical reactions can produce electricity.	

Fig. 5.6 Jessie's post-test extract on curricular saliency

Suggested concepts and sequence	Reasons
1. Electricity can be used to produce a chemical reaction.	1 and 2 are concepts of equal weighting. They are basically the reason for studying electrochemistry – either to use/generate electricity, to drive chemical reactions i.e. this is the Big Picture. 2 is the basis of why you always need two different reactions, and why they need to both happen at the same time. Important for galvanic cells. It is important for learners to make links with real life uses
2. Energy from chemical reactions can produce electricity.	
3. Oxidation and reduction occur simultaneously.	

Fig. 5.7 Annah's post-test extract on curricular saliency

ture of a topic based on most core concepts. For example, she mentioned oxidation and reduction happening simultaneously at the respective electrodes (see Fig. 5.7) and her rationale made reference to the need for learners to see the big picture in a topic. While Annah's response is, however, short of clear links between the core concepts and their specific subordinate concepts, it is different to her pretest in that her rationale for sequencing was found to be sound.

Sipho's list of core concepts (see Fig. 5.8) remained similar to that in his pretest without any evidence of a developing vision for the network of connections between the core concepts and their corresponding subordinate ideas.

Suggested concepts and sequence	Reasons
1. Energy from chemical reactions can produce electricity.	Electrochemistry is mainly based on energy that is converted from chemical to electrical and other way around. It is important that learners know what is important in everyday life where electrochemistry is applied.
2. Electricity can be used to produce a chemical reaction.	
3. Electrochemistry has important applications in everyday life.	

Fig. 5.8 Sipho's post-test extract on curricular saliency

His response also identified the reciprocal relationship between chemical and electrical energies as the fundamental important concept in electrochemistry. His inclusion of the uses of electrochemistry in everyday life as a core concept remained broad, while his reasons lacked the needed rationale with specific uses linked to the specific concepts to be taught.

The key difference in the responses, across the three responses, lies in the observed differentiated extent of success in visioning content structure that highlights the core concepts to be taught in the topic, that is, seeing a topic structure that is sequenced to show the core and subordinate concepts from the perspective of teaching and learning. Such a vision is important in the development of pPCK in the topic as it confirms absorption, or not, of important features of the shared knowledge for teaching the topic presented and discussed in the course as cPCK in electrochemistry. The responses provided by the three pre-service teachers reflect the differentiated extent of their developing pPCK in the topic. Sipho's personal belief about the importance of explicit links of topic content knowledge to real-life examples was evident in his vision of the content structure. The belief is observed in 5.5 and 5.8 where his inclusion of this view is placed in the same level of consideration (i.e. amplified) as the key core concepts of the topic.

Developing Teacher Identification Beyond Learners' Doubt: Seeing Connections and Disconnections

The analysis identifying this factor was based on the responses the pre-service teachers provided for a test item that required a teacher to provide an explanation to a learner who expressed doubt in her understanding about the oxidation/reduction at the different electrodes across the two kinds of electrochemistry cells. The responses provided by all three respective participants in the pre-intervention topic-specific PCK tool reflected good conceptual understanding of the concepts, as they could all affirm accurate content knowledge to the learner in response to her question. Their confirmations are however visibly brief, all starting with a direct confirmation, e.g. Jessie, *Yes, your understanding is correct*; or Sipho, *Yes, you are right*. During the intervention in a class tutorial, the pre-service teachers were asked to describe their thoughts about the source of the problem expressed by the learner in the same test item. The responses from the sub-sample reflected striking differences. Jessie and Annah interestingly described the learner's problem as situated within the oxidation/reduction processes and the naming of the anode/cathode electrodes accordingly. For example, Jessie's comment:

> I think the fog in the learner's understanding is about connecting the processes of oxidation and reduction to respective electrodes by name, and their charge signage across the different electrochemical cells.

On the other hand, Sipho's response simply focussed on a single concept, the signage of the electrodes, as shown.

> The learner is confused about signage of the electrodes in the electrolytic cell

In the post-intervention topic-specific PCK test, the participants' response continued to reflect correct understanding of content knowledge on electrochemistry cells; however, the response differed in depth and the emphasis in their explanations. Jessie and Annah's response demonstrated an explicit awareness of the need to draw the learner's attention to the importance of understanding the oxidation/reduction processes as a major concept in both galvanic and electrolytic cells. For example, Jessie's response:

> The most important concept is the location of the loss or gain for electrons. The loss is known as oxidation and gain as reduction. In both the galvanic and electrolytic cells, oxidation takes place at the anode, while reduction always takes place at the cathode. Note the difference in the signage of the anode and cathode across the different cells, they are only due to signage of the terminals of the battery.

Jessie's response interestingly provided descriptions of both these processes with reference to the gain and loss of electrons, and affirmation of the consistency of their location indicated by the use of the word *always* in the above example. Annah's response in 5.9 similarly provided descriptions of the oxidation process with reference to 'loss' of electrons at the anode and also affirmed the consistency of its location at the anode. She further added the use of a representation that displayed electrons movement at both electrodes. Her response, as shown in Fig. 5.9, demonstrated a dynamic interaction among the learner prior knowledge and representations components of her pPCK in electrochemistry. However, Annah's description is provided only for one electrode, leaving the learner to work out the second electrode (cathode).

Both Jessie and Annah ranked the issue of the signage of the electrodes as low priority in the rank order of what is important to understand in the topic. Annah added the importance of establishing conceptual understanding versus memorisation as a more desirable way of learning science. In contrast, the response by Sipho has a selective focus on the issue of signage of the electrodes. For example, he suggests:

Fig. 5.9 Annah's post-test extract on learner prior knowledge

It is important to understand oxidation as the loss of electrons, and that the anode always "loses" electrons in order to be able to deduce the difference with cathode, and figure out the charge signage of the terminals, as shown in the figure below.

This is the more scientific way of understanding than simply memorizing the positive/negative

In an electrolytic cell the anode is connected to the positive terminal of the electrical source and therefore will have a positive charge. Oxidation still takes place at the anode

Furthermore, Sipho mentions the process of oxidation without reference to loss or gain of electrons, and the explanation is placed at the end of the description as an affirmation of the consistency of the location of oxidation at the anode.

The key difference across these three responses is the extent to which the process of analysis of learner's doubt occurred. The degree of analysis influenced the varying depth of coverage and the location of emphasis in the respective teacher responses. Both responses from Jessie and Annah demonstrated awareness of the learner's problem, but went beyond the source to interpret the learner's conceptual doubt in terms of connections and disconnections to the most important concepts. Sipho's response on the other hand, while also accurate, is limited as it lacked depth in making connections to the main concept of oxidation and reduction in each cell. An analysis that identified the source of learner's doubt and stopped there seems to have resulted in a brief isolated response with no acknowledgement of the connections or disconnections to the most core concepts to be understood (curricular saliency). However, an analysis that identified the source of the learner's doubt and went beyond to interpret it, in terms of connections and disconnections in the learners understanding, influenced the depth and the visibility of the location of emphasis in the provided explanations, which is a signal of a developing pPCK at the topic level.

The Use of Representations Interactively with Other Topic-Specific PCK Components

This test item required the pre-service teachers to choose a representation from a provided list and show how they would use it in teaching the concepts in the topic.

In the pre-intervention topic-specific PCK test, Jessie and Annah chose the same representation (2) that has a potential to explain concepts at a sub-microscopic level; see Fig. 5.10.

Sipho chose a simplistic representation with less sub-microscopic detail but presents both electrochemistry cells at the same time. All three participants, despite their choice, referred to a simplistic use of the representation in a lesson, with a focus placed more on the description of the representation than the targeted concepts. For example, Sipho responded: *I will use the simple representation not to confuse learners and explain each component*. Jessie also had a similar response as shown in her extract:

> I will use the galvanic cell representation and explain the components of the representation. Then introduce the representation for electrolytic cell and do the same.

In the class tutorial during the intervention, Jessie and Annah indicate a developing consideration for *sequence and timing*, as well as *being strategic* in their use of a representation, specifically the introduction of the different levels (macroscopic, symbolic and sub-microscopic). For example, Annah said:

> It is important for learners to be familiar with the representation use. I think one should be strategic about when to introduce the different levels, especially macroscopic.

In contrast, during his tutorial Sipho showed little ability to consider different aspects, other than using a representation to establish a big picture and retaining a linear sequence where the representation is presented first with little interaction with content. In the post-intervention topic-specific PCK test for electrochemistry, Jessie's description reflected a considerable shift from her pre-intervention statement. She refers to a stepwise sequence for introducing different levels of the representation interactively with specific concepts. She first suggested a physical demonstration, which would be a macroscopic presentation of the concepts, and then increasingly

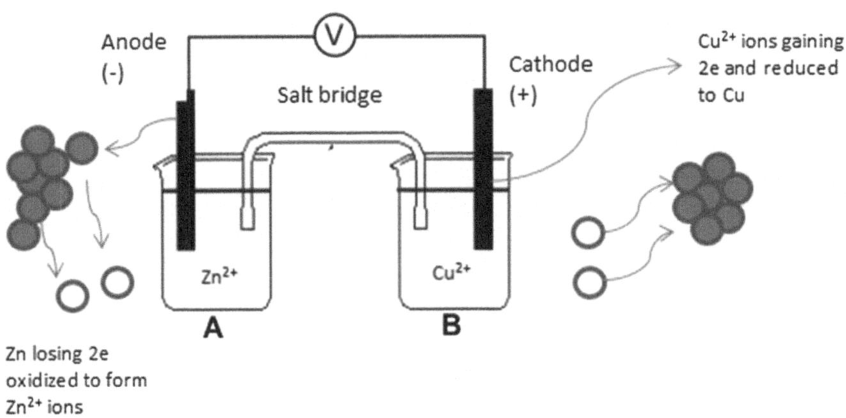

Fig. 5.10 Typical representation used by Jessie and Annah

scaffolded learner understanding with interactive use of different levels of the representation with specific concepts. For example, she suggested the use of symbolic representations to follow her explanation of the oxidation and reduction half-cell reactions. These were preceded by an explanation of how electrons flow through the cell. See Jessie's extract:

> After setting up the physical demonstration of a galvanic cell, I will use the representation to explain how electrons flow, while not showing the electrode extrapolations and signage. I will symbolically explain the electrode oxidation and reduction using half-cell reactions, after which I will then show the electrodes microscopic insert. I will then discuss the use of the salt bridge, and learners can set the see the symbolic representations of the salt bridge. I will conclude by explaining the cell as a whole entity.

Jessie provided insights into her deliberate decision not to discuss a specific component of her representation, leaving it out of her teaching sequence at a certain point, then picking it up again later and emphasising it by visibly linking it to the introduction of a specific subordinate concept—half-reactions. Annah's extract seen below, although less detailed than Jessie's, reflected a similar understanding of the value of the interactive use of representations of the same concept that are at different macroscopic/microscopic levels and placed strategically at different points of her explanations.

> I would first show learners the real-set up, and let then engage with it. Then I would explain it in more detail using the schematic part, leaving out the microscopic representations. I would add the label lines and explain terminology. Finally, I would then introduce the microscopic view.

Jessie and Annah's demonstration of the strategic interactive use of the components of topic-specific PCK, when planning to teach the topic of electrochemistry, indicates the beneficial extent of their ePCK at a topic-specific level in planning. As alluded to earlier, their ePCK in the topic is drawn from their pPCK in the topic learned during the structured methodology course that taught cPCK in electrochemistry. On the other hand, Sipho retained his thinking from his learning experiences during the intervention, applying it to his planning where he decided to first provide learners with explanations of all the components in the representation of both galvanic and electrochemical cells, then followed by explanations of the charges at the electrodes as shown below.

> I will then indicate all the components and charges of the electrode. I will then be able to explain how each cell operates

His post-intervention extract, however, lacked insight into how those explanations could be constructed. The key difference between the three pre-service teachers was in their developing ability to use representations strategically and interactively with other topic-specific PCK components, which is the key evidence of the presence of desirable quality of ePCK.

Discussion and Conclusion

Two major purposes were cited at the beginning of this chapter. The first was aligned to the collective objective of this book to demonstrate the use of the RCM in empirical studies of PCK. This chapter aimed at contributing to this objective by articulating the position of topic-specific PCK as a grain size in the continuum of PCK found within the three realms of PCK, namely ePCK, pPCK and ePCK. A discussion that revisited the conceptualisation of topic-specific PCK from previous separate studies was undertaken to provide a rationale behind the suggested composition of topic-specific PCK with five content-specific components. As with PCK, Mavhunga and Rollnick (2013) argued that PCK at the topic level offers the key benefit of providing a means for pedagogically transforming the content knowledge of a specific topic to versions that are accessible for learner understanding. The interactions among the content-specific components of topic-specific PCK are regarded as the key emphasis in pre-service programmes along with developing knowledge of the individual components (Aydin et al., 2014a). It is important to note that the suggested description of content-specific components of PCK in this chapter is not conclusive, but offers a reasoned starting point for further research work on the construct, and more excitingly in conceptualising the other grain size PCK constructs such as discipline- and concept-specific PCK.

The second purpose of this chapter was to demonstrate through an existing empirical study how the development of topic-specific PCK could also be tracked across the collective–personal–enacted realms of PCK as depicted in the RCM. It was shown in this particular study how pre-service teachers drew from a course based on collective knowledge (cPCK) for teaching the topic to develop their pPCK and ePCK in the topic, and the findings indicated that all 16 participating pre-service teachers exposed to the intervention experienced some improvement in their quality of pPCK in the topic. Their improvement was, however, at differentiated extents despite having experienced the same intervention. This differentiated development of their pPCK was influenced by factors mentioned below.

The pre-service teachers were required to draw on their differentiated topic-specific pPCK to demonstrate their ePCK in planning for teaching the topic. It was at this point that factors amplifying or filtering the development of their pPCK for the electrochemistry topic were visible in three of the five components of topic-specific PCK, namely curricular saliency, learner prior knowledge and representations. Personal beliefs emerged as a hindering factor or filter for Sipho when structuring the topic through identification of core concepts. His beliefs about linking concepts to real-life examples saw him place this view at the same level as core concepts. This outcome had the effect of blurring evidence of his ability to identify the core concepts in the specific topic as his statements were very broad. The next factor identified as influencing the development of pPCK at the topic-specific level was the ability to read beyond learners' misconceptions and to identify connections and disconnections with the core concepts considered most important for understanding in a topic. This ability was seen to promote or amplify the depth and the placement of emphasis in the teacher's responses to teaching tasks in the topic of electrochemistry. Lastly, the emergence of a seamless and interactive use of the content-specific components of topic-specific PCK in the use of the component of representations by several of the pre-service teachers illustrated desirable and sophisticated ePCK in planning, i.e. amplified ePCK development. Both Jessie and Annah demonstrated interactive use of the curricular saliency and what is difficult to understand components in their use of representations. While it is acknowledged that such an exemplary level of ePCK in the topic was not seen in all the pre-service teachers, the value of organising and implementing cPCK for pPCK and ePCK development topic by topic in pre-service teachers is promising.

The acknowledgement of the multidimensional nature of PCK in the RCM challenges science educators to break away from ambiguous references to PCK. It challenges us to speak with specificity of the grain size and the realm of PCK being explored. Such clarity allows the reader and the science education community at large to establish and pinpoint the extent and the limitations of the professional knowledge developed at given times. This study had a limitation in that it focused only on the development of ePCK in planning for a specific topic, thereby excluding experiences in actual teaching and reflection. On the other hand, the acknowledgement of such gaps indicates the next steps for progressive research in the field.

References

Abell, S. A. (2007). Research on science teacher knowledge. In S. A. Abell & N. G. Lederman (Eds.), *Handbook of research in science education* (pp. 1105–1150). Mahwah: Lawrence Erlbaum.

Abell, S. A. (2008). Twenty years later: Does pedagogical content knowledge remain a useful idea? *International Journal of Science Education, 30*(10), 1405–1416.

Aydeniz, M., & Kirbulut, Z. D. (2014). Exploring challenges of assessing pre-service science teachers' pedagogical content knowledge (PCK). *Asia-Pacific Journal of Teacher Education, 42*(2), 147–166. https://doi.org/10.1080/1359866x.2014.890696.

Aydin, S., Friedrichsen, P. M., Bozc, Y., & Hanuscinb, D. L. (2014a). Examination of the topic-specific nature of pedagogical content knowledge in teaching electrochemical cells and nuclear reactions. *Chemistry Education Research and Practice, 15*(4), 658–674.

Aydin, S., Friedrichsen, P. M., Bozc, Y., & Hanuscinb, D. L. (2014b). Examination of the topic-specific nature of pedagogical content knowledge in teaching electrochemical cells and nuclear reactions. *Chemistry Education Research and Practice,, 15*(4), 658–674.

Aydin, S., Demirdogen, B., Atkin, F. N., Uzuntiryaki-Kondakci, E., & Tarkin, A. (2015). The nature and development of interaction among components of pedagogical content knowledge in practicum. *Teaching and Teacher Education, 46*, 37–50.

Darling-Hammond, L., & Richardson, N. (2009). Teacher learning: What matters? *How Teachers Learn, 66*(5), 46–53.

Davidowitz, B., & Potgieter, M. (2016). Use of the Rasch measurement model to explore the relationship between content knowledge and topic-specific pedagogical content knowledge for organic chemistry. *International Journal of Science Education, 38*(9), 1483–1503. https://doi.org/10.1080/09500693.2016.1196843.

Geddis, A. N., Onslow, B., Beynon, C., & Oesch, J. (1993). Transforming content knowledge: Learning to teach about isotopes. *Science Education, 77*, 575–591.

Geddis, A. N., & Wood, E. (1997). Transforming subject matter and managing dilemmas: A case study in teacher education. *Teaching and Teacher Education, 13*(6), 611–626.

Gess-Newsome, J. (2015). A model of teacher professional knowledge and skill including PCK: Results of the thinking from the PCK summit. In A. Berry, P. Friedrichsen, & J. Loughran (Eds.), *Re-examining pedagogical content knowledge in science education* (pp. 28–42). London: Routledge.

Huang, K., Lubin, I. A., & Ge, X. (2011). Situated learning in an educational technology course for pre-service teachers. *Teaching and Teacher Education: An International Journal of Research and Studies, 27*(8), 1200–1212.

Loughran, J. J., Berry, A., & Mulhall, P. (2004). In search of pedagogical content knowledge in science: Developing ways of articulating and documenting professional practice. *Journal of Research in Science Teaching, 41*(4), 370–391.

Mavhunga, E., & Rollnick, M. (2013). Improving PCK of chemical equilibrium in pre-service teachers. *African Journal of Research in Mathematics, Science and Technology Education, 17*(1–2), 113–125.

Nezvalová, D. (2011). Researching science teacher pedagogical content knowledge. *Problems of Education in the 21st Century*.

Park, S., & Chen, Y. (2012). Mapping out the integration of the components of pedagogical content knowledge (PCK): Examples from high school biology classrooms. *Journal of Research in Science Teaching, 49*(7), 922–941. https://doi.org/10.1002/tea.21022.

Patton, M. Q. (2002). *Qualitative Research and Evaluation Methods*. Thousand Oaks, CA: 3rd Sage Publications.

Shulman, L. S. (1986). Those who understand: Knowledge growth in teaching. *Educational Researcher, 15*(2), 4–14.

Shulman, L. S. (1987). Knowledge and teaching: Foundations of the new reform. *Harvard Educational Review, 57*(1), 1–22.

Van Driel, J. H., Verloop, N., & de Vos, W. (1998). Developing science teachers' pedagogical content knowledge. *Journal of Research in Science Teaching, 35*(6), 673–695.

Veal, W. R., & MaKinster, J. G. (1999). Pedagogical content knowledge taxonomies. *Electronic Journal of Science Education, 3*(4).

Chapter 6
The Development of Science Teachers' Professional Competence

Stefan Sorge, Anita Stender, and Knut Neumann

Abstract On re-examining some of our earlier research into secondary science teachers' PCK, as we reposition this work within the Refined Consensus Model (RCM) of PCK, we uncover the RCM is not only a model of PCK. We argue the RCM is also a model of science teachers' professional competence and its development since the model identifies other elements of science teachers' professional competence, including the role played by broader knowledge bases as well as amplifiers and filters moderating exchanges between knowledge bases. To support our argument, in this chapter, we utilise data from two earlier studies that investigated exchanges between knowledge bases as secondary science teachers develop professional competence. Re-examining this data through the interpretive lens of the RCM, the first study utilised paper–pencil-tests that assessed pre-service physics teachers' content knowledge (CK), collective PCK (cPCK) and pedagogical knowledge (PK). The analyses reveal a stronger correlation between PK and cPCK in the first half and stronger correlation between CK and cPCK in the second half of teacher education. Again from a RCM perspective, the second study used the same instrument to assess cPCK, plus instructional planning vignettes to assess physics teachers' personal/enacted PCK (pPCK/ePCK) and standardised paper–pencil questionnaires to examine selected amplifiers and filters. The results suggest an increased influence of cPCK on pPCK/ePCK for more experienced physics teachers, moderated by motivational orientations. This retrospective treatment of earlier research data reveals, as the RCM implies, the development of cPCK is informed by broader

The original version of this chapter was revised: Reference year and location were changed. The correction to this chapter is available at https://doi.org/10.1007/978-981-13-5898-2_15

S. Sorge (✉) · K. Neumann
Leibniz Institute for Science and Mathematics Education, Kiel University, Kiel, Germany
e-mail: sorge@ipn.uni-kiel.de

K. Neumann
e-mail: neumann@ipn.uni-kiel.de

A. Stender
Department of Physics Education, University Duisburg-Essen, Duisburg, Germany
e-mail: anita.stender@uni-due.de

A. Hume et al. (eds.), *Repositioning Pedagogical Content Knowledge in Teachers' Knowledge for Teaching Science*, https://doi.org/10.1007/978-981-13-5898-2_6

professional knowledge bases, whereas cPCK plays a major role in the development of pPCK/ePCK.

Introduction

The expectations teachers are confronted with in their professional lives are manifold. Teachers are expected, for example, to organise learning opportunities that support students in developing competence in domains such as science (National Research Council [NRC], 2014), to develop a collaborative relationship with their colleagues (Darling-Hammond et al., 2005) or to counsel parents effectively (Hertel, Bruder, Jude and Steinert, 2013). In order to meet these expectations, teachers require a broad spectrum of knowledge and skills, as well as motivational orientations and beliefs. The knowledge and skills, as well as the motivational orientations and beliefs needed to successfully act in their professional lives, is what constitutes teachers' professional competence (see Baumert and Kunter, 2013; Weinert, 2001).

There has been a multitude of efforts to describe the different elements of teachers' professional competence, how they develop and how they play out in teachers' professional lives (for an overview see Abell, 2007 or Fischer, Borowski, and Tepner, 2012). One of the most influential has been Shulman's (1986) proposal of pedagogical content knowledge (PCK) as the unique province of teachers lying between subject matter or content knowledge (CK) and general pedagogical knowledge (PK). This triplet of professional knowledge bases has served as a theoretical framework in many different studies investigating science teachers' professional competence (e.g., Kirschner, Borowski, Fischer, Gess-Newsome and von Aufschnaiter, 2016), how it develops (e.g. Rollnick, 2017) or how it plays out in teaching (e.g., Park, Jang, Chen and Jung, 2011). However, the knowledge base of PCK has been subject to varying conceptualisations (e.g., Kind, 2009). In the field of science teacher education and research, the Refined Consensus Model (RCM) of PCK, presented in Chap. 2, represents the latest and most elaborate effort in proposing a unified model of science teachers' professional competence. The model incorporates three distinct conceptualisations (realms) of PCK: the specialised professional knowledge of a teacher of science that he/she shares with multiple other professionals in the field and develops outside of his/her own classroom in formal science teacher education or professional development courses—such knowledge is derived from the collective canonical knowledge of science education (collective or cPCK); the personalised professional knowledge and skills related to an individual science teacher's classroom context (personal or pPCK); and the unique subset of knowledge and skills that a teacher of science draws on and uses to engage in pedagogical reasoning and actions during the planning of, teaching of and reflecting on a specific lesson (enacted or ePCK). The RCM also recognises that broader professional knowledge bases, such as CK and PK, are foundational to the development of these three realms of PCK. Furthermore, the RCM highlights the relevance of different amplifiers and filters, such as motivational orientations, to exchange between the three realms of PCK as teachers of science develop professional competence throughout their careers.

The purpose of this chapter in re-examining data from two previous studies on physics teachers' professional competence (namely Sorge, Kröger, Petersen and Neumann, 2017 and Stender, Brückmacnn and Neumann, 2017) was to learn more about the exchanges between the broader professional knowledge bases and the three realms of PCK. More specifically, we will analyse the data to obtain information about (a) the relationship between CK, PK and cPCK at different stages of physics teacher education and thus the role of CK and PK in the development of cPCK, and (b) the relationship between teachers' cPCK in physics and their pPCK/ePCK at different stages of teachers' professional career and thus the role of cPCK in the development of pPCK/ePCK.

Teachers' Professional Competence and Its Development in Science

The RCM depicts science teachers' professional competence as a series of interdependent knowledge bases that teachers develop throughout their career and draw on in their planning, performance and reflection of instruction (see Chap. 2). In the model, the outer rim of teacher professional competence in science comprises five broader professional knowledge bases other than PCK: pedagogical knowledge (PK), knowledge of students, curricular knowledge, assessment knowledge and, most importantly, content knowledge (CK). However, the inner three realms of PCK form the core of teacher professional competence in science. The RCM arranges these three realms on a continuum from publicly shared PCK of a group of professionals in science teaching to privately held knowledge of an individual science teacher. The most public realm of PCK is named collective PCK (cPCK) as it represents the knowledge related to teaching a particular science subject, topic or concept shared amongst a group of professionals (e.g. teachers, teacher educators and teacher education researchers). Such knowledge is generated through science education research or best practice examples developed or approved by a professional community (see the idea of topic specific professional knowledge [TSPK] in Gess-Newsome, 2015). That is, to our understanding, canonical knowledge communicated in teacher education or professionalisation. An individual teachers' cPCK in science is his/her knowledge of this canonical knowledge that is not derived from but potentially reinforced by his/her own teaching experiences. The realm that represents more private knowledge situated in the classroom context is personal PCK (pPCK), which is the cumulative knowledge and skills of an individual teacher of science that reflects the teacher's own teaching and learning experiences. The most private realm of PCK, the very core of teacher professional competence, is enacted PCK (ePCK)—the knowledge and skills used by an individual teacher of science in a specific context for specific students to achieve a specific goal (for details see Chap. 2).

The RCM goes beyond other models of teachers' professional competence (e.g., Baumert and Kunter, 2013) in specifying the relationship between the broader professional knowledge bases and teachers' PCK in science and amongst the three realms of PCK as one of exchange. For example, the RCM identifies an exchange between the broader professional knowledge bases and cPCK, which is a relationship that has been repeatedly confirmed by previous research (e.g., Baumert and Kunter, 2013; Großschedl, Harms, Kleickmann and Glowinski, 2015; Kirschner et al., 2016; Riese and Reinhold, 2010). Kirschner et al. (2016), for example, report a medium correlation between CK and (c)PCK and a small correlation between PK and (c)PCK in investigations of physics teachers' professional knowledge. However, according to Chap. 2, the exchange between the broader professional knowledge bases and cPCK in science should be ongoing throughout a teacher's professional career. Yet, there is little research on how the relationship between the broader professional knowledge bases changes as science teachers' professional competence develops (for an exception, see Krauss et al., 2008). The RCM also assumes an exchange between cPCK and pPCK and ePCK. That is, pPCK plays an important role in the RCM as it mediates exchanges with cPCK as well as ePCK (e.g., Alonzo, Berry and Nilsson, 2018). However, little is known about the relationship between science teachers' cPCK and their pPCK/ePCK. Last but not least, the RCM assumes exchange between the professional knowledge bases (i.e., the broader professional knowledge bases and the three realms of PCK) to be moderated by so-called amplifiers and filters. These amplifiers and filters include goals, beliefs and values, motivational orientations and self-regulatory skills (Woolfolk-Hoy, Davis and Pape, 2006). However, evidence about the role of individual amplifiers and filters in the exchange between professional knowledge base as teachers of science develop professional competence is largely missing.

The development of teachers' professional competence spans their whole professional life. Kunter, Kleickmann, Klusmann and Richter (2013a) accordingly distinguish teachers' learning opportunities in three categories: school education, teacher education and professional experience. The effects of the respective learning opportunities on the development of professional competence are mediated by teachers use (i.e., uptake) of learning opportunities according to their personal characteristics and dependent on the learning context (i.e., the teacher education system). We can, however, generally assume that students develop initial knowledge with respect to several professional knowledge bases during school education, where future teachers develop, for example, initial content knowledge (e.g., Sadler and Tai, 2001), and likely also some naïve intuitive PCK that they might fall back on during their teaching career (Lortie, 1975). In secondary science teacher education (since pre-service teachers attend classes in the subjects, general pedagogy and subject education) teachers should develop knowledge in the area of the broader professional knowledge bases as well as cPCK in science (see Chap. 2). However, there may also be possibilities to develop pPCK and ePCK for science teaching through the planning, performance and reflection of instruction in school practicums. The scope of possibilities, however, varies considerably across teacher education systems (Pedersen, Isozaki and Hirano, 2017). In-service teachers have possibilities to plan, teach and reflect every day, to the effect that with increasing professional experience, in-service teachers develop more extensive pPCK and ePCK. Since professional development experiences, again depending on the specifics of the respective education system, are likely less frequent, teachers often have fewer opportunities to develop their broader professional knowledge bases as well as their cPCK. Thus, the different foci of learning opportunities should lead to a differential development of the broader professional knowledge bases and the three realms of PCK in science throughout a teacher's career.

In our research, we set out to learn about the development of secondary science teacher professional competence throughout their career. More specifically, we were interested in the relationship of CK and PK as two broader professional knowledge bases and pre-service physics teachers' cPCK at different stages of teacher education and how physics teachers' cPCK relates to their pPCK/ePCK across different stages of their career. Based on previous research (e.g., Kirschner et al., 2016; van Driel, de Jong and Verloop, 2002), we expected future physics teachers' CK, PK and cPCK to be related, but we were interested in differences in this relationship at different stages of science teacher education to derive conclusions about the role of CK and PK in the development of cPCK. In the same way, we expected physics teachers' cPCK to be related to their pPCK and ePCK, but we aimed to learn more about the specifics of this relationship—in particular considering the role of amplifiers and filters—throughout physics teachers' professional lives.

The Role of CK and PK in the Development of cPCK for Physics Teaching

In order to investigate the relationship between content knowledge (CK) and pedagogical knowledge (PK), as two broader professional knowledge bases, and the realm of collective PCK (cPCK) at different stages of science teacher education, we use data previously collected from a study of pre-service secondary physics teachers at 12 major teacher education universities in Germany (Sorge et al., 2017). Teacher education in Germany is organised in two stages. The first stage consists of a five-year pre-service teacher education programme at a university, and the second stage of a two-year in-service training programme at a school. Pre-service teacher education programmes include classes in two subjects (e.g., physics and mathematics), respective subject education (i.e., physics education and mathematics education) and educational sciences, but little practical experience (for details, see Neumann, Härtig, Harms and Parchmann, 2017). That is, by design, the German teacher education system can be said to focus on CK, cPCK (viewed as canonical PCK) and PK.

To ensure that we covered the CK taught in subject courses and the cPCK taught in subject education courses in their full breadth and depth, we began our instrument development with an analysis of the curriculum for physics teacher education of 16 teacher education institutes and an in-depth literature review. Based on the findings from this analysis, we reviewed existing instruments at the college level. In addition, we invited experts in physics and physics education to develop new items. The process resulted in a pool of items that underwent a pilot study, subsequent expert rating and a think-aloud study. During the think-aloud study $N = 10$, pre-service physics teachers were asked to continuously verbalise their own thoughts as they worked through selected items. This process allowed us to see if the participants of the study used relevant strategies to solve the items. Based on the results from the pilot study, the expert rating and the think-aloud study, we selected 59 CK and 39 cPCK items for inclusion in the final instrument (for more details, see Sorge et al., 2017). The CK items covered the following content areas: Mechanics, Electrodynamics, Optics, Thermodynamics, Solid State Physics, Atomic and Nuclear Physics and Quantum Mechanics. The cPCK items covered four aspects of PCK identified by Magnusson, Krajcik and Borko (1999): knowledge about science curriculum, knowledge about students' understanding of specific science topics, knowledge about assessment in science and knowledge about instructional strategies (see also Park and Oliver, 2008). While the majority of the CK items were multiple-choice, 18 of the cPCK items were open-ended, 15 were multiple-choice and the rest true-false, matching or short-answer. The open-ended items were scored dichotomously by one rater based on a detailed coding manual. To ensure reliable scoring, a second rater scored a random subset of 50 booklets resulting in a good overall inter-rater agreement ($\kappa_{cPCK} = 0.73$). Figure 6.1 provides a sample item for CK from the area of Mechanics, and a sample cPCK item assessing knowledge of students' understanding of Mechanics (for details on the PK instrument see Hohenstein, Kleickmann, Zimmermann, Köller and Möller, 2017).

The final test instrument was administered to $N = 201$ pre-service physics teachers. Since we had to exclude one participant due to invalid data, we obtained data from a final sample of $N = 200$ pre-service physics teachers. The average age of the participants was 23.7 (SD = 3.0) years, and on average they were enrolled in their 2.9 (SD = 1.3) years of teacher education. To investigate the relationship between CK, PK and cPCK at different stages of teacher education, we split the sample into two groups: with three or less years of experience in teacher education (i.e. students enrolled in a bachelor programme) and students with more than three years of experience (i.e. students enrolled in a master programme). This way, we obtained one group of $N_1 = 91$ beginning and $N_2 = 109$ advanced pre-service teachers.

In a first step of analysis, we examined the relationship between CK, PK and cPCK for all participants using structural equation modelling. The results revealed a strong correlation between CK and cPCK ($r = 0.78, p < 0.001$), an equally strong correlation between cPCK and PK ($r = 0.81, p < 0.001$) and a considerably smaller correlation between CK and PK ($r = 0.54$, $p < 0.001$). In principle, these results support the relationships identified in the RCM that the broader professional knowledge bases CK and PK influence the cPCK of science teachers. The results for the beginning and more advanced pre-service physics teachers, however, reveal a differential pattern (see Fig. 6.2). Whereas there is a strong correlation between PK and cPCK for the beginning pre-service physics teachers, that correlation is substantially lower for the

Fig. 6.1 Sample items for content knowledge (CK, top) and collective pedagogical content knowledge (cPCK, bottom)

Different masses are attached to a beam as shown in the picture.

For which value of the mass m is the beam balanced?

□ 2.0 kg
□ 4.5 kg
þ 5.5 kg
□ 6.0 kg
□ 7.8 kg

Students often struggle in correctly applying Newton's laws to solve concrete problems. Consider the following situation: A small object with mass m moves without friction and constant velocity $\vec{v}$ to the right. There are three forces acting on the body, two are shown in the figure. You ask your students to draw the third force.

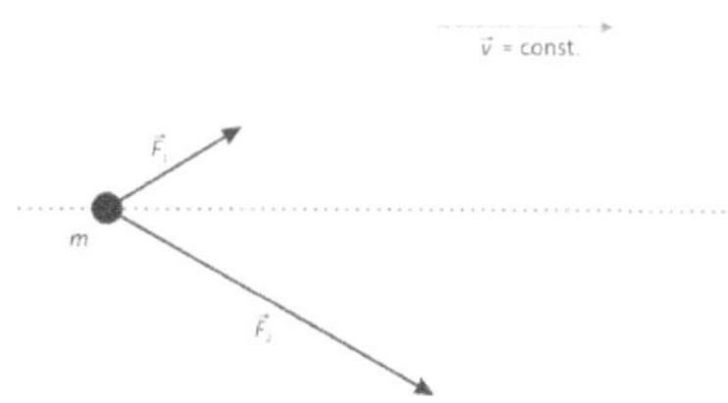

What physically incorrect answer would you expect? Draw your answer and briefly describe, based on which (incorrect) conception a student could have come to this answer.

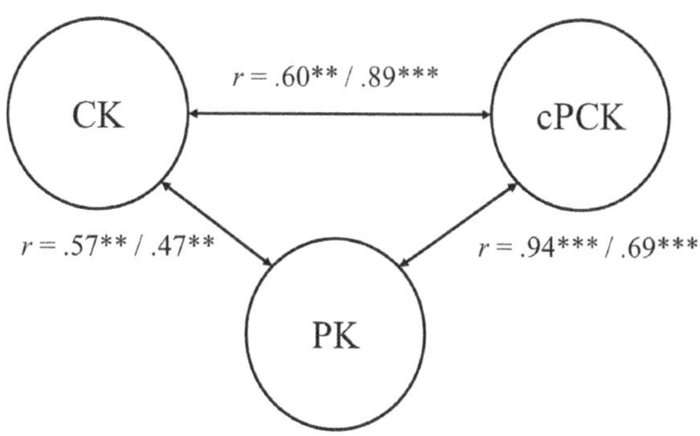

Fig. 6.2 Relationship amongst the broader professional knowledge bases and cPCK for beginning/advanced pre-service teachers. $^{**}p < 0.01$, $^{***}p < 0.001$

advanced pre-service physics teachers. The advanced pre-service physics teachers, on the other hand, exhibit a considerably higher correlation between CK and cPCK.

The Role of Physics Teachers' cPCK in the Development of PPCK and EPCK

To obtain insights into the relationship between cPCK and pPCK/ePCK and, more importantly, the role of different amplifiers and filters in this relationship at different stages in science teachers' professional careers, we draw on data from a study by Stender et al. (2017). This study investigated physics teachers' collective PCK (cPCK) utilising items developed as part of the study presented in the previous section, as well as teachers' pPCK/ePCK by means of teaching scripts in order to obtain insights into the role of cPCK for the development of pPCK/ePCK.

In the RCM, science teachers' cPCK is conceptualised as the specialised knowledge base for teaching that can be articulated and shared amongst professionals (see Chap. 2), and this realm of PCK may be considered to exist mainly in the form of propositions (Shulman, 1987, p. 10). Such knowledge is not easily retrieved for interactive decision-making during instruction (Anderson, 1983; Stender et al., 2017). Teachers' pPCK/ePCK, on the contrary, is conceptualised as the knowledge and skills that a teacher can draw/draws upon and actions during the practice of teaching science (see Chap. 2). These two realms must therefore be represented in terms of mental structures other than propositional ones (Schank and Abelson, 1977). Shavelson (1986) assumes that parts of such knowledge are stored as teaching scripts [see also Bishop and Whitfield (1972)]. Teaching scripts can be regarded as mental representations about students and sequences of events in classroom settings that are acquired with teaching experience (Bishop and Whitfield, 1972; Borko, Roberts and Shavelson, 2008; Henze and van Driel, 2015; Shavelson, 1986). In terms of the RCM, a single teaching script becomes visible when engaging in the practice of teaching a group of students in a specific science-teaching situation. The instantiation of a single teaching script can be understood as parts of a science teacher's enacted PCK (ePCK), whereas the cumulative amount of teaching scripts of a science teacher represents parts of personal PCK (pPCK).

Based on our assumptions of how cPCK and pPCK/ePCK of science teachers are represented in different knowledge structures, we conceptualised the relationship between cPCK and pPCK/ePCK as a complex transformation process of propositional knowledge into procedural knowledge, driven by the process of the planning, performance and reflection of instruction (for details see Stender et al., 2017). Based on this conceptualisation, we assumed that a science teacher, when they start to teach has little pPCK/ePCK and will draw mainly on their cPCK when planning instruction. Through performance and reflection, elements of the planned instruction are reinforced—if they worked well in the eyes of the teacher—or adapted—if they did not. The elements are thus increasingly refined and internalised by the science teacher; that is, they become teaching scripts. As the reinforcement or refinement of instructional plans depends on a teacher's beliefs, motivation and self-regulatory skills (Shavelson and Stern, 1981; Stender et al., 2017), these dispositions of a teacher should be expected to play a key role in the development of teaching scripts; that is, in the transformation of cPCK into pPCK/ePCK, respectively.

The teaching scripts developed and tested in real classroom situations should enable experienced science teachers to outperform novice teachers in noticing meaningful instructional details, interpreting them and identifying alternative strategies for solving problems during instruction (Berliner, 1986; Borko and Livingston, 1989; Livingston and Borko, 1989). For domains other than teaching, this capability is based on scripts which are easily retrievable, show high dependency of decisions and high concreteness (Bransford, Brown and Cocking, 2000). Based on these findings, we assume that these formal features of teaching scripts (retrievability, dependency of decisions and concreteness) also predict to what extent teaching scripts enable teachers to routinely act in teaching situations. However, routine does not necessarily indicate quality instruction. Rather, the teaching scripts should fulfil particular functional features. For example, appropriateness of decisions to antecedent conditions is shown as a feature-predicting instructional quality (Shavelson and Stern, 1981). Coherence of decisions for instruction is seen as a second feature (Seidel, Rimmele and Prenzel, 2005). The third feature characterises the potential of the script to cognitively activate the students during classroom instruction, an aspect that also has to be considered during lesson planning (Kunter, Klusmann, Baumert, Richter, Voss and Hachfeld, 2013b; Lipowsky et al., 2009). It is thus assumed that these three functional features describe whether or not teaching scripts can lead to high instructional quality.

The activation of (teaching) scripts for teaching in given contexts should be stimulated by the recognition of typical situations (Rumelhart, 1980; Schank and Abelson, 1977; Schnotz, 1994). Hence, in order to capture physics teachers' teaching scripts, we developed an instrument based on three planning vignettes. Each vignette presented a task in the planning of a different lesson from a lesson sequence in the field of mechanics, more specifically, the introduction of the force concept and Newton's laws (9th grade level). Figure 6.3 shows a sample planning vignette.

In order to ensure a certain level of standardisation in the analysis of teachers' performance on the planning tasks, the teachers were guided through each task by a series of questions similar to a semi-structured interview. The sequence and selection of questions depended on teachers' responses to the actual questions. The flowchart displayed in Fig. 6.4 shows the adaptivity in the sequence and selection of questions (for a detailed description, see Stender et al., 2017). In order to achieve this level of adaptivity and to be able to capture teachers' responses online, the vignettes were implemented in an online-survey platform (LimeSurvey Project Team, 2015).

In order to assess the quality of physics teachers' teaching scripts, we focused on the formal features (i.e. retrievability, high dependency and concreteness) and functional features (i.e. appropriateness, coherence and activation) of the teaching scripts. As shown in Fig. 6.4, two different response types were used in our questionnaire: a closed-response format and an open-response format. To assess the retrievability and the dependency of decisions, we used the questions 2 and 4a or b with the closed format and the answer options of 'yes' or 'no'. To assess the concreteness, the appropriateness, the coherence and the activation of a teaching script, we used the answer to the questions with the open-response format (Questions 3a, 7 and 8). The answers were rated on a scale from 'does not apply' (0 points), 'partially applies' (1 point),

Fig. 6.3 Planning vignette for an experimental lesson (Stender et al., 2017)

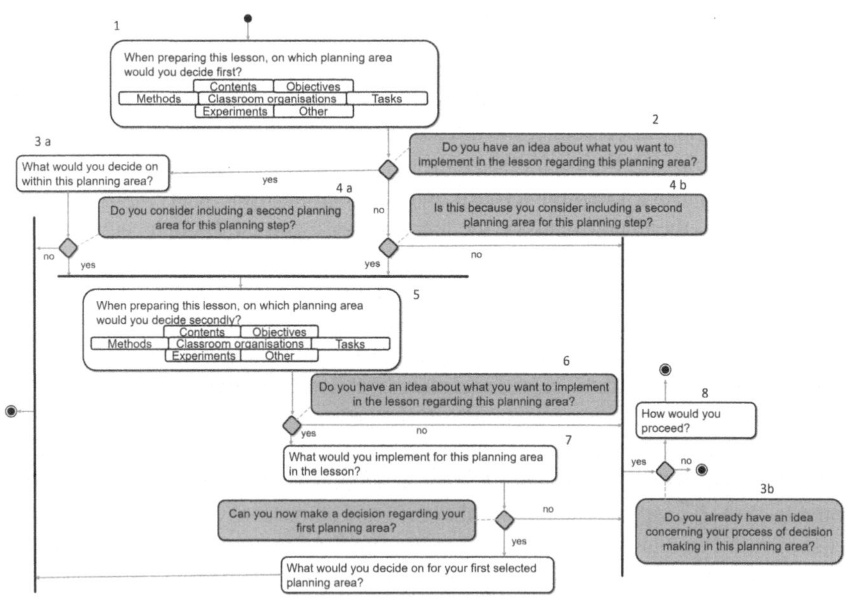

Fig. 6.4 Flowchart of the questionnaire (Stender et al., 2017)

'applies' (2 points) to the feature. The objectivity of the rating was checked by double coding 20% of the data set. The ordinal-scaled data meant the weighted Kappa and Spearman's ρ were computed for inter-rater agreement (Concreteness ($\kappa = 0.64$), appropriateness ($\kappa = 0.71$), coherence ($\kappa = 0.66$) and activation ($\kappa = 0.66$)). According to Landis and Koch (1977), values of Kappa above 0.75 represent excellent agreement, and values between 0.40 and 0.75 represent fair to good agreement, suggesting that our instrument allows for a reliable assessment of the quality of a single teaching script of a physics teacher. The internal consistency of the rating of the six features across all three vignettes was also found to yield sufficiently high values (retrievability: $\alpha = 0.76$; dependency: $\alpha = 0.63$; concreteness: $\alpha = 0.62$; appropriateness: $\alpha = 0.\ 69$; coherence: $\alpha = 0.62$; activation: $\alpha = 0.70$); suggesting a reliable assessment of the quality of teaching scripts of a teacher (in the domain of mechanics). In terms of the RCM, the instrument allows a reliable assessment of an individual physics teachers' ePCK (in the case of a single teaching script) and a physics teacher's pPCK (across multiple teaching scripts).

In order to examine the exchange between cPCK and pPCK/ePCK, we administered the developed instrument together with an instrument utilising items developed as part of the study presented in the previous section to physics teachers with different professional experience ($n = 51$ pre-service physics teachers, $n = 48$ trainee physics teachers, $n = 26$ beginning physics teachers (<10 years of teaching experience), $n = 23$ advanced physics teachers ($\geq$10 years of teaching experience). In line with inherent assumptions of the RCM related to filters and amplifiers, we surveyed the teachers' self-efficacy and enthusiasm (Baumert, 2009), constructive beliefs (Seidel, Prenzel, Duit and Lehrke, 2006) as well as self-regulatory skills (Schwarzer and Jerusalem, 1999).

Since the development of the quality of teaching scripts underpins an acquisition of expertise as well as a transformation of science teachers' cPCK into teaching scripts (as one facet of pPCK/ePCK), we expected (1) the quality of teaching scripts to develop with expertise in science teaching and (2) the influence of cPCK on the quality of teaching scripts to be higher with a higher expertise in science teaching. In order to test the first expectation, we analysed the quality of teaching scripts across the different groups of physics teachers using analysis of variances with the assumption of a linear trend (ANOVA). The results suggested that the formal features retrievability (F (2134) $= 10.71$, $p < 0.01$, $\omega = 0.24$) and dependence (F (2134) $=$ 17.55, $p < 0.01$, $\omega = 0.32$) and the functional feature activation (F (2117) $= 5.87$, $p < 0.05$, $\omega = 0.18$) increase with experience. For the other features, no such pattern was observed. In order to test the second expectation, we used Multi-Group—Partial Least Square structural equation modelling for the different groups of physics teachers at different stages of their career (Hair, Hult, Ringle and Sarstedt, 2014). The results suggested that pre-service physics teachers' cPCK has no direct impact on the formal and functional quality of teaching scripts, whereas cPCK of experienced physics teachers has an impact on the functional quality in terms of coherence and activation (direct effect: $\gamma = 0.48$, $p < 0.01$, $f^2 = 0.07$). The influence of cPCK on the quality of teaching scripts is not only—as assumed—higher with more experience; it only becomes significant in the group of experienced physics teachers (for details, see Stender et al., 2017). Based on our theoretical model of the exchange between science teachers' cPCK and pPCK/ePCK, an interpretation of this result can be that the influence of cPCK on the quality of teaching scripts developed over time through the process of planning, implementing and reflecting of science instruction. The results also confirm the assumption that motivational aspects and self-regulatory skills moderate this influence (interaction effect: $\gamma = 0.21$, $p < 0.01$, $f^2 = 0.01$). The direct effect of cPCK and the interaction effect of motivational aspects with cPCK explain together 26% of the variance of the functional quality of teaching scripts. So, the effect of cPCK on the formal and functional quality of scripts was higher if the physics teachers additionally had a higher motivation and higher self-regulatory skills, while, in our study, constructive beliefs had neither a direct nor a moderating effects (for details see Stender et al., 2017).

Discussion and Conclusion

The Consensus Model (CM) of Teacher Professional Knowledge and Skill including PCK (Gess-Newsome, 2015) presented an initial attempt to delineate the complexity of science teachers' PCK as a construct and to identify other aspects of teacher professional competence. The RCM aims to address concerns that have been raised and discussed in the community after publication of the initial CM. Amongst these were concerns that the initial CM identifies only one specific aspect of science teachers' professional knowledge as PCK, that is, teachers' personal PCK. However, there are other realms of PCK that a teacher of science can possess such as the PCK shared by a community of professionals, which may be taught in teacher education, or the PCK driving teachers' decisions in the classroom, which is a subset of a teachers' personal PCK only. In the context of science teacher research and education, the RCM addresses these concerns by distinguishing between three different realms of PCK—collective PCK (cPCK), personal PCK (pPCK) and enacted PCK (ePCK) and, in addition, broader professional knowledge bases such as content knowledge (CK) or pedagogical knowledge (PK). The model envisions these knowledge bases and the different realms of PCK to be interrelated in a specific way: the broader knowledge bases are expected to be related via a two-way knowledge exchange to a teachers' cPCK in science, which is in turn related to a teachers' pPCK via a similar two-way knowledge exchange, and finally the reciprocal relationship between pPCK and ePCK. Each of these knowledge exchange relationships is moderated or mediated by amplifiers and filters such as teachers' beliefs, motivational orientations or self-regulatory skills.

The next step after publication of the RCM is to test out its specific assumptions and, in particular, obtain further information about the exchanges between knowledge bases as teachers of science develop competence throughout their career—with the aim to better understand the development and optimise it. In this chapter, we attempted to test out the model and obtain further insight into the relationship between two broader professional knowledge bases, CK and PK and cPCK of pre-service physics teachers at different stages of teacher education in order to learn about the role of the broader professional knowledge bases in the development of cPCK. We also attempted to obtain further insights into the relationship between cPCK and pPCK/ePCK in order to learn about the role of cPCK in the development of pPCK/ePCK. To do so, we used data from two studies that we had previously carried out on physics teachers' professional competence and its development (e.g., Sorge et al., 2017; Stender et al., 2017).

Our analyses confirm the relevance of broader knowledge bases such as CK and PK for the development of cPCK for science teaching. More specifically, the results showed that for beginning pre-service physics teachers, cPCK was more strongly related to PK, whereas for the advanced pre-service physics teachers cPCK was more strongly related to CK. These results suggest that CK plays an increasing role in the development of pre-service secondary teachers' cPCK in science as they progress through the teacher education programme. Our analyses also confirmed an apparent exchange between physics teachers' cPCK and pPCK/ePCK. Interestingly, we found that this exchange seems to be moderated largely by teaching experience as there was no relationship between cPCK and pPCK/ePCK for the pre-service teachers (with negligible teaching experience) but a notable relationship for the experienced teachers. We also found that the relationship is moderated by motivational aspects and self-regulatory skills, but not by beliefs.

The results are in line with our assumption that through science-teaching experience (i.e., the planning, performance and reflection of instruction), cPCK is transformed into pPCK/ePCK and that this process depends on aspects of teacher professional competence other than their knowledge. Our findings suggest that the strong focus on knowledge in research about science teacher professional competence should be reconsidered (for further evidence on the important role of motivational aspects for teacher professional competence, see Keller, Neumann and Fischer, 2017 or Sorge, Keller, Neumann and Möller, 2018).

In summary, our findings support the RCM in science teacher education and research in that they confirm expected relationships between the broader professional knowledge bases and cPCK, and the relationship between cPCK and pPCK/ePCK for science teaching. Furthermore, the comparison of relationships between knowledge bases across teachers at different stages in their professional career provides some indications of how teachers develop professional competence. We find CK plays an increasing role in (the development of) pre-service physics teachers' cPCK, and we find the role of cPCK for (the development of) teachers' pPCK/ePCK to develop with teachers' teaching experiences—moderated by motivational aspects and self-regulatory skills. The nature of our data (i.e. quantitative and cross-sectional) meant we cannot draw conclusions about the nature or processes of the exchanges between knowledge bases in the development of teacher professional competence. Further studies are needed that take a more longitudinal, potentially qualitative approach that also includes more information about (the actual quality of) learning opportunities.

References

Abell, S. K. (2007). Research on science teacher knowledge. In S. K. Abell & N. G. Lederman (Eds.), *Handbook of research on science education* (pp. 1105–1149). Mahwah, NJ: Lawrence Erlbaum.

Alonzo, A. C., Berry, A., & Nilsson, P. (2019). Capturing and representing the complexity of PCK in action. In A. Hume, R. Cooper, & A. Borowski (Eds.), *Repositioning pedagogical content knowledge in teachers' professional knowledge*. pp. 271–286, Singapore:Springer.

Anderson, J. R. (1983). *The architecture of cognition*. Cambridge, MA: Harvard University Press.

Baumert, J. (2009). *Professionswissen von Lehrkräften, kognitiv aktivierender Mathematikunterricht und die Entwicklung von mathematischer Kompetenz (COACTIV): Dokumentation der Erhebungsinstrumente*. Berlin: Max-Planck-Institut für Bildungsforschung.

Baumert, J., & Kunter, M. (2013). The COACTIV model of teachers' professional competence. In M. Kunter, J. Baumert, W. Blum, U. Klusmann, S. Krauss, & M. Neubrand (Eds.), *Cognitive activation in the mathematics classroom and professional competence of teachers* (pp. 25–48). Springer US. https://doi.org/10.1007/978-1-4614-5149-5_2.

Berliner, D. C. (1986). In pursuit of the expert pedagogue. *Educational Researcher, 15*(7), 5–13. https://doi.org/10.3102/0013189X015007007.

Bishop, A. J., & Whitfield, R. C. (1972). *Situations in teaching*. McGraw-Hill.

Borko, H., & Livingston, C. (1989). Cognition and improvisation: Differences in mathematics instruction by expert and novice teachers. *American Educational Research Journal, 26*(4), 473–498. https://doi.org/10.3102/00028312026004473.

Borko, H., Roberts, S. A., & Shavelson, R. J. (2008). Teachers' decision making: From Alan J. Boshop to today. In A. J. Bishop, P. C. Clarkson, & N. C. Presmeg (Eds.), *Critical issues in mathematics education. Major contributions of Alan Bishop* (pp. 37–67). Springer.

Bransford, J., Brown, A. L., & Cocking, R. R. (2000). *How people learn: Brain, mind, experience, and school*. Washington, DC: National Academy Press.

Darling-Hammond, L., Banks, J., Zumwalt, K., Gomez, L., Sherin, M. G., Griesdorn, J., et al. (2005). Educational goals and purposes: Developing a curricular vision for teaching. In L. Darling-Hammond & J. Bransford (Eds.), *Preparing teachers for a changing world. What teachers should learn and be able to do* (pp. 169–200). San Francisco, CA: Jossey-Bass.

Fischer, H. E., Borowski, A., & Tepner, O. (2012). Professional knowledge of science teachers. In B. Fraser, K. Tobin, & C. J. McRobbie (Eds.), *Second international handbook of science education* (pp. 435–448). Dordrecht: Springer.

Gess-Newsome, J. (2015). A model of teacher professional knowledge and skill including PCK. In A. Berry, P. J. Friedrichsen, & J. Loughran (Eds.), *Teaching and learning in science series. Re-examining pedagogical content knowledge in science education* (1st ed., pp. 28–41). New York, NY: Routledge.

Großschedl, J., Harms, U., Kleickmann, T., & Glowinski, I. (2015). Preservice biology teachers' professional knowledge: Structure and learning opportunities. *Journal of Science Teacher Education, 26,* 291–318.

Hair, J. F., Hult, G. M., Ringle, C., & Sarstedt, C. (2014). *A primer on partial least squares structural equation modelling (PLS-SEM)*. Thousand Oaks, CA: Sage.

Henze, I., & van Driel, J. H. (2015). Towards a more comprehensive way to capture PCK in its complexity. In A. Berry, P. J. Friedrichsen, & J. Loughran (Eds.), *Teaching and learning in science series. Re-examining pedagogical content knowledge in science education* (pp. 120–134). New York, NY: Routledge.

Hertel, S., Bruder, S., Jude, N., & Steinert, B. (2013). Parental counselling at secondary schools. *ZEITSCHRIFT FUR PADAGOGIK*, 40–62.

Hohenstein, F., Kleickmann, T., Zimmermann, F., Köller, O., & Möller, J. (2017). Erfassung von pädagogischem und psychologischem Wissen in der Lehramtsausbildung: Entwicklung eines Messinstruments. *Zeitschrift für Pädagogik, H, 1,* 91–113.

Keller, M. M., Neumann, K., & Fischer, H. E. (2017). The impact of physics teachers' pedagogical content knowledge and motivation on students' achievement and interest. *Journal of Research in Science Teaching, 54*(5), 586–614. doi:10.1002/tea.21378

Kind, V. (2009). Pedagogical content knowledge in science education: Perspectives and potential for progress. *Studies in Science Education, 45*(2), 169–204.

Kirschner, S., Borowski, A., Fischer, H. E., Gess-Newsome, J., & von Aufschnaiter, C. (2016). Developing and evaluating a paper-and-pencil test to assess components of physics teachers' pedagogical content knowledge. *International Journal of Science Education, 38*(8), 1343–1372.

Krauss, S., Brunner, M., Kunter, M., Baumert, J., Blum, W., Neubrand, M., et al. (2008). Pedagogical content knowledge and content knowledge of secondary mathematics teachers. *Journal of Educational Psychology, 100,* 716–725.

Kunter, M., Kleickmann, T., Klusmann, U., & Richter, D. (2013a). The development of teachers' professional competence. In M. Kunter, J. Baumert, W. Blum, U. Klusmann, S. Krauss, & M. Neubrand (Eds.), *Cognitive activation in the mathematics classroom and professional competence of teachers* (pp. 63–77). US: Springer.

Kunter, M., Klusmann, U., Baumert, J., Richter, D., Voss, T., & Hachfeld, A. (2013b). Professional competence of teachers: Effects on instructional quality and student development. *Journal of Educational Psychology, 105*(3), 805–820. https://doi.org/10.1037/a0032583.

Landis, R., & Koch, G. (1977). The measurement of observer agreement for categorical data. *Biometrics, 33,* 159–174. https://doi.org/10.2307/2529310.

LimeSurvey Project Team. (2015). LimeSurvey: An open source survey tool. Retrieved from http://www.limesurvey.org.

Lipowsky, F., Rakoczy, K., Pauli, C., Drollinger-Vetter, B., Klieme, E., & Reusser, K. (2009). Quality of geometry instruction and its short-term impact on students' understanding of the Pythagorean theorem. *Learning and Instruction, 19*(6), 527–537.

Livingston, C., & Borko, H. (1989). Expert-Novice Differences in Teaching: A Cognitive Analysis and Implications for Teacher Education. *Journal of Teacher Education, 40*(4), 36–42. https://doi.org/10.1177/002248718904000407.

Lortie, D. C. (1975). *Schoolteacher: A sociological study.* Chicago: University of Chicago Press.

Magnusson, S., Krajcik, J., & Borko, H. (1999). Nature, sources, and development of pedagogical content knowledge for science teaching. In J. Gess-Newsome & N. G. Lederman (Eds.), *Examining pedagogical content knowledge* (pp. 95–132). Dordrecht: Kluver.

National Research Council [NRC]. (2014). *Developing assessments for the next generation science standards.* Washington, D.C: The National Academies Press.

Neumann, K., Härtig, H., Harms, U., & Parchmann, I. (2017). Science teacher preparation in Germany. In J. Pedersen, T. Isozaki, & T. Hirano (Eds.), *Model science teacher preparation programs* (pp. 29–52). Charlotte, NC: Information Age Publishing.

Park, S., Jang, J.-Y., Chen, Y.-C., & Jung, J. (2011). Is pedagogical content knowledge (PCK) necessary for reformed science teaching? Evidence from an empirical study. *Research in Science Education, 41,* 245–260.

Park, S., & Oliver, J. S. (2008). Revisiting the conceptualization of pedagogical content knowledge (PCK): PCK as a conceptual tool to understand teachers as professionals. *Research in Science Education, 38*(3), 261–284.

Pedersen, J., Isozaki, T., & Hirano, T. (2017). *Model science teacher preparation programs.* Charlotte, NC: Information Age Publishing.

Riese, J., & Reinhold, P. (2010). Empirische Erkenntnisse zur Struktur professioneller Handlungskompetenz von angehenden Physiklehrkräfte. *Zeitschrift für Didaktik der Naturwissenschaften, 16,* 167–187.

Rollnick, M. (2017). Learning about semi conductors for teaching—The role played by content knowledge in pedagogical content knowledge (PCK) development. *Research in Science Teaching, 47,* 833–868.

Rumelhart, D. (1980). Schemata: The building blocks of cognition. In R. Spiro, B. Bruce, & W. Brewer (Eds.), *Theoretical issues in reading comprehension* (pp. 33–58). Hillsdale, NJ: Erlbaum.

Sadler, P. M., & Tai, R. H. (2001). Success in introductory college physics: The role of high school preparation. *Science Education, 85*(2), 111–136.
Schank, R., & Abelson, R. (1977). *Scripts, plans, goals, and understanding*. Hillsdale, NJ: Erlbaum.
Schnotz, W. (1994). *Aufbau von Wissensstrukturen: Untersuchungen zur Kohärenzbildung beim Wissenserwerb mit Texten [Development of knowledge structures: Studies on the formation of coherence in knowledge acquisition with texts]*. Weinheim: Beltz.
Schwarzer, R., & Jerusalem, M. (1999). *Skalen zur Erfassung von Lehrer- und Schülermerkmalen: Dokumentation der psychometrischen Verfahren im Rahmen der wissenschaftlichen Begleitung des Modellversuchs Selbstwirksame Schulen*. Berlin: Freie Universität Berlin.
Seidel, T., Prenzel, M., Duit, R., & Lehrke, M. (2006). *Technischer Bericht zur Videostudie "Lehr-Lern-Prozesse im Physikunterricht" [Technical report of the IPN video study]*. Kiel: IPN.
Seidel, T., Rimmele, R., & Prenzel, M. (2005). Clarity and coherence of lesson goals as a scaffold for student learning. *Learning and Instruction, 15*, 539–556. https://doi.org/10.1016/j.learninstruc.2005.08.004.
Shavelson, R. J. (1986). Toma de decision interactiva: Algunas reflexiones sobre los procesos cognoscitivos de los profesores [Interactive decision making: Some thoughts on teacher cognition]. In L. M. V. Angulo (Ed.), *Pensiamentos de los profesores y toma de decisiones* (pp. 164–184). Servicio de Publicaciones.
Shavelson, R. J., & Stern, P. (1981). Research on teachers' pedagogical thoughts, judgments, decisions, and behaviour. *Review of Educational Research, 51*(4), 455–498. https://doi.org/10.3102/00346543051004455.
Shulman, L. S. (1986). Those who understand: Knowledge growth in teaching. *Educational Researcher, 15*(2), 4–14.
Shulman, L. (1987). Knowledge and teaching: Foundations of the new reform. *Harvard Educational Review, 57*, 1–23. https://doi.org/10.17763/haer.57.1.j463w79r56455411.
Sorge, S., Keller, M. M., Neumann, K., & Möller, J. (2018). Investigating the relationship between pre-service physics teachers' professional knowledge, self-concept and interest (Manuscript submitted for publication).
Sorge, S., Kröger, J., Petersen, S., & Neumann, K. (2017). Structure and development of pre-service physics teachers' professional knowledge. *International Journal of Science Education*. Online first https://doi.org/10.1080/09500693.2017.1346326.
Stender, A., Brückmann, M., & Neumann, K. (2017). Transformation of topic-specific professional knowledge into personal pedagogical content knowledge through lesson planning. *International Journal of Science Education, 39*(12), 1690–1714. https://doi.org/10.1080/09500693.2017.1351645.
van Driel, J. H., de Jong, O., & Verloop, N. (2002). The development of preservice chemistry teachers' PCK. *Science Education, 86*(4), 572–590.
Weinert, F. E. (2001). Concept of competence: A conceptual clarification. In D. S. Rychen & L. H. Salganik (Eds.), *Defining and selecting key competencies* (pp. 45–65). Göttingen: Hogrefe & Huber.
Woolfolk-Hoy, A., Davis, H., & Pape, S. (2006). Teachers' knowledge, beliefs, and thinking. In P. A. Alexander & P. H. Winne (Eds.), *Handbook of educational psychology* (pp. 715–737). Mahwah, USA: Erlbaum.

Chapter 7
Illustrating and Developing Science Teachers' Pedagogical Content Knowledge Through Learning Study

Rebecca M. Schneider

Abstract Using a teacher educator's perspective to study pedagogical content knowledge (PCK) as described in the Refined Consensus Model (RCM) of PCK, this chapter explores an approach to uncover teacher thinking that is grounded in the complex work of teaching and learning about teaching. PCK is described in the RCM as the knowledge that supports science teachers' pedagogical reasoning during teaching. Stated from a teacher educator's perspective, PCK is the knowledge used and developed by science teachers in their teaching practice. The complexity of teaching, however, creates a challenge for researchers and teacher educators interested in gathering evidence to better understand and document science teachers' developing PCK. An approach to supporting science teacher learning that is embedded in the facets of teaching—planning, enacting, and reflecting—is learning study. Learning study engages teachers in cycles of describing phenomenon-based tasks, anticipating students' ideas, and analysing learning. Each of these phases in the study of learning draws on science teachers' pedagogical reasoning in ways that appear to be aligned with descriptions of collective PCK (cPCK) and distinguishes qualitative differences between individual teacher's ePCK. In the context of graduate teacher education, this chapter describes the potential of learning study to enable researchers and teacher educators to capture, unpack, and refine our ideas about the features of PCK that guide science teachers' thinking within the different facets of their teaching.

Introduction

As a construct that promises to be helpful in describing what science teachers need to learn, pedagogical content knowledge (PCK) can be an interesting idea for teacher educators. The challenge has been how to incorporate ideas about PCK into designs for science teacher education and, in turn, demonstrate teachers' learning. The work of the Second (2nd) PCK Summit was to refine our model of PCK in ways that would better illustrate the group's thinking and to explore methods of gathering data

R. M. Schneider (✉)
University of Toledo, Toledo, OH, USA
e-mail: Rebecca.Schneider@utoledo.edu

A. Hume et al. (eds.), *Repositioning Pedagogical Content Knowledge in Teachers' Knowledge for Teaching Science*, https://doi.org/10.1007/978-981-13-5898-2_7

on science teachers' PCK. How to use this Refined Consensus Model (RCM) of PCK to frame research that identifies science teachers' PCK from empirical data is an interesting challenge for both teacher educators and researchers. The research work described in this chapter is approached from a teacher educator's perspective. The idea is to investigate how PCK, as conceptualised in the RCM, might be useful in guiding the design of experiences intended for science teachers' learning while at the same time illustrating teachers' PCK. This chapter describes how teacher-developed learning studies might be an approach to observe teachers' developing PCK in ways that can inform our thinking about how to model and research PCK.

Studying PCK in Science Teacher Education

Studying PCK from a teacher educator's perspective means thinking about PCK as a tool for describing what teachers need to learn about teaching science. Teaching is an incredibly complex and dynamic activity that takes time and effort to learn well (Lampert, 2001). It is a multifaceted activity that involves planning, enacting, and reflecting around tasks intended for student learning. To support teacher learning, teacher educators are challenged to design tasks for teachers—preservice and continuing—that will develop teachers' thinking about teaching and that document their progress (Grossman, 2005). A model for science teacher knowledge that represents the complexity of knowing about science teaching could guide teacher educators in developing teacher-educative tasks and assessing teacher learning.

The RCM of PCK described in Chap. 2 of this book is a significant step in creating a model that represents what teachers know about teaching subject matter. Based on years of thoughtful work and multiple conversations, the refined model represents the thinking of the participants in the second international summit on PCK. This model provides a framework in which to situate studies of PCK for science teaching. The predictive power of the model now needs to be tested, for its ability to be helpful in describing teachers' thinking in ways that predict outcomes for students. For teacher educators, having predictive power means being helpful in designing tasks that support and assess teachers' progress in learning about teaching subject matter. To do this work well, a set of methods for gathering evidence that describes science teachers' PCK in meaningful ways is needed.

With this goal in mind, I have been exploring a method to investigate science teachers' PCK within a framework of learning about teaching. In teacher education, PCK is conceptualised as the knowledge of teaching subject matter (in this case, science) that is used and developed within practice (Feiman-Nemser, 2008). Thus, tasks intended for teacher learning should engage teachers in the activities of teaching while learning about teaching (Hammerness et al., 2005). It is also important for teachers to engage in thinking about student learning while learning about teaching (Sykes, 1999). These outcomes require the development of teacher-educative tasks that emphasise the study of student thinking in connection with plans, enactments, and reflections. In other words, teacher-educative tasks should highlight teachers' pedagogical reasoning behind the act of science teaching. An approach that is embedded within the complex work of science teaching, with the potential to both develop and illustrate teachers' thinking, is learning study.

Learning Study to Support Teacher Learning

Learning study is a recent proposal for supporting teacher learning that builds upon the well-known lesson study approach to professional development (Cheng & Ling, 2013; Tan & Nashon, 2013). In contrast to lesson study, learning study is not focused on lesson plans per se, but rather on plans for engaging and uncovering students' thinking in order to learn about learning (Wood, 2015). It is important to note that planning is a complex activity of teaching that requires sophisticated thinking across multiple timeframes from within a class period to across weeks (Calderhead, 1996). In learning study, teachers focus on the complexity of understanding subject matter and how students express their thinking. Teachers are asked to select an object of learning. In other words, teachers select a phenomenon for students to investigate and develop explanations around. A well-chosen object will lead to explanations that require and develop subject matter thinking. Learning study requires teachers to construct complex tasks that will enable students to illustrate their thinking in multiple ways. Teachers anticipate and then analyse students' thinking while students develop and revise the products of their work (e.g., investigation plans or reports, annotated diagrams or models, written explanations, and oral presentation of reasoning). In this way, learning study illustrates what teachers know and are learning about their students' interactions with subject matter ideas. By embedding the study of PCK within tasks for science teacher learning, researchers can get closer to uncovering the reasoning that illustrates teachers' developing PCK.

Learning Study and PCK for Science Teaching

Learning study and PCK are approaches to understanding the development of teacher knowledge and skills that are both squarely situated with the practice of teaching. It is reasonable to think that by engaging in the structured study of science learning, science teachers will use and develop PCK. And like teaching science, the processes of studying learning and developing PCK are both dynamic and complex. The RCM attempts to untangle this complexity for PCK by describing three realms of PCK from the professional knowledge of a community of science teachers and educators, to that of an individual teacher, to the ideas used to inform and the actions taken in an instance of teaching (see Chap. 2). In a parallel fashion, learning study is framed by the professional community's knowledge of teaching and aims to develop each teacher's ideas in ways that will enable him or her to skilfully teach in specific cases. The RCM also describes the interplay of levels of PCK in ways that align with the dynamic work of studying learning. Teachers necessarily engage in planning, enacting, and reflecting with a particular setting, student(s), and learning goals in mind in order to add to their own understanding, which can in turn contribute to the community's understanding of science learning and teaching. The next step in this line of thinking is to unpack how the RCM can guide the design of a learning study assignment and how the assignment artefacts can inform the RCM.

From a teacher educator's perspective, the design of a learning study as a task for teacher learning should be framed by the broader ideas about teaching science while being situated in a specific case of teaching science. Although teacher learning about teaching is considered to be embedded in specific instances of planning, enacting, and reflecting, teachers will need the knowledge and skills to reason about many instances of teaching science. Teacher educators are charged with preparing science teachers for multiple settings and sets of learning goals for students at different stages. The RCM identifies this broader or community-based knowledge as collective PCK (cPCK). Similarly, enacted PCK (ePCK) is described as the knowledge and skills used by a science teacher when they are engaged in the practice of teaching in a particular setting with a particular learning goal for particular student(s). Situating the work of science teacher education within the RCM implies using the realm of cPCK to frame the design of tasks for teacher learning while using ePCK to frame the analysis of a teachers' work within these tasks.

Studying science teachers' ideas about learning and teaching subject matter within the context of a learning study is a twofold, intertwined task of developing and analysing science teachers' PCK. One phase of the work is to integrate ideas about PCK in order to develop and describe the experience for teachers and in turn support their learning. Another phase is to use ideas about PCK to analyse teachers' responses in ways that enable the qualitative features of their thinking to be described and documented. Framed by the RCM, one phase requires the broader framework of cPCK, while the other phase illustrates instances of ePCK. Thus, the questions guiding this exploration into how PCK might be studied in teacher education could be framed as:

One: In what way can cPCK guide the design of learning study as a task for science teacher learning?
Two: In what ways do teacher-developed learning studies illustrate science teachers' ePCK?

As a teacher educator, my examination of an approach to use and study PCK is embedded within tasks developed for teacher learning. In this case, a learning study approach was used to design an assignment to serve two purposes. One was to support teachers in learning about teaching science, while the other was to explore how the assignment could uncover teachers' developing ideas about teaching science. Developing the assignment and examining teachers' work was an iterative process across several semesters of a graduate course in curriculum and instruction. For clarity, this work is presented here as two phases—designing the learning study and illustrating teachers' ideas—and uses examples from one cohort.

Designing the Learning Study Assignment

The learning study was developed as an assignment in a graduate course in curriculum and instruction. This course is a regular offering that is not necessary for teachers only, but since it is a core course in curriculum and instruction, most students who enrol are licensed teachers. The course topics include subject matter for teaching (Grossman, Schoenfeld, & Lee, 2005), learning progressions where students' thinking about science becomes more sophisticated over broad spans of time (Corcoran, Mosher, & Rogat, 2009), and ambitious teaching that considers inquiry and discourse essential to developing all students' scientific thinking (Windschitl, 2008, 2013). As a course assignment, the graduate students' study of learning is informed by reading and discussion around each of these perspectives for thinking about learning and teaching science.

Learning Study Plans

The learning study assignment in this course is focused on the planning phase of a learning study. The process of planning in learning study generally includes selecting subject matter, identifying an object of learning, and considering patterns of variation (Wood, 2015). Translated to an assignment, graduate students select and justify a subject matter idea for learning, identify and describe a cognitive task centred on a phenomenon, and create an "anticipation guide" describing on-target and off-target student thinking. Since the learning study is not a lesson plan, details about materials and student activities are not emphasised. It is also important to point out that learning study goes beyond planning to include the examination of student work using the anticipation guides. Then, based on their students' work, teachers refine their ideas about student thinking and plan future instructional tasks. This second phase of learning study is part of a subsequent graduate course for teachers.

Defining cPCK for the Learning Study Assignment

To inform the design of the learning study assignment, components of PCK were identified by reviewing the literature on science teacher PCK and through empirical work with science teachers (Park & Oliver, 2008; Schneider, 2015). The five components of cPCK used as a guide for this assignment are described below.

- *Orientations* to teaching science. Teachers' ideas about: (a) *nature of learning and teaching* science, (b) *goals* of teaching science, and (c) *purpose* of teaching science.
- *Science curriculum* prepared for teacher and student thinking. Teachers' ideas about: (a) *scope* of science ideas that are important and worth learning, (b) *stan-*

dards as guides for planning and assessing, and (c) *sequence* of science ideas organised for learning.
- *Frameworks* for science teaching. Teachers' ideas about: (a) *inquiry* science learning environments that characterise science and (b) *discourse* in science, both oral and written.
- *Student thinking* about science. Teachers' ideas about: (a) students' *initial* science ideas and experiences, (b) *development* of students' science ideas, (c) how students *express* science ideas, (d) *challenging* science ideas for students, including why the ideas are challenging, and (e) appropriate *level* of science understanding.
- *Instructional strategies* for science topics. Teachers' ideas about: (a) *natural phenomena* experiences and (b) *assessment* of science learning.

Aligning Learning Study and cPCK

The first question for this work asks in what way can cPCK guide the design of learning study as a task for science teacher learning? To answer this question, the development of the learning study assignment was informed by both the components of learning study and the components of cPCK. The task of developing a learning study as an assignment involved creating clear and helpful directions for how, exactly, teachers should plan a learning study. To uncover teachers' pedagogical reasoning, the assignment was designed to prompt their reasoning about teaching science. In addition, the framework of the learning study needed to be consistent with the work of planning and the directions to prompt teachers' pedagogical reasoning had to fit with the purpose of the pedagogical task.

Thinking about cPCK did indeed improve the description of this assignment by supporting the addition of descriptive details for the directions (see Appendix 1). For example, rather than ask teachers to simply identify a target science idea (i.e., learning objective), the directions guide teachers in how to identify a "big idea" and then support their decision. Informed by thinking about specific cPCK components of purpose, scope, and goals, the directions were refined to have teachers select a high impact idea that is worthwhile and meaningful for students and appropriate for students across multiple grade levels. Based on the cPCK concept of sequence (see part c of the Science Curriculum component above), the "big idea" is one where students can develop increasingly more sophisticated thinking over broad spans of time. Similarly, directions for identifying a phenomenon and describing a task were refined when thinking about inquiry, discourse, and expressing ideas, while directions for anticipating student thinking were refined by thinking about initial and challenging ideas for students. The complete alignment between the components of the learning study assignment and cPCK is outlined in Table 7.1.

Table 7.1 Alignment of learning study and collective PCK components

Aspects of learning study assignment	Aspects of collective PCK
High impact idea: teachers select a science idea for student learning that will have a high pay-off for students in understanding science and is appropriate for students across multiple grade levels	*Orientations*: *purpose* of teaching science *Science curriculum*: *standards* as guides for planning and assessing
Sophisticated idea: teachers describe how students can develop increasingly more sophisticated thinking regarding this science idea over broad spans of time	*Science curriculum*: *sequence* of science ideas organising for learning
Worthwhile: teachers describe how the science idea is of value to *science*	*Science curriculum*: *scope* of science ideas that are important and worth learning
Meaningful: teachers describe how the science idea is of value for students *outside* of academic tasks	*Orientations*: *goals* of teaching science
Cognitive task: teachers describe what students will be asked to think about and do that is complex and cognitively demanding	*Instructional strategies*: *natural phenomena* experiences *Frameworks*: *inquiry* science learning environments
Artefact: teachers describe the artefact students will create (write, draw, present, etc.) while engaged in the task *Frameworks*: *discourse* in science both oral and written including argumentation and technical writing	*Instructional strategies*: *assessment* of science learning
Student thinking: teachers describe how this task and artefact will make student thinking visible	*Student thinking*: how students *express* science ideas
Target-level artefact: teacher creates an example of an on-target artefact to illustrate goal for student performance	*Student thinking*: appropriate *level* of science understanding
Describing on-target ideas: teachers describe, list, or illustrate what they anticipate that student will say or do or draw that unpacks complex or sophisticated thinking. Teacher creates a checklist or other method that makes sense for the task	*Student thinking*: *development* of students' science ideas *Student thinking*: how students *express* science ideas
Not on-target ideas: teachers describe, list, or illustrate what they anticipate that student will say or do or draw that illustrate initial or challenges. Teacher creates a checklist or other method that makes sense for the task	*Student thinking*: students' *initial* science ideas and experiences *Student thinking*: *challenging* science ideas and why the ideas are challenging
Role of the teacher: teachers describe their role during this task. What they will do, pay attention to, record, examine, interpret, and revise	*Frameworks*: *inquiry* science learning environments *Frameworks*: *discourse* in science

(continued)

Table 7.1 (continued)

Aspects of learning study assignment	Aspects of collective PCK
Rationale: teachers describe how their plan is an illustration of how they frame their thinking about teaching and learning science	*Frameworks*: *inquiry* science learning environments *Frameworks*: *discourse* in science
Teacher learning: teachers describe what they are learning about teaching and learning	*Orientations*: *nature of learning and teaching* science

Illustrating Teachers' ePCK

This particular learning study assignment was part of an introductory graduate-level course in curriculum and instruction. Ohio science teachers enrolled in the course as part of a programme to prepare current high school teachers to teach introductory-level college content in their high school classrooms. The study group included 19 high school chemistry teachers across multiple course sections in the same semester, and as Ohio teachers, all were using the same state-provided content standards for chemistry. These teachers developed learning study plans as part of the course.

In order to investigate possible differences in their enacted PCK (ePCK), teachers were identified as new (1–3 years of experience), some experience (4–10 years), or much experience (11 or more years). Their content knowledge background was described as excellent (content major with high grades in area), good (content major with lower grades or non-major with high grades in area), or developing (non-major with modest grades area). The examples presented here were selected from three chemistry teachers who focused their work on atomic models. This selection of the same teaching topic meant qualitative differences in ePCK could be highlighted. Teacher A was a new teacher with a good background in chemistry, while Teacher B also had a good background in chemistry but more teaching experience (some). Teacher C was a new teacher but had an excellent background in chemistry. With different levels of experience and chemistry background, the work of these three teachers tests the learning study as a task to uncover differences in teachers' ePCK.

Describing ePCK

The second question for this work asks in what ways do teacher-developed learning studies illustrate teachers' ePCK? To answer this question, teachers' responses to components of the assignment were examined in relation to the corresponding components of cPCK. In other words, the cPCK component determined to be most aligned with each component of the learning study (Table 7.1) was used to guide the review of that aspect of a teacher's response. The intention was to develop qualitative descriptions of teachers' ideas. For example, when a teacher describes how the subject matter idea they have selected is meaningful for students outside of academic tasks, his/her response is examined for ideas about goals of teaching science. To explore whether the learning study responses were helpful in illustrating differences in teachers' ePCK, responses from teachers with different levels of experience were compared. To determine whether this task was illustrating ePCK separately from content knowledge, teachers with different levels of chemistry background were compared. Tables 7.2, 7.3 and 7.4 include sample responses selected from these three teachers' learning studies in order to demonstrate how ePCK is illustrated.

To illustrate ePCK for teachers with different levels of teaching experience, responses from Teacher A (new teacher with a good background in chemistry) and Teacher B (some teaching experience with a good background in chemistry) were compared (see Table 7.2). For example, Teacher A describes the selected subject matter idea (atomic model) as meaningful for students because all matter is made of atoms. In comparison, Teacher B does not directly describe how this idea is meaningful but does mention the need to understand the viewpoint of students. Both responses begin to illustrate the teachers' ideas about goals for teaching science. Although it is premature to suggest one response is more advanced or correct than the other, differences based on experience with students are suggested.

Comparing responses from Teacher B (some experience with a good background in chemistry) and Teacher C (new teacher with an excellent background in chemistry) explores possible differences in ePCK based on teaching experience for teachers who also have different levels of content background (Table 7.3). In one example, Teacher B describes student thinking by stating that students will use arrows to represent movement, but it is not clear why these ideas are challenging for students. On the other hand, Teacher C describes student thinking by stating that students' drawings will show their thinking, but it is more specific in describing how students will misunderstand ideas about models and elements. Again, these responses illustrate differences that might begin to uncover ePCK.

A third set of comparisons highlights two new teachers with different levels of content background and limited teaching experience. Teacher A (new teacher with a good background in chemistry) and Teacher C (new teacher with excellent background in chemistry) are both novice teachers, but one has more chemistry background (Table 7.4). In this case, both teachers describe the role of the teacher in an inquiry and discourse environment as encouraging students to investigate or collaborate, but do not have specific ideas about how to do so. This response is reasonable for new teachers. Their responses also differ in that Teacher A mentions feedback, while Teacher C is more specific about the chemistry ideas students will explain. These responses might be indicating similar ePCK, but differences in content knowledge.

Discussion

Using the RCM (i.e. cPCK, pPCK, and ePCK) as a guide, this chapter describes how a learning study was designed and teachers' responses were examined. It makes sense that PCK, as a construct that is intended to describe what teachers know about

Table 7.2 Teacher A and Teacher B: responses to components of the learning study assignment and ePCK illustrated

Teacher A	Teacher B
Orientations, purpose: to prepare students for the next level of schooling in chemistry	*Orientations, purpose*: to prepare students to understand further chemistry
Science curriculum, standards: clear linking of standards across grades	*Science curriculum, standards*: not focused on standards
High impact idea: the big idea I have selected for my learning study is the atomic model. The atomic model is a big idea that is built upon throughout students' education. In the Ohio model curriculum, the idea that all matter is composed of atoms is presented in the elementary grades. Later, in middle school, they are to understand that these atoms are made up of subatomic particles and a model of this atom can be created based on current scientific evidence. At the high school level, different models of the atom are presented	*High impact idea*: this learning plan deals with the formation of ions in order to increase atomic stability. This idea builds on the knowledge of atomic structure and leads to understanding the formation of bonds and chemical reactions
Science curriculum, sequence: focused on components of the atomic model, mentions these are useful (more below)	*Science curriculum, sequence*: focused on how the model explains bonds and reactions
Sophisticated idea: the atomic model is a big idea that is built upon throughout students' education. In the Ohio model curriculum, the idea that all matter is composed of atoms is presented in the elementary grades. Later, in middle school, they are to understand that these atoms are made up of subatomic particles and a model of this atom can be created based on current scientific evidence. At the high school level, different models of the atom are presented. The two most useful models include the Bohr model and the quantum mechanical model	*Sophisticated idea*: this idea builds on the knowledge of atomic structure and leads to understanding the formation of bonds and chemical reactions…. after we have learned the structure of atoms and their stability and have begun to work with ionisation
Science curriculum, scope: describes detail about how this idea will help students think about chemistry	*Science curriculum, scope*: describe that structure relates to function
Worthwhile: knowing the atomic model is worthwhile because having a deep understanding of the atomic structure is a key to all topics covered in chemistry. For example, the trends seen in the periodic table can be explained by understanding how the protons and electrons within an atom are arranged. The more advanced quantum mechanical model helps to describe exactly how the electrons are arranged, which gives rise to the properties of elements and compounds. Chemical bonding also relies on the atomic orbitals becoming hybridized, and this gives rise to molecular geometry, molecular polarity, and many other concepts	*Worthwhile*: this idea builds on the knowledge of atomic structure and leads to understanding the formation of bonds and chemical reactions (did not give a distinct response for worthwhile)

(continued)

Table 7.2 (continued)

Teacher A	Teacher B
Orientations, goals: to help students think about properties and models, closer to classroom than outside of the classroom	*Orientations, goals*: to match students' perspective, no details yet on what that is
Meaningful: this idea is meaningful because all matter is made of atoms. If one is able to understand the structure of the atom, one can begin to make sense of the physical and chemical properties of the materials they encounter in their everyday experiences. This idea is also meaningful because it exemplifies the use of a model in science. Models can be used to show things at the very macroscopic and very microscopic levels and are used to visualizse abstract ideas	*Meaningful*: by analysis of the results, I would hope to better understand the viewpoint of my children (did not have a response for meaningful)

teaching subject matter, would be helpful in designing tasks for teacher learning. It also makes sense that evidence collected from artefacts of teaching (in this case, planning), would make teachers' ideas visible in ways that can illustration their personal PCK (pPCK). Descriptions of whether and in what ways this is, indeed, the case are needed. The descriptions provided here are from a teacher educator's perspective, exploring this potential approach as a means of illustrating teachers' ideas while supporting teachers in learning about teaching.

Designing Tasks for Teacher Education

The RCM identifies three realms of PCK, each of which describes PCK at a different level from a community's knowledge, to an individual teacher, to a subset of ideas and actions used in a particular instance of teaching. It turns out, quite reasonably, that the different realms of the RCM were useful for thinking about PCK in different situations. To design a task for teacher learning, cPCK was a helpful framework for thinking about what ideas should frame a teaching-based task. The RCM, in and of itself, did not have the detail needed to guide the design of the learning study as a planning task. However, thinking about cPCK as a community-based knowledge is consistent with existing thinking about a collective understanding of the components of PCK. The science education research community, for example, has been describing and researching ideas to identify a set of PCK components for some time (Schneider & Plasman, 2011). The RCM helps to clarify how to use the components of PCK most often described in the literature. In this work, the components of cPCK were helpful in developing the learning study as an assignment for teachers. The specific components were a guide in adding specific details and directions to guide teachers in thinking deeply about learning that otherwise might

Table 7.3 Teacher B and Teacher C: responses to components of the learning study assignment and ePCK illustrated

Teacher B	Teacher C
Instructional strategies, phenomena: focused on one model to connect to bonding, link to phenomenon not implied	*Instructional strategies, phenomena*: focused on two specific atoms to compare using the model and to connect to bonding, link to phenomenon explicit
Frameworks, inquiry environments: one guiding question	*Frameworks, inquiry environments*: several guiding questions
Cognitive task: students will look at the Bohr model of a standard atom and determine how it will become stable. "Would this cause these two atoms to bond?"	*Cognitive task*: the student will have to draw the Bohr model for a francium atom and a fluorine atom and describe why they believe each element to be either reactive or unreactive and why. They will be asked to think about what happens to the size of things when you continually add more to it and to consider the charges of all types of the particles
Instructional strategies, assessment: students use model to begin to predict. Students draw and label, but do not describe what the drawing represents	*Instructional strategies, assessment*: students draw, label, and describe what the drawing represents
Frameworks, discourse: focus on representation and notation only	*Frameworks, discourse*: focus on representation and description in their own words about their ideas
Artefact: draw a second Bohr model showing the resulting atom/ion. Indicate the atomic number, mass number, and ionic charge for each model. Finally, use arrow to show how electrons move from one atom to another	*Artefact*: students will illustrate the atomic structure (Bohr model) for a francium and fluorine atom and describe why you believe each element to be reactive or unreactive. (Number of energy levels and electrons not necessary but shown on the artefact.)
Student thinking, express science ideas: students draw and use notation to represent thinking	*Student thinking, express* science ideas: students' descriptions and aspects of their drawings illustrate their thinking
Student thinking: the task involves having students determine electron stability and drawing a model of the ion that is formed. This artefact will show common misconceptions such as how many electrons and atom tends to gain or lose	*Student thinking*: through their descriptions of the reactivity of the atom, I can determine what they understand about how electrons are gained and lost through attractive forces of the nucleus and how that impacts the reactivity of the atom. I would be able to determine if the students understand that similar charges repel, where like charges attract. I would also be able to determine how they understand the changes in electrical attraction or repulsion as the distances between the particles changes and how that distance affects the reactivity of the element and its ability to form an ion

(continued)

Table 7.3 (continued)

Teacher B	Teacher C
Student thinking, appropriate *level*: using diagram to represent simple bonding, using arrows to represent movement	*Student thinking*, appropriate *level*: using model to contrast elements on several dimensions
Target-level artefact: 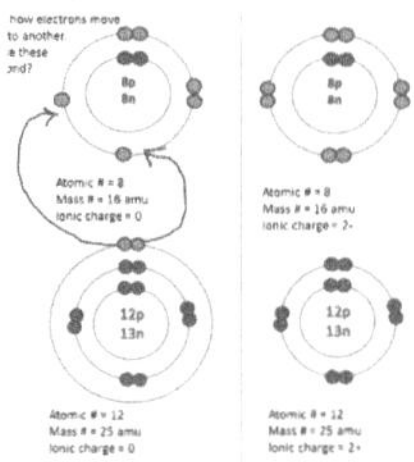 	*Target-level artefact*: 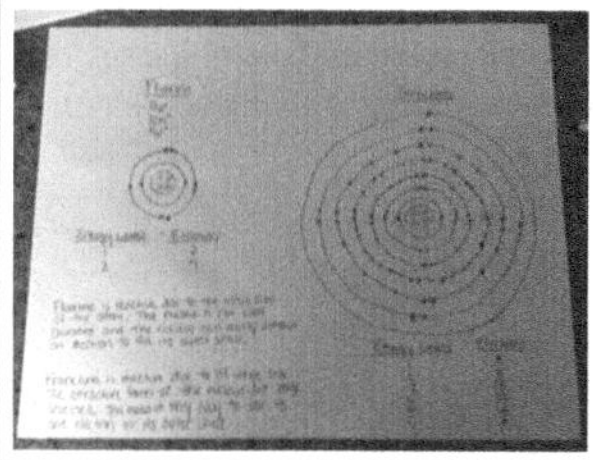
Student thinking: *development* of science ideas—students learn the components	*Student thinking*: *development* of science ideas—students learn by contrasting features
Student thinking: *express* science ideas—students draw the components	*Student thinking*: *express* science ideas—student model the components and explain in their words
Describing on-target ideas	*Describing on-target ideas*
Concepts: • Outer shell is full (8) • Atomic number (protons) does not change • Mass number does not change • Number of neutrons does not change • Electrons form pairs • Only outer shell is affected • Ionic charge = number of protons − number of electrons • Electrons transfer from one atom to another into appropriate places	Model (excerpt) Correct ratio in scale size (francium larger, fluorine smaller). Due to the addition of energy levels (and electrons), it will continue to make the atom larger Description (excerpt) ____Francium is larger in size due to its number of energy levels and electrons ____Fluorine is smaller in size due to the limited number of energy levels. This means the attractive forces can pull in and hold the electrons very easily
Student thinking: *initial* science ideas and experiences—student do not know the components	*Student thinking*: *initial* science ideas and experiences—students misunderstand the interactions
Student thinking: *challenging* science ideas— superficial ideas about what is challenging	*Student thinking*: *challenging* science ideas—students misunderstand the connections such as size or forces
Not on-target ideas	*Not on-target ideas*

(continued)

Table 7.3 (continued)

Teacher B	Teacher C
Misconcepts: • Outer shell is not full • Number of protons changes • Number of neutrons changes • Mass number changes • Electrons added/removed from inner shells • Gaining electrons instead of losing electrons or vice versa • Ionic charge is incorrect • Electrons are not transferred	Models (excerpt) ____: Models drawn the same size or fluorine drawn bigger Used to seeing computer pictures of the Bohr models and thinks they are all the same size. Does not understand the pull of the positively charged nucleus or that adding more energy levels increases the diameter of the atom Description ____Describes the elements as unreactive for various reasons ____Size is not used in the explanation for either element or is not used correctly ____The attractive forces of the nucleus to the electrons are not mentioned in the descriptions correctly

have been overlooked. Because the assumptions underlying the RCM of PCK and the ideas about teacher learning were based on the same fundamental ideas about teaching and learning, it was possible to align each of the components of cPCK with a corresponding component of the learning study task.

The design of the learning study assignment described here suggests that cPCK can be a useful guide for designing educative tasks for teachers. In this way, the RCM of PCK can be helpful in strengthening the education of teachers. Teacher education is frequently the focus of critique with some reformers recommending more robust programmes with stronger links to classroom-based experiences, while others advocate for reducing formal teacher education (Darling-Hammond, Holtzman, Gatlin, & Heilig, 2005). The learning study assignment is a carefully designed task that is anchored in well-thought-out ideas about teacher knowledge that supports teachers in thinking about their own students. If more assignments in teacher education were carefully constructed based on ideas about what and how teachers learn, this practice would not only make these experiences more powerful, it may also aid teacher educators in describing the importance of teacher education in ways that could inform policy for teacher education.

Observing ePCK

Learning study does appear to be a useful approach to illustrate teachers' ePCK for planning. The learning study assignment is consistent with the work of planning and is closely linked to teachers' pedagogical reasoning around specific subject matter for specific students. The examples above suggest that teachers with similar content

Table 7.4 Teacher A and Teacher C: responses to components of the learning study assignment and ePCK illustrated

Teacher A	Teacher C
Frameworks, inquiry environments: low inquiry environment, unclear about experiences	*Frameworks, inquiry* environments: students gather information and collaborate, draw conclusions, reasoning with explanation
Frameworks, discourse: visual representations, teachers provide feedback	*Frameworks, discourse*: students collaborate and discuss ideas, explain ideas
Role of the teacher: the teacher must investigate what knowledge the students have in order to determine where to begin. … then they can provide students with experience that will build upon their prior knowledge. In this task, the prior knowledge would be the Bohr model. Once students are engaged in their learning, the teacher must observe and monitor how students are incorporating the new material by having them create visual representation s of their understanding during the learning process. Once students have shown how they understand the new material, the teacher must examine their work and provide them with constructive feedback letting them know how successful they were with the task	*Role of the teacher*: I will allow the students to investigate and draw conclusions on their own. I will encourage the students to use their periodic table as a guide and investigate the subatomic particles themselves to determine where they should be located. Proper materials will also be provided so if a student wants to use a compass to make uniform circles so they can accurately represent the size of the atoms, it will be possible. I will encourage cooperative learning amongst students and student discussion of ideas. I will pay attention to how they draw their conclusions about reactivity of each of the elements and if they used the size and valence electrons in their reasoning. I will not just be looking for if they know it is reactive or not, but their ability to be able to explain why it is reactive
Frameworks, inquiry environments: translating from a simpler model to a more sophisticated model, visual, teacher feedback	*Frameworks, inquiry* environments: ideas build on previous ideas; student create and explain; use extremes examples first
Frameworks, discourse: NA	*Frameworks, discourse*: NA
Rationale: this plan builds upon the less sophisticated atomic model, the Bohr model, to help students understand the more sophisticated quantum model. Students find the Bohr model easy to create, while the quantum model tends to be more challenging to understand. I believe by relating the two, students will more clearly see the relationship from one model to the next. The artefact created by student during this task also allows the teacher to visually determine a student's understanding of the quantum model. The teacher can then more easily provide feedback to students to help them with their learning, as well as determine the success of the instruction based on student learning	*Rationale*: this plan illustrates subject matter knowledge for teaching as it builds on previous background knowledge the students have and provides a chance to reinforce that material along with building upon it. This gives the students an opportunity to create something to explain their thinking. I have also used the smallest halogen and largest alkali metal as they are two of the most reactive elements on the periodic table. If students understand these elements, then we could explore deeper into the other elements and use those same principles to discuss the reactivity of more complex elements
Orientations, nature of learning and teaching science: teachers need to monitor learning, students need to think about their own understanding, ideas develop in sophistication	*Orientations, nature of learning and teaching* science: learning requires verbal communication; question identify areas to work on and misconceptions; students learn differently

(continued)

Table 7.4 (continued)

Teacher A	Teacher C
Teacher learning: I learned that monitoring students learning by making their thinking visual could be a powerful tool to help the teacher guide their students to the desired learning outcome and to give students the opportunity to think about their own understanding. I also learned that the path to learning could be aided by building upon prior, less sophisticated knowledge to develop a more sophisticated way of thinking	*Teacher learning*: this artefact will support my learning about learning as I will be able to identify the students' thought processes, not only through the artefact itself, but also through verbal communication. The questions students will ask can help me identify the areas of weakness in the material and allow me to determine if it is an individual weakness or a class weakness. If it is a class weakness, and they all have the same misconception, there may be an experience in previous learning that created that misconception. I could then communicate with previous teachers to help sort out that misconception for future students. Through the students' ideas, I may also be able to determine how to best present the information initially, to give them a more solid understanding of the information. Each student learns a bit differently, so over time I may be able to compile an assortment of methods and allow the students to pick how they would like to learn a topic

preparation may illustrate different types of ePCK. This finding needs to be explored in more depth, but initial indications are that learning study may be illustrating more than content knowledge. The nature of the differences observed for the chemistry teachers appears to reflect differences related to teaching experience separately from differences related to chemistry background. Although not included in the samples provided in this chapter, learning studies from teachers working outside of their expertise (e.g., biology teachers planning for physical science) indicated that these teachers struggled to a greater degree with this planning task. Although this learning study assignment was focused on only the planning aspect of teaching, ePCK ideas suggested by the teachers were representative of the components of cPCK used to design this planning task. It is reasonable to predict that when teachers collect and examine student artefacts in the next stage of the learning study, their ePCK ideas will be further illustrated. Perhaps their ideas about inquiry and discourse environments, in particular, will be better illustrated.

As an approach proposed to capture (i.e. assess) teachers' ideas, it is important to think about the validity of learning study as an assessment tool. Learning study can be thought of as a performance assessment, and, as such, factors of validity for performance assessments should be considered (Messick, 1994). This learning study assignment has a relative low consequence in that it will not be used to determine anything more than a single grade in a course. However, some other factors worth keeping in mind are content coverage, cognitive complexity, and meaningfulness.

The components of cPCK (notably science curriculum, frameworks, and student thinking) were represented in the components of the learning study assignment. Learning study is embedded in teachers' work of planning for their students and, thus, should be a meaningful task outside of the course assignment. This task also reflects the complexity of teaching and learning about teaching. Learning study appears to be a fruitful path to pursue with assessment design in mind.

Mapping Trajectories

As an approach to thinking about assessing ePCK in ways to infer teachers' personal PCK (pPCK), it is interesting to think about mapping trajectories to describe how teachers' learning progresses. Learning progression is a framework for thinking about how learners (in this case teachers) develop increasing more sophisticated ways of thinking over broad spans of time and in connection with instruction and assessment (Heritage, 2008). Measuring progressions is a complex task that involves construct mapping (Wilson, 2009), that is, mapping the layers of increasingly sophisticated ideas for the construct, in this case cPCK. Based on a well-thought-out construct map, artefacts illustrating teachers' thinking are analysed to suggest a trajectory or path of learning progress. Instruction and assessment become an iterative process in the uncovering of trajectories. This process matches that described here in this chapter, that is, where a construct is used to design instruction and assess learning. This type of work is a step towards describing trajectories for teachers' pPCK.

Conclusion

Overall, learning study as an approach to develop and illustrate teachers' pedagogical content knowledge shows promise. The RCM as a model for thinking about PCK in realms or layers was helpful to situate and parse the work of design and research associated with teacher education. Perhaps most interesting is the potential to begin mapping the PCK construct and teachers' learning trajectories in conjunction with learning study. Learning study is a complex and meaningful task for teachers. It is also a more efficient or concise approach than lesson planning. This quality is an important consideration for an assessment tool. Learning study, however, does require teachers to learn about learning study. While teachers are accustomed to being asked to plan lessons, learning study planning is more focused and complex and can push their thinking in new ways. These are the features, though, that make learning study valuable. This thinking is the type of work that is needed to advance our understanding of how to design and demonstrate excellence in teacher education.

Appendix 1: Learning Study Assignment Directions

Subject Matter Idea Description: Describe the big or high impact idea you have selected for your learning study. Describe how this idea(s) is meaningful (of value outside of academic tasks), worthwhile (of value to science), and of high impact for students (will make a big difference for students). Identify a core idea that will require increasingly sophisticated thinking over time. Use and revise your description of the high impact ideas that you posted for your journal.

Cognitive Task and Artefact Description: Describe the cognitive task and artefact that will help you uncover students' thinking. Describe what students will be asked to think about and do. Describe what the artefact students would create (write, draw, present, etc.). Describe how this task and artefact will make student thinking visible to you and them. Think about a task that is substantive so that students can participate and develop ideas. Create a task that is cognitively demanding, has multiple ways for students to participate (a complex task), and results in an artefact.

Sample Artefact: Mock up a sample of what you expect students to create. This should be an on-target example. It does not need to be actual student work. It can be an extract or sample of the key aspects of the artefact. This might include essays, diagrams, and illustrations.

Assignment Directions for Anticipation of Student Thinking: Describe or illustrate and list what you anticipate that students will say or do or draw, etc., including pieces that are on target and not on target. What do you anticipate as student thinking about the target ideas? Include these ideas in a checklist or other method as makes sense for your task. For example, if the task is to draw a representation of a molecule, what features would you look for in the drawing that would tell you what they are thinking?

Reflecting on the Study of Learning: (a) *Role of the teacher as learner*: What role will a teacher (i.e. you) have that will make this a study of learning. What will you do, pay attention to, record, examine, interpret, and revise as you complete your study of learning? How will this support your learning about learning? Be specific to this learning study, the artefact, and anticipation of student thinking. (b) *Theoretical underpinnings*: How does your plan reflect theoretical frameworks for thinking about curriculum (specifically learning progressions, subject matter knowledge for teaching (teacher knowledge; PCK), and learning studies)? Be explicit in linking your plan to these ideas. (c) *Planning to learn*: What did you learn by creating this plan? How would you create another plan to study learning?

References

Calderhead, J. (1996). Teachers: Beliefs and knowledge. In D. C. Berliner & R. C. Calfee (Eds.), *Handbook of educational psychology* (pp. 709–725). New York: Macmillan.

Cheng, E. C. K., & Ling, L. M. (2013). *The approach of learning study: Its origin and implications*. Retrieved from http://www.oecd.org/edu/ceri/innovativelearningenvironments.htm.

Corcoran, T., Mosher, F. A., & Rogat, A. (2009). *Learning progressions in science: An evidence-based approach to reform*. Retrieved from New York.

Darling-Hammond, L., Holtzman, L., Gatlin, S. J., & Heilig, J. V. (2005). Does teacher preparation matter? Evidence about teacher certification, teach for America, and teacher effectiveness. *Educational Policy Analysis Archives, 13*(42). Retrieved from http://epaa.asu.edu/epaa/v13n42/.

Feiman-Nemser, S. (2008). Teacher learning: How do teachers learn to teach? In M. Cochran-Smith, S. Feiman-Nemser, D. J. McIntyre, & K. Demers (Eds.), *Handbook of research on teacher education: Enduring questions in changing contexts* (pp. 697–705). New York: Routledge.

Grossman, P. (2005). Research on pedagogical approaches in teacher education. In K. Zeichner (Ed.), *Studying teacher education* (pp. 425–476). Mahwah, NJ: Lawrence Erlbaum Associates.

Grossman, P., Schoenfeld, A. H., & Lee, C. (2005). Teaching subject matter. In L. Darling-Hammond & J. D. Bransford (Eds.), *Preparing teachers for a changing world: What teachers should learn and be able to do*. San Francisco, CA: Jossey-Bass.

Hammerness, K., Darling-Hammond, L., Bransford, J., Berliner, D., Cochran-Smith, M., McDonald, M., et al. (2005). How teacher learn and develop. In L. Darling-Hammond & J. Bransford (Eds.), *Preparing teachers for a changing world* (pp. 358–389). Hoboken, NJ: Jossey-Bass.

Heritage, M. (2008). *Learning progressions: Supporting instruction and formative assessment*. Retrieved from Washington, DC.

Lampert, M. (2001). *Teaching problems and the problems of teaching*. New Haven, CT: Yale University Press.

Messick, S. (1994). The interplay of evidence and consequences in the validation of performance assessments. *Educational Researcher, 23*(2), 13–23.

Park, S., & Oliver, J. S. (2008). Revisiting the conceptualization of pedagogical content knowledge (PCK): PCK as a conceptual tool to understand teachers as professionals. *Research in Science Education, 38*(3), 261–284.

Schneider, R. M. (2015). Pedagogical content knowledge reconsidered: A teacher educator's perspective. In A. Berry, P. Friedrichsen, & J. Loughran (Eds.), *Re-examining pedagogical content knowledge in science education* (pp. 162–177). London: Routledge.

Schneider, R. M., & Plasman, K. (2011). Science teacher learning progressions: A review of science teachers' pedagogical content knowledge development. *Review of Educational Research, 81*(4), 530–565.

Sykes, G. (1999). Teacher and student learning: Strengthening their connection. In L. Darling-Hammond & G. Sykes (Eds.), *Teaching as the learning profession: Handbook of policy and practice* (pp. 151–179). San Francisco, CA: Jossey-Bass.

Tan, Y. S. M., & Nashon, S. M. (2013). Promoting teacher learning through learning study discourse: The case of science teachers in Singapore. *Journal of Science Teacher Education, 24*, 859–877.

Wilson, M. (2009). Measuring progressions: Assessment structures underlying a learning progression. *Journal of Research in Science Teaching, 46*(6), 716–730.

Windschitl, M. (2008). How novice science teachers appropriate epistemic discourses around model-based inquiry for use in classrooms. *Cognition and Instruction, 26*(3), 310–378.

Windschitl, M. (2013). *The vision of ambitious science teaching*. Retrieved from https://ambitiousscienceteaching.org/get-started/.

Wood, K. (2015). Deepening learning through lesson and learning study. In K. Wood & S. Sithamparam (Eds.), *Realizing learning: Teachers' professional development through lesson and learning study* (pp. 1–24). New York: Routledge.

Chapter 8
The PCK Map Approach to Capturing the Complexity of Enacted PCK (ePCK) and Pedagogical Reasoning in Science Teaching

Soonhye Park and Jee Kyung Suh

Abstract This chapter focuses on how the Refined Consensus Model (RCM) of PCK for teaching science provides a useful conceptual framework for informing a methodological approach to PCK research called "PCK mapping". The PCK map approach, as it is known, was originally designed to identify and illustrate interactions among PCK constituent components through visualisation and quantification. In this chapter, we first describe and discuss aspects of the PCK map approach as they relate to the RCM. These aspects include (1) the theoretical underpinnings and assumptions of the approach, (2) step-by-step procedures of the approach, (3) its applications and usefulness to PCK research, and (4) contributions of the approach to advancing research on PCK. We also illustrate how repositioning the PCK map approach within the RCM enabled us to critique methodologies in two previous science education studies where we utilised the PCK map approach in different ways, while identifying and addressing methodological issues. Finally, we highlight the potential of the PCK map approach as a methodological tool to capture the essence of science teachers' enacted PCK (ePCK) throughout the pedagogical cycle of planning, enactment, and reflection, and the knowledge exchanges occurring between the realms of pPCK and ePCK.

Introduction

In an effort to capture and display the abstract and complex construct of PCK in a more explicit and concrete manner, the PCK map approach was first created as an analysis method that sought to quantify and visualise interactions among PCK

S. Park (✉)
Department of STEM Education, North Carolina State University, 326 K Poe Hall, Campus Box 7801, Raleigh, NC 27695, USA
e-mail: spark26@ncsu.edu

J. K. Suh
Curriculum and Instruction, The University of Alabama, 3403 Science and Engineering Complex, Box 870232, Tuscaloosa, AL 35401, USA
e-mail: jksuh@ua.edu

A. Hume et al. (eds.), *Repositioning Pedagogical Content Knowledge in Teachers' Knowledge for Teaching Science*, https://doi.org/10.1007/978-981-13-5898-2_8

constituent components (Park & Chen, 2012). The PCK map approach is not only theoretically based on the pentagon model of PCK (Park & Oliver, 2008), but also utilises that model as an analytic device. In this chapter, we discuss how the PCK map approach can be situated and utilised in the context of the Refined Consensus Model (RCM) of PCK for teaching science. First, we introduce theoretical assumptions underlying the mapping approach, followed by a description of the step-by-step procedures the approach employs. Next, by comparing and contrasting two research studies in science teacher education that employed the approach in different ways, we demonstrate how the PCK map approach can be applied to serve varying research purposes. In addition, repositioning the PCK map approach within the RCM, we identify and address methodological issues associated with the PCK map approach of the two studies. Finally, we explore how the RCM provides a useful conceptual framework for situating the PCK map approach as a viable and powerful means of capturing a science teacher's PCK in classroom actions, i.e. enacted PCK (ePCK) and complex interactions of personal PCK (pPCK) and ePCK throughout a full pedagogical cycle of planning, enactment, and reflection.

Theoretical Framework Underlying the PCK Map Approach

To explicate interplay among the components constituting PCK, various approaches have been used in science education research. For example, Henze and her colleagues (2008) drew maps visualising the relationship among the components in order to investigate how two different types of teacher orientations (type A PCK vs. type B PCK), knowledge of learner, assessment, and instructional strategy were integrated while teaching science. Kaya (2009) utilised a quantitative approach featuring a PCK scoring rubric to explore intra-relationships between preservice science teachers' PCK components. Later, Padilla and van Driel (2011) combined quantitative and qualitative methods to analyse relationships between PCK components of six university instructors teaching chemistry using an adapted version of the PCK model of Magnusson, Krajcik, and Borko (1999). As a means of explicating interplay among PCK components, the PCK map approach has demonstrated its usefulness as a data analysis method for capturing the interconnectedness and integration of PCK components (Aydin & Boz, 2013; Park & Chen, 2012).

The PCK map approach was grounded in the pentagon model of PCK (Park & Oliver, 2008) in which PCK is defined as an integration of five components: (1) Orientations toward Teaching Science (OTS), (2) Knowledge of Students' Understanding in Science (KSU), (3) Knowledge of Science Curriculum (KSC), (4) Knowledge of Instructional Strategies and Representations (KISR), and (5) Knowledge of Assessment of Science Learning (KAs) (see Park & Oliver, 2008 for descriptions of the five components). Although the pentagon model involves the same five components as the PCK model of Magnusson et al. (1999), it places emphasis on the reciprocal interactions across the components, while the Magnusson et al. model emphasises hierarchical interactions between the orientations towards teaching science and the other four components (Park & Chen, 2012). Furthermore, the pentagon model stresses the importance of coherent and synergistic interactions among the PCK components to the quality of PCK. In other words, advancing to higher levels of PCK requires teachers to strengthen the coherence between the components in addition to improving individual components. Those coherent connections between the components are accomplished through the complementary and ongoing readjustment occurring through both reflection in action and reflection on action processes that are conceptualised as "pedagogical reasoning" in the RCM of PCK (see Chap. 2; Park & Oliver, 2008).

The notion that a teacher's PCK level depends on the degree of integration and coherence among the components resonates with the work of many PCK scholars (e.g., Friedrichsen et al., 2009; Hashweh, 2005; Krauss et al., 2008; van Driel, De Jong, & Verloop, 2002). In this regard, research on the nature and development of PCK needs to pay more attention to interactions among PCK components than to each component as a silo. This shift in focus calls for a new and yet robust research method to capture the interactions, which the PCK map approach can provide. For example, there is a great deal of uncertainty about what types of interactions among PCK components provide leverage for promoting PCK as an entirety and how those interactions work. This uncertainty can be largely attributed to difficulties in capturing the dynamics of interactions. Thus, it is imperative to develop sound analytic methods that enable researchers to explicitly identify and represent the interactions occurring among PCK components. With this goal in mind, the PCK map approach was created as an attempt to visually represent identified interactions by presenting them in a pentagon form of PCK.

Step-by-Step Procedures of the PCK Map Approach

When the PCK map approach is situated within the RCM, the approach enables us to capture the integration and interaction of the five PCK components as revealed in a science teacher's ePCK and pedagogical reasoning throughout a cycle of planning, enactment, and reflection. In order to use the PCK map approach, data sources must include classroom observations and interviews in combination with each observation, i.e., pre-/post-observation interviews encompassing the full pedagogical cycle. A pre-observation interview focuses on a teacher's planning of the lesson to be observed. This interview involves defining goals of the lesson, identifying curriculum consideration, designing science activities and resources, and planning assessments. A post-observation interview conducted after each observation is not only to elicit a teacher's pedagogical reasoning for instructional decisions, but also to provide the teacher opportunities to reflect on various classroom incidents and instructional actions. Such opportunities for reflection will enable the teacher to tease out a complex interaction of knowledge and skills during acts of teaching that involve processes of pedagogical reasoning and acts of teaching influenced by various contextual and affective factors.

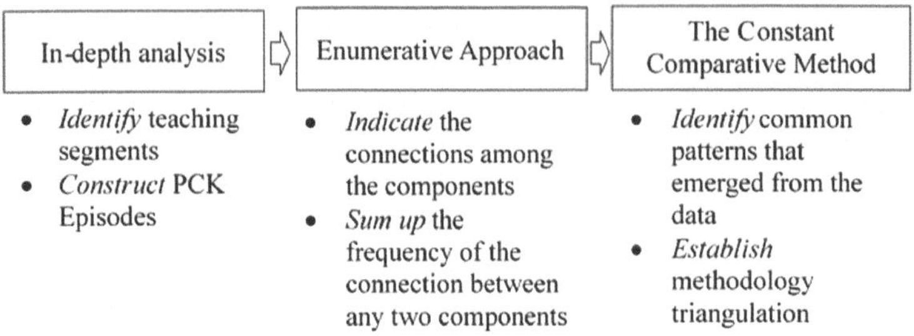

Fig. 8.1 Steps of the PCK map approach

Data analysis, using the PCK map approach, is carried out through several steps: (1) in-depth analysis of explicit PCK, (2) the enumerative approach, and (3) the constant comparative method (Strauss & Corbin, 1990). The steps are shown in Fig. 8.1 and explained in full below.

In-Depth Analysis of Explicit PCK

The main purpose of the in-depth analysis step is to identify PCK components integrated into a science teacher's ePCK in a specific teaching segment within a given context. In this step, teaching segments representing ePCK are first identified from videotaped instructional sessions. The RCM defines ePCK as "the specific knowledge and skills utilised by an individual teacher in a particular setting, with a particular student or group of students, with a goal for those students to learn a particular concept, collection of concepts, or a particular aspect of the discipline" (see Chap. 2). However, for the PCK map approach to depict ePCK as interactions of PCK constituent components, the RCM needs to specify components that constitute ePCK, which it does not. In this regard, it seems methodologically reasonable to use the five components from the pentagon model as a means to capture ePCK through the PCK map approach. Selecting the teaching segments involving ePCK can then be guided by an operational definition of ePCK drawn from the pentagon model of PCK; that is, PCK is an integration of *two* or *more* components in the pentagon model. Once a science teaching segment that indicates the presence of two or more components of ePCK is identified, the teaching segment is labelled as a PCK Episode. Three features of the PCK Episode are then described in detail: (1) what the teacher and students did, (2) what PCK components were integrated, and (3) the evidence that identifies the presence of the components. The description is derived primarily from observations, but complemented by interviews, documents, and artefacts related to the PCK Episode.

The Enumerative Approach

In this step, the PCK components identified in a particular PCK Episode and the connections among the components are indicated using the pentagon model as an analytic device—this representation constitutes a PCK map. Specifically, connections between any two identified components are determined based on two assumptions: (1) there must be at least one connection between any two of the identified components, and (2) every connection has the same strength (Park & Chen, 2012). For example, if three components like Orientations toward Teaching Science (OTS), Knowledge of Students' Understanding in Science (KSU), and Knowledge of Instructional Strategies and Representations (KISR) are identified as working components in a particular PCK Episode, one connection is recorded between any two of the three as depicted in Fig. 8.2. Although individual connections between component pairs could be different, the same strength for each connection is assumed for analytic convenience. Each PCK Episode results in one PCK map, and the same procedure is repeated for the other PCK Episodes identified in the same instructional session. Then, the frequency of the connection between any two components must be summed up across all PCK Episodes from the particular instructional session being studied and indicated in the pentagon model.

Evidence that supports the presence of the PCK components identified in a PCK Episode	PCK map
In the post interview, Mr. K said his students would have misconceptions about the concept of energy because it is new and it is a little bit abstract to them. [KSU] He stated that he would not look to correct any of those misconceptions since it is still a kind of formative understanding of what energy is. [KISR] Also, he said "I've seen plenty of times in teaching, when a student doesn't understand, if I just step in and in an authoritative way say 'Nope this is it', the connection isn't made. It is stopped. It stunts the growth process in the way that they form their understanding of it." [OTS]	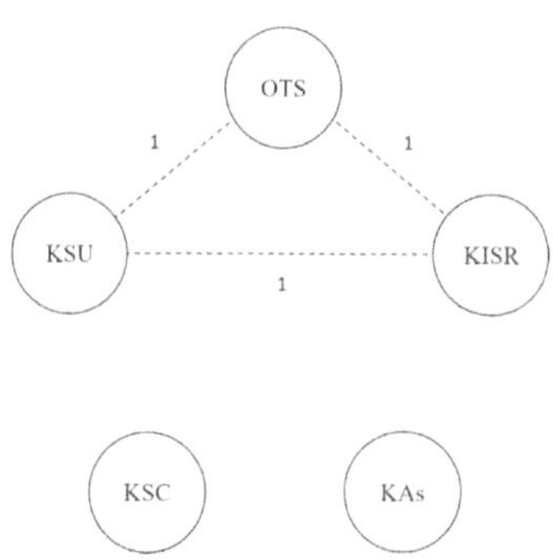

Fig. 8.2 Example of the first step of the enumerative approach

The Constant Comparative Method

Using the constant comparative method, the same data is analysed to identify common patterns emerging from the data in terms of interactions among the components. The results from the constant comparative method are then compared and contrasted with PCK maps. They are then created through the in-depth analysis and the enumerative approach for the purpose of methodological triangulation. Table 8.1 summarises all steps of the PCK map approach.

Applications of the PCK Map Approach in Two Example Studies

The PCK map approach has been used in two of our science education research projects on PCK. The two research studies reported here were exclusively concerned about the interactions of the PCK components identified in a science teacher's ePCK and pedagogical reasoning behind the teacher's utilisation of ePCK. In this section, we compare and contrast the two previous studies where we used the PCK map approach in different ways in order to show the approach's applications and usefulness to PCK research. Building on the discussion in this section, we address methodological issues associated with the approach in the next section by repositioning the PCK map approach within the RCM. The first study conducted by Park and Chen (2012) utilised the PCK map approach to examine the nature of the integration of the five PCK components that affected teaching practice in high school biology classrooms. Later, Suh and Park (2017) used a modified version of the PCK map approach to investigate common patterns in the interactions among PCK components related to science teachers' sustained implementation of an argument-based inquiry (ABI) approach.

Although both studies were grounded in the pentagon model, each made some modifications to the model in order to correspond to its respective research purpose. Park and Chen (2012) used the pentagon model that Park and Oliver (2008) proposed but they rearranged the components in the original pentagon model and added abbreviations for the components. On the other hand, in our study (Suh & Park, 2017) we modified Park and Chen's model to make it more compatible with the notion of the ABI approach. With the ABI approach, students have opportunities to engage in scientific argumentation based on evidence and reasoning which require students' learning to use language to construct and critique scientific arguments. This practice requires teachers to shift their epistemological orientations to be more aligned with the main assumptions of this approach in terms of how students learn, language use in science, and what science is. Acknowledging this shift in our 2017 study, we redefined the "Orientations to Teaching Science" (OTS) component and named it OTS-A (Orientation to Teaching Science specialised for ABI). At a more specific level, in the Park and Chen 2012 model, the OTS component comprises three major

Table 8.1 Summary of the steps of the PCK map approach

Steps	Data sources	Analysis procedures	Outcomes
In-depth analysis of PCK	• Observations (videotaped instruction sessions)	• Identify a teaching segment that indicate the presence of two or more PCK components • Repeat the procedure to identify more teaching segments	• Teaching segments
	• A teaching segment • Observations, interviews, documents, and artefacts related to the teaching segment identified	• Create a detailed description of the PCK Episode • Repeat the procedure to create multiple PCK Episodes	• PCK Episodes
Enumerative approach	• PCK Episodes	• Identify connections among the components of PCK in a particular PCK Episode • Repeat the procedure for other PCK Episodes	• Frequency of the connection between any two components of PCK identified in PCK Episodes
	• A number of PCK connections between any two components	• Sum up the frequency of the connection between any two components • Draw a PCK map	• Total number of the connection that one component makes with the other components • PCK maps
Constant comparative method	• Interview, observations, documents, and artefacts • PCK Episodes • PCK maps	• Identify common patterns that emerged from the data • Compare and contrast the results from the PCK maps	• Common patterns/themes • Methodological triangulation

Table 8.2 Comparison of two studies utilising the PCK Map approach

Studies	Park and Chen (2012)	Suh and Park (2017)
Participants	Four high school biology teachers	Three experienced fifth-grade teachers who were purposefully selected
Major data sources	(1) Classroom observations (2) Teacher interviews (three semi-structured interviews) (3) Other sources (lesson plans, instructional materials, and students' work samples)	(1) Classroom observations (2) Teacher interviews (two semi-structured and two video-stimulated recall interviews) (3) Other sources (instructional materials and students' work samples)
Topics taught at the time of the study	Photosynthesis and Heredity	Force and motion
# of lessons selected for data analysis	Two lessons on each topic for each teacher	Eight lessons for each teacher
PCK Episodes identified for each teacher	12–24 episodes	22–24 episodes
Patterns in interactions of the PCK components emerging through the data analysis	(1) The integration of the components was idiosyncratic and topic-specific (2) KSU and KISR were central in the integration (3) KSC had the most limited connection with other components (4) KAs were more often connected with KSU and KISR than OTS and KSC (5) Didactic OTS directed KISR inhibiting its connection with other components	(1) The connections between OTS-A, KSU, and KISR were most strongly connected to each other. This triangular relationship was closely related to the teachers' sustained implementation of the ABI approach (2) OTS-A was a critical driver for significant changes in PCK and practice to adopt ABI (3) KSC and KAs interacted least with other components

sub-components: beliefs about the purpose of learning science, decision-making in teaching, and beliefs about the nature of science. However, in our 2017 study, we revised the orientation component to include three aspects of epistemological orientations key to implementing the ABI approach: learning theory, the nature of science, and language use in science classroom (Hand, 2008) as shown in Fig. 8.3.

A detailed comparison of these two example studies is now provided. Table 8.2 summarises the applications of the PCK map approach in the two example studies.

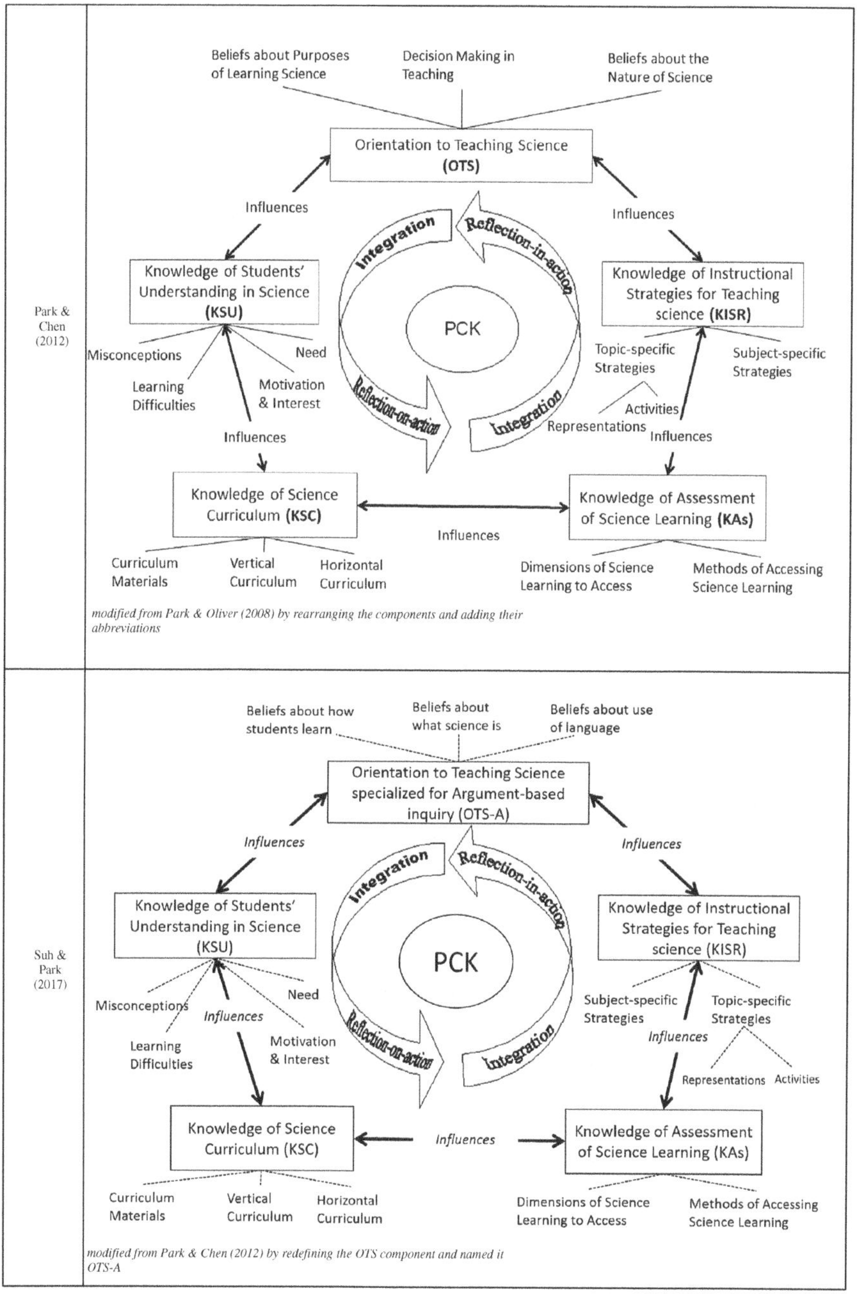

Fig. 8.3 Comparisons of the pentagon model of PCK employed in two studies

Participants in the Two PCK Map Studies

Four biology teachers working in a high school were selected in the first study while three fifth-grade teachers, who had been implementing an ABI, were purposefully selected in the second study. In both studies, the topic specificity and the context specificity of PCK were taken into consideration. In the first study, the four biology teachers were working at the same high school and teaching the topics of photosynthesis and heredity with the same curricular materials. In the second study, three experienced teachers were teaching the same science topic, force and motion, at the same grade level. In addition, the three teachers were selected based on two criteria: (1) teachers must have been implementing an ABI approach for at least five years in their science teaching, with at least two years without professional development facilitation, and (2) teachers must have demonstrated a high level of implementation of the ABI approach, as measured by the modified Reformed Teaching Observation Protocol (RTOP) (Martin & Hand, 2009; Sawada et al., 2002).

Major Data Sources

In both studies, major data sources included classroom observations, teacher interviews, and teaching artefacts. As a point of difference, the second study replaced the post-observation interview with video-stimulated recall (VSR) interviews. Such interviews are considered an effective technique for examining teachers' thoughts, decisions, and reasons for actions (i.e., their pedagogical reasoning). The interviews enabled researchers to explore and evaluate teachers' reflections and commentaries regarding their actions while allowing teachers to relive an episode of science teaching by providing a retrospective, accurate, and verbalised account of their thought processes (Calderhead, 1981). Before conducting VSR interviews, eight videotaped instructional sessions were purposefully selected for each teacher given the purpose of the study. Participants were asked to watch the videotaped teaching episodes and answer interview questions based on recollections of their thoughts and actions. While watching a video clip, the teachers could pause the video by themselves; they were then allowed to end the interview or prolong it as they wished. The VSR interviews focused on the pedagogical reasoning behind their instructional decisions and practices, which we regard as a key data collection method for researching the pedagogical reasoning underpinning the utilisation of ePCK during acts of science teaching.

Emerging Patterns in Interactions Among the PCK Components

In each study, to identify the patterns of interactions among the PCK components, PCK maps were constantly compared and contrasted to each other and to the results from the constant comparative method. The first study revealed that the interplay of the components was idiosyncratic and specific to the biology topic. It also found that knowledge of student understanding and knowledge of instructional strategies and representations were central in the interaction, which suggests that the stability and coherence of ePCK are related to the strength of the connection between these two components. The second study highlighted the interactions between orientations to teaching science, knowledge of student understanding, and knowledge of instructional strategies and representations. These findings suggest that OTS-A was a critical driver for significant changes in ePCK and practice to adopt ABI. Knowledge of curriculum and assessment had the least connection with other components in both studies. While the PCK map approach appears an effective means for capturing complex interactions among the components, it lacks the capability to recognise contextual and affective factors influencing teacher ePCK and the interplay between ePCK and pPCK. This limitation is discussed in depth in the following section.

Contributions of the Approach to Advancing PCK Research in the Context of the Refined Consensus Model of PCK

As demonstrated in the aforementioned studies, the PCK map approach is an effective method for analysing interactions among PCK components underpinning pedagogical reasoning and ePCK in science teaching. By modifying or refining each component to fit varying research purposes, the approach is potentially applicable to the answering of a wide range of research questions about ePCK and knowledge exchange between pPCK and ePCK. An important strength of this approach is its ability to make the abstract and complex structure of PCK more visible, explicit, and accessible. This feature is significant, especially in the context of the RCM as it provides a useful tool to examine the pedagogical reasoning and action that drives knowledge exchange between pPCK and ePCK. For example, the PCK map approach can be used to explicate a process of the pedagogical reasoning required for the translation of pPCK into ePCK. In addition to a data analysis method, the PCK map approach can be used as a reflective tool for science teachers and assist them in identifying which components and connections they need to improve for teaching a particular science topic more effectively (Park & Chen, 2012).

Repositioning the PCK approach in the RCM enabled us to identify some methodological issues associated with the approach. As Aydin and Boz (2013) pointed out, the PCK map approach assumes the same strength to every connection for analytic convenience, but is unable to differentiate between qualities and strengths of individual connections. Such differentiation is essential for more precise representations of ePCK. Given that limitation, Aydin and Boz (2013) used a scoring rubric to gauge the strength of each connection and found their method useful for quantifying the level of PCK that can be used for further quantitative analyses. We believe that this modification to the PCK map approach can make it possible to score individual science teachers' PCK map(s) as a means of assessing ePCK. Furthermore, individual teachers' PCK scores produced by quantifying their PCK maps with scoring rubrics will allow researchers to examine the relationships between PCK and variables associated with student learning through quantitative analyses, such as correlational analysis between science teacher PCK scores and student achievement scores on a standardised science test. That line of research could consequently contribute to revealing and articulating the "student outcomes" and "student contributions" aspects within the RCM.

Another limitation is that the PCK map approach focuses on individual science teachers' cognitive processes, represented by interactions of the PCK components, and thus pays insufficient attention to contextual factors influencing PCK. The RCM underscores learning context as an amplifier and a filter of teacher PCK as well as a mediator for teacher actions. As a result, we suggest modifying the mapping procedures and visual representation of PCK to explicitly involve contextual factors associated with a science teacher's PCK. For example, adding a field to the PCK Episode description template (created during the in-depth analysis of PCK in the first step of the approach), in which contextual factors influencing a particular PCK Episode can be identified and described, would be an appropriate modification. The representation of these factors in relation to the PCK map using arrows, circles, or boxes will help better interpret ePCK encapsulated with the PCK map. Taken together, the PCK map approach certainly has some methodological benefits and limitations. However, we think that the PCK map approach has great potential to help advance PCK research in science education with careful adjustments, especially within the context of the RCM.

References

Aydin, S., & Boz, Y. (2013). The nature of integration among PCK components: A case study of two experienced chemistry teachers. *Chemistry Education Research and Practice, 14*(4), 615–624.

Calderhead, J. (1981). Stimulated recall: A method for research on teaching. *British Journal of Educational Psychology, 51*(2), 211–217.

Friedrichsen, P., Abell, S. K., Pareja, E. M., Brown, P. L., Lankford, D. M., & Volkmann, M. J. (2009). Does teaching experience matter? Examining biology teachers' prior knowledge for teaching in an alternative certification program. *Journal of Research in Science Teaching, 46*(4), 357–383.

Hand, B. (2008). Introducing the science writing heuristic approach. In B. Hand (Ed.), *Science inquiry, argument and language: A case for the science writing heuristic*. Rotterdam, The Netherlands: Sense Publishers.

Hashweh, M. Z. (2005). Teacher pedagogical constructions: A reconfiguration of pedagogical content knowledge. *Teachers and Teaching: Theory and Practice, 11*(3), 273–292.

Henze, I., Van Driel, J. H., & Verloop, N. (2008). The development of experienced science teachers' pedagogical content knowledge of models of the solar system and the universe. *International Journal of Science Education, 30*(10), 1321–1342.

Kaya, O. N. (2009). The nature of relationships among the components of pedagogical content knowledge of preservice science teachers:'Ozone layer depletion'as an example. *International Journal of Science Education, 31*(7), 961–988.

Krauss, S., Brunner, M., Kunter, M., Baumert, J., Blum, W., Neubrand, M., et al. (2008). Pedagogical content knowledge and content knowledge of secondary mathematics teachers. *Journal of Educational Psychology, 100*(3), 716.

Magnusson, S., Krajcik, L., & Borko, H. (1999). Nature, sources and development of pedagogical content knowledge. In J. Gess-Newsome & N. G. Lederman (Eds.), Examining pedagogical content knowledge (pp. 95–132). Dordrecht: Kluwer.

Martin, A. M., & Hand, B. (2009). Factors affecting the implementation of argument in the elementary science classroom. A longitudinal case study. *Research in Science Education, 39*(1), 17–38.

Padilla, K., & Van Driel, J. (2011). The relationships between PCK components: The case of quantum chemistry professors. *Chemistry Education Research and Practice, 12*(3), 367–378.

Park, S., & Chen, Y.-C. (2012). Mapping out the integration of the components of pedagogical content knowledge (PCK) for teaching photosynthesis and heredity. *Journal of Research in Science Teaching, 49*(7), 922–941.

Park, S., & Oliver, J. S. (2008). Revisiting the conceptualisation of pedagogical content knowledge (PCK): PCK as a conceptual tool to understand teachers as professionals. *Research in Science Education, 38*(3), 261–284.

Sawada, D., Piburn, M. D., Judson, E., Turley, J., Falconer, K., Benford, R., et al. (2002). Measuring reform practices in science and mathematics classrooms: The reformed teaching observation protocol. *School Science and Mathematics, 102*(6), 245–253.

Strauss, A., & Corbin, J. (1990). *Basics of qualitative research* (Vol. 15). Newbury Park, CA: Sage.

Suh, J., & Park, S. (2017). Exploring the relationship between pedagogical content knowledge (PCK) and sustainability of an innovative science teaching approach. *Teaching and Teacher Education, 64*, 246–259.

Van Driel, J. H., Jong, O. D., & Verloop, N. (2002). The development of preservice chemistry teachers' pedagogical content knowledge. *Science Education, 86*(4), 572–590.

Part III
New Approaches to PCK Research in Science

Anne Hume

Overview

This part offers new methodologies, tools and/or perspectives on PCK measurement that are aimed at providing a broader view of PCK for science teaching, as depicted in the Refined Consensus Model (RCM) of PCK. Each of the following chapters reveals how the authors are currently using the new model to inform and advance their thinking around PCK research in science teacher education.

In Chap. 9, Ineke Henze and Erik Barendsen investigate mechanisms by which student science teachers develop personal pedagogical content knowledge (pPCK) in a methodology featuring purpose-built data-gathering tools. Using PCK-oriented templates developed by the researchers, student teachers create authentic artefacts (lesson plans, evaluations and reflections) during their teacher training programme that are used as data sources for tracking their pPCK development. The student teachers' development of pPCK is examined in terms of developmental steps, pPCK components and moderating personal factors. Combining analytical frameworks from three related models in teacher education allows the authors to characterise individual differences within the student science teachers' pPCK development. The authors argue the findings can inform new ways of tailored scaffolding of pPCK development in student science teachers and give some insights into the nature and role of amplifiers and filters in the RCM of PCK.

Chapter 10 features a methodology informed by the RCM that uses a research tool in a novel way as a professional development intervention to enhance science teachers' topic-specific personal and enacted PCK (pPCK and ePCK) within a school-based professional learning community. Jared Carpendale and Anne Hume reveal how collaborative Content Representation (CoRe) design among members of a school science department can be used to access, share and collate aspects of individual teachers' topic-specific pPCK into a collective form of topic-specific PCK i.e., cPCK. By capitalising on in-house expertise in this way, the pPCK and ePCK of science teachers less knowledgeable in the topic can be significantly enhanced, as evidenced by data gathered of classroom teaching by these teachers using an observation protocol designed specifically for the study.

Kennedy Chan, Marissa Rollnick and Julie Gess-Newsome explore the possibilities of a 'grand science PCK rubric' in Chap. 11 for measuring science teachers' PCK in all its various forms as depicted in the RCM. The authors recount how PCK that is widely agreed upon and generated by research and/or collective wisdom of practice is referred to as canonical PCK, i.e., a form of collective PCK (cPCK), and how rubrics are increasingly being used in science PCK studies to evaluate this canonical PCK. They argue that well-designed rubrics have the potential to make a significant contribution to the establishment of international standards for articulating not only canonical PCK for a number of commonly taught science topics but also for pPCK and ePCK at the topic level. The chapter reviews and critiques the characteristics of current rubrics and in an interesting approach uses the forum of an expert discussion group to examine and analyse the critical considerations in the construction of a grand rubric. The result is a grand science rubric proposal, generic in nature that can be customised for use with different content topics as well as for measurement of specific variants of PCK in the RCM.

Finally in Chap. 12, Amanda Berry, Pernilla Nilsson and Alicia Alonzo turn attention to personal and enacted PCK (pPCK and ePCK) in science teaching and expand upon the interrelationship between these two critical realms of PCK. They identify and highlight the role of the plan–teach–reflect cycle at the heart of the model as the mechanism by which pPCK is transformed to ePCK and vice versa. This elaboration of the processes underpinning the specific knowledge and skills science teachers use and/or develop in their classroom teaching enhances, and challenges, the potential of the RCM to further PCK research in the all-important arena of the classroom.

Chapter 9
Unravelling Student Science Teachers' pPCK Development and the Influence of Personal Factors Using Authentic Data Sources

Ineke Henze and Erik Barendsen

Abstract In this chapter, we discuss a methodology to analyse student teachers' personal pedagogical content knowledge (pPCK) development in a chemistry teacher education programme using PCK-oriented forms for lesson planning, evaluation and reflection. We unravel the student science teachers' development of pPCK in terms of (1) developmental steps, (2) pPCK components and (3) moderating personal factors. Our method relies on authentic data sources only, namely student teachers' products based on assignments in their teacher training programme. By combining three analytical frameworks, we were able to characterise individual differences within the student teachers' pPCK development for chemistry teaching. Such results can inform new ways of tailored scaffolding of this development in student teachers.

Introduction

As part of their teacher education programme, student teachers are expected to plan and carry out instruction without being able to draw upon substantial relevant prior knowledge and experience in the classroom. This expectation makes these teachers an interesting group for pedagogical content knowledge (PCK) research, particularly in light of the cognitivist–constructivist point of view on knowledge development (Greeno, Collins, & Resnick, 1996) that has been the basis of most educational research since the 1980s. The essence of this perspective is the assumption that teacher knowledge (including PCK) is built upon prior knowledge and beliefs from different domains and upon experiences from practice, among which are interactions with students and colleagues (Clarke & Hollingsworth, 2002; Van Driel, 2010).

I. Henze (✉)
Delft University of Technology, Delft, Netherlands
e-mail: f.a.henze-rietveld@tudelft.nl

E. Barendsen
Radboud University, Nijmegen, Netherlands
e-mail: e.barendsen@cs.ru.nl

Open University of the Netherlands, Heerlen, Netherlands

A. Hume et al. (eds.), *Repositioning Pedagogical Content Knowledge in Teachers' Knowledge for Teaching Science*, https://doi.org/10.1007/978-981-13-5898-2_9

From this perspective, it is natural to conceptualise PCK as a collection of private and personal pedagogical *constructions*, gradually built up 'as a result of repeated planning and teaching of, and reflection on the teaching of the most regularly taught topics' (Hashweh, 2005, p. 277).

These private and personal aspects of PCK are reflected in the realm of knowledge known as personal PCK (pPCK) in the Refined Consensus Model (RCM) of PCK (see Chap. 2). From now on, we will use 'pPCK' to refer to this particular personal knowledge and use PCK for the generic construct (Shulman, 1986). Note the above-mentioned repeated planning, teaching and reflecting activities, so necessary to the building of pPCK, appear in the RCM as the so-called *pedagogical cycle* (i.e., planning-enactment-reflection cycle).

For their lesson planning, beginning student teachers necessarily draw upon external knowledge sources (with respect to topic level pedagogy) together with their personal beliefs, their own learning experiences and their general pedagogical knowledge. The subsequent execution of the plan and reflection close the first cycle of pPCK development. In the research on which this chapter is based, we aimed to study these student teachers' initial development tracks in the context of science teacher education.

Among teacher educators, it is commonly known that there are substantial individual differences between student teachers with respect to rate and focus of the development of the pedagogical constructions constituting their pPCK. Indications of such differences can be visible in several ways. For example, lesson preparations remain superficial for some student teachers, whereas others quickly integrate pedagogical and content aspects of the topic they are about to teach. Moreover, during teacher training, most student teachers shift their attention from teacher activities to more pupil-oriented reasoning before, during and after their teaching. For some student teachers, however, this transition is delayed or does not appear at all.

Challenges in Dutch Science Teacher Training Programmes

This chapter is based on an exploratory multiple case study with seven chemistry student teachers who were enrolled in a teaching methodology course in the second half of their one-year teacher training programme at Delft University of Technology (TU Delft). In the Netherlands, such a compact programme is typically carried out after completion of a content-specific (chemistry, in this case) Bachelor's and Master's programme, so the student teachers can be expected to have a strong content knowledge background. This feature of Dutch pre-service teacher education is important when considering PCK development, as PCK is the intersection of content knowledge and pedagogical knowledge specific to a content area (Shulman, 1986).

For this particular group of student science teachers, we can identify four aspects of the teacher training programme challenging their initial pPCK development. Firstly, when starting the programme student science teachers *suddenly enter the 'new' field of social sciences* and find learning about human interaction through an interpretivist paradigm contrasts with their study background in natural sciences characterised by a more positivist tradition in which objectivity, predictability and 'exactness' prevail. This transition tends to cause some initial confusion in the student teachers, obstructing or delaying their professional growth, including the construction of pPCK.

Secondly, there is *the early start to their internship at school* in which student teachers enrol at the very beginning of the training programme. Although practical experience is one important source for pPCK development, at this early stage of their training the student teachers lack relevant prior knowledge about teaching as another important resource for building their initial pPCK.

The third challenging aspect is that *of combining two different roles*. Throughout their programme, the student teachers spend 3–4 days a week as a trainee in a school and 1–2 days at university. Combining the roles of *teacher* (internship) and *student* (teacher education), they have to adjust to both the rules in school and the requirements of the teacher training programme. One way to handle this situation is to switch into 'survival mode', in which science student teachers tend to focus on (gaining competences for) class management and teaching strategies, as opposed to pedagogy for specific topic levels (Wehrmann & Henze, 2016).

The fourth and crucial challenge of the teacher training programme for pPCK development is *its short span* in combination with teacher educators' (high) expectations of *the student teachers' ability to take a critical perspective on their teaching practice.* As argued before, reflection on the teaching of specific content is recognised as central for pPCK development (cf. Schön, 1983). However, it might be difficult for student teachers, especially when they are young adults, to make sense of their teaching experiences in an advanced way (c.f., Korthagen & Vasalos, 2005). Indeed, the position of many of them in relation to their ability to give meaning to their professional role as a teacher can be recognised as Stage 3 in Kegan's model of (adult) cognitive development, which defines five stages of mental complexity or 'orders of mind' (Kegan, 1982, 1994; Wehrmann & Henze, 2016) (see 9.1). In this third stage of 'socialised mind', student teachers' sense of self is *socially determined* based on the real or imagined expectations of pupils, colleagues and supervisors. At this stage of personal development, they tend to seek approval and feedback from their surroundings, as a measure of how well they are doing (Rodgers & Scott, 2008). As a consequence, student teachers' reflections often concern their classroom *behaviour* rather than personal knowledge and beliefs, as they feel that their *performance* is the aspect of their learning they are 'judged' on.

In Table 9.1, we summarise Kegan's model of cognitive development in relation to teachers' capacity to take a perspective on their teaching. We will exclude Stage 1 (concerning childhood) and Stage 5 (usually not achieved until middle age or later).

A one-year teacher training programme offers little opportunity to facilitate student teachers' capacity to 'find their voice' and take the authority needed to shape their own professional path (Stage 4 in Table 9.1), that is, to take a perspective on information from experiences, evaluate it and then decide how to act upon it, according to their *own internal standards*.

Table 9.1 Kegan's (1982, 1994) stages of adult cognitive development in relation to teacher role (Rodgers & Scott, 2008)

Level of cognitive development	Teacher capacity to take a perspective on oneself (sense-making of professional role)
Stage 1: Impulsive mind (early childhood)	–
Stage 2: Instrumental mind (adolescence, 6% of the adult population)	Teachers have a naïve and concrete conception of teacher role; they view experiences in 'black and white'
Stage 3: Socialised mind (58% of the adult population)	Teachers enact a role that has been defined by the system (school context and teacher education programme); they are able to report on feelings and emotions concerned with teaching
Stage 4: Self-authoring mind (35% of the adult population)	Teachers are self-critical, and self-author their experiences (regarding teacher role) according to their inner voice; they see how relationships with others impact upon teaching
Stage 5: Self-transforming mind (1% of the adult population)	–

About This Chapter

In this chapter, we discuss the methodological aspects of a study seeking to identify individual differences in TU Delft student teachers' pPCK development for teaching chemistry. Instead of 'measuring' the development *result* by comparing student teachers' pPCK before and after teaching, we analysed the *process* of their pPCK development, that is, their planning, enactment and reflection activities through which they build their pedagogical constructions (Hashweh, 2005). Our methodology is designed to yield detailed insight into this development so that it enables educators and supervisors to identify problematic (or positive) aspects of a student science teacher's pPCK development and apply suitable interventions (or choose no intervention for the moment).

Instead of relying on traditional PCK research instruments, we chose to use 'non-obtrusive' methods for capturing pPCK, that is, we did not bother the student teachers in their short and busy programme with instruments other than the formats for lesson planning and reflection they were using already. In other words, we aimed to analyse the student teachers' pPCK development using *authentic* data only.

The remainder of this chapter is structured as follows. First we discuss the theoretical basis of our research, including three models serving as frameworks to unravel (1) the components of students' pPCK for science teaching, (2) its stepwise development and (3) personal factors influencing this development. Then, we formulate our research questions in terms of these frameworks and describe the assignments that will serve as authentic data sources. Then we specify our methods for data collection and analysis, followed by examples taken from the results of the study, illustrating the promise of the analytic approach. We conclude with some reflections on the study and the Second (2nd) PCK Summit (held in 2016 in Leiden, The Netherlands).

Theoretical Background

Components of Pedagogical Content Knowledge

By definition, teachers' PCK is content-specific knowledge. This 'content' can be understood in various ways. The RCM, (see Chap. 2) refers to the grain size (or levels) of science content knowledge and recognises it can be specified along a continuum of broad to narrow ideas in all realms of PCK. Thus it can refer to a topic or a scientific concept (c.f., Shulman, 1986), and we argue in a broader sense it can also refer to a specific problem (e.g., using scientific knowledge in reasoning about socio-scientific dilemmas), a practice field (e.g., design, research) or even a discipline.

The notion of PCK (in the general sense) can be considered as an integrated or holistic construct ('amalgam') of content and pedagogy, but there are ways of subdividing this knowledge domain into meaningful components. In this respect, the most cited PCK model in the science education field is that of Magnusson, Krajcik and Borko (1999), which distinguishes five components of PCK. Four of these correspond to the following aspects of content-specific pedagogy (which we refer to as M1 to M4, respectively): goals and objectives for teaching specific content in the curriculum (M1); students' understanding of this content (M2); instructional strategies concerning this content (M3); and ways to assess students' understanding of this content (M4). These four components are universal in the sense that they appear in a variety of general pedagogical models in literature and in teacher education materials (e.g., Van Gelder, Peters, Oudkerk Pool, & Sixma, 1973). Their pedagogical completeness and simplicity convinced us to use Magnusson et al.'s M1 to M4 components to characterise aspects of content-specific pedagogy and components of the corresponding PCK. Note Magnusson et al. (1999) propose 'teacher orientations to teaching science' as a fifth PCK component. As we consider orientations to be less content-specific than the other components, we do not include these in our analysis. Besides, Magnusson et al. themselves present orientations as an underlying influence on the components M1 to M4.

Within their PCK model, Magnusson et al. (1999) do not relate the different components with each other. In our opinion, however, connections between these components (i.e., internal coherence or logic) are crucial (c.f., Van Driel & Henze, 2012). Indeed, M1 to M4 need to be interconnected to enable effective scaffolding of students' learning (c.f., Park & Chen, 2012).

Thus apart from the extent of the respective components, the strength of the interconnections is an indicator for the quality of a teacher's personal pedagogical constructions (in the sense of Hashweh, 2005) and hence of their pPCK, that is, strong connections contribute to strong pPCK. In this, we respect Shulman's (1987, p. 15) model of *pedagogical reasoning and action*, implying that teachers' knowledge of students' understanding of a specific topic, in a specific context, provides grounds for choices and actions with regard to goals, instructional strategies and ways to assess students' understanding that *correlate with each other* and *with the specific situation*. Although experienced teachers use pPCK as one integrated knowledge source, the four components (M1 to M4) above can be investigated separately (Henze, Van Driel, & Verloop, 2008).

Stepwise Development of pPCK

There is ample evidence that teachers develop their pPCK in various ways. Clarke and Hollingsworth (2002, p. 951) designed a non-linear, empirically grounded model, which we regard as a useful means of augmenting and refining aspects of the RCM. The *interconnected model for teachers' professional growth* consists of four interacting domains: (1) the Personal Domain, which contains teachers' knowledge, beliefs and attitudes; (2) the External Domain containing external sources of information and stimuli; (3) the Domain of Practice, which involves professional classroom experimentation and (4) the Domain of Consequence, containing salient outcomes related to classroom practice. The model can be used to describe and analyse teachers' professional growth in terms of changes within the domains through the specific mechanisms of *enactment* and *reflection*, see Fig. 9.1.

The interconnected structure of the model enables the identification of particular sequences of enactment and reflection steps (so-called *pathways*), which may differ in complexity. For example, a teacher might try out an idea for an instructional strategy in class (enactment step between Personal Domain and Domain of Practice) and reflect on the experience (reflection step between Domain of Practice and Personal Domain), thus extending his/her knowledge (Personal Domain). Alternatively, the observed learning outcomes might be involved, resulting in a sequence containing reflection on the Domain of Consequence.

From the perspective of the RCM (see Chap. 2), the model by Clarke and Hollingsworth can be seen as a refinement of the enactment and reflection parts of the pedagogical cycle 'planning-enactment-reflection' in the RCM because it differentiates between forms of subsequent enactment and reflection, and between the domains (as defined by Clarke and Hollingsworth, 2002) involved in these enactment and reflection steps. We can use the interconnected model for teachers' professional growth to discuss different types of PCK as represented in the RCM. Science teacher's pPCK is situated within the Personal Domain. Teachers may use external sources (External Domain) on content-specific pedagogies such as books, courses or colleagues to inform their pPCK and the knowledge represented in these sources can be

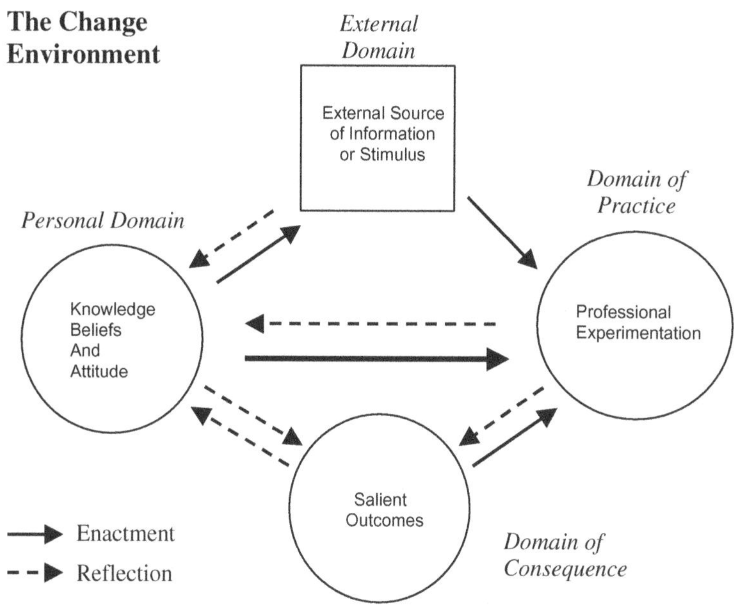

Fig. 9.1 Interconnected model for teachers' professional growth (Clarke & Hollingsworth, 2002, p. 951)

regarded as a form of PCK, although it represents collective rather than individual knowledge. Therefore, this type of PCK is referred to as *collective PCK* (cPCK) in the RCM. Incorporation of cPCK into pPCK (i.e., knowledge exchange) is one mechanism of development ('growth') of this pPCK.

We can also refine the notion of pPCK by looking at the process of the *use and exchange* of knowledge during the science teachers' enactment and reflection steps. When preparing a lesson, a teacher draws upon his/her pPCK, which can be seen as the teachers' complete repertoire for teaching specific science content. *Planning* involves using this pPCK to select a specific part (goals, expected student factors, instructional strategy and assessment methods) of the teacher's pPCK for the lesson in question. This part then becomes active during enactment of the planned activity in the Domain of Practice. The RCM refers to this 'active part' of pPCK as *enacted PCK* (ePCK). During a *reflection* step, the teacher draws upon the experience in the Domain of Practice gained by using this particular ePCK to enhance or refine his/her pPCK in the Personal Domain. We regard ePCK as a *dynamic* construct defined by the part of pPCK that is 'active' at a certain moment during teaching practice. Indeed, decision-making during enactment involves going through one reflection-enactment cycle and is likely to change the part of personal pPCK that is currently 'in use', that is, there is knowledge exchange back to the realm of pPCK as depicted in the RCM. The dynamic aspect of ePCK makes it difficult to elicit this type of teacher knowledge. Observations of classroom practice (e.g., Barendsen & Henze,

2017) and stimulated recall interviews based on recordings of classroom situations are possible methods to capture this ePCK. Note the mechanisms for extending pPCK and exchanging knowledge between pPCK and ePCK that occur within these pathways of professional growth can be moderated by factors called *amplifiers and filters* in the RCM (see Chap. 2). From empirical research, we concluded that pathways as complex growth networks, rather than more simple change sequences, including the Domain of Consequence (see Fig. 9.1) are important for the simultaneous development of PCK for science teaching *and* classroom practice (Wongsopawiro, Zwart, & Van Driel, 2016).

Amplifiers and Filters in pPCK Development

The ways teachers develop and use their PCK is influenced by personal as well as 'extra-personal' (i.e., outside the individual's direct influence) factors. Personal factors are part of a teacher's personal and professional identities. Extra-personal factors include elements of the teaching situation, such as curriculum, classroom context and the school's atmosphere among others. We view these factors as amplifiers and filters, which act on teachers' pedagogical reasoning in their decision-making processes, before, during and after teaching and consequently on their actions.

Regarding decision-making processes in teaching, one can distinguish between long-term decision-making (during planning) and short-term decision-making (during enactment of a plan). This latter within-lesson decision-making is sometimes called 'decision-making-on-the-spot' (c.f., Bishop & Whitfield, 1972; Borko, Roberts, & Shavelson, 2008). Long-term decision-making involves pPCK, whereas short-term decision-making affects ePCK.

This decision-making appears to be shaped by both personal factors (such as beliefs and values) and situational factors, see Fig. 9.2 (based on Bishop & Whitfield, 1972; Borko et al., 2008).

With respect to *personal* factors, many scholars have claimed that teaching is highly charged with feeling, aroused by and directed towards not just people, but

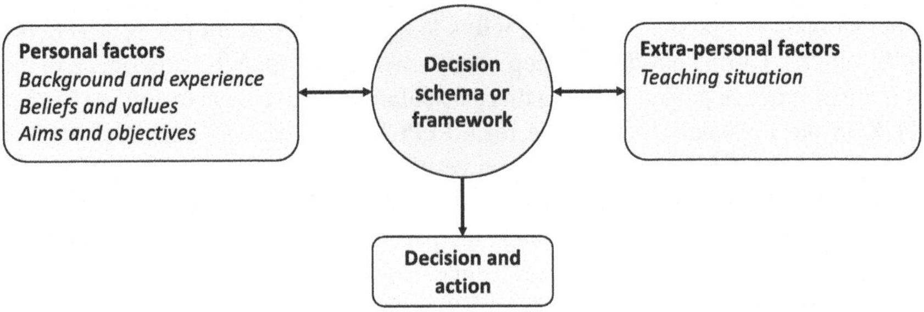

Fig. 9.2 Decision-making framework (based on Bishop & Whitfield, 1972; Borko et al. 2008)

also values and ideals. In this respect, McCaughtry (2004) and Zembylas (2007) explicitly pointed to the interrelations between PCK and 'emotional knowledge' in teaching and learning. Park and Oliver (2008) introduced 'teacher efficacy' as an affective component *within* their hexagon model of pedagogical content knowledge for science teaching (Park & Oliver, 2008, p. 279). Also Garritz (2010) suggested *incorporating* an affective domain *within* PCK, related to the teachers' interests, attitudes and emotions about their own ways of teaching. This affective PCK domain would include, among other things, teachers' interest and value beliefs, self-concept, self-efficacy, self-esteem and control beliefs. However, just as 'orientations to teaching science', we see these aspects not as a part of pPCK, but rather as factors influencing its development, that is, amplifiers and filters as in the RCM.

Hong (2010) defined a model in which personal factors such as *value, efficacy, commitment, emotion, confidence regarding power and control in the school organization* (micro-politics), and *personal knowledge (including PCK) and beliefs* are interrelated to each other. These aspects of the model can be explained as follows: (1) *value*: the interest and enjoyment the teacher gets from a specific activity, and the importance of doing well on a given task, (2) *self-efficacy*: judgment of own capabilities to work as a successful teacher, (3) *commitment*: commitment to becoming a teacher or to working as a teacher, (4) *emotions*: stress and emotions the teacher experience through interactions with students, other teachers, or administrators, (5) *knowledge and beliefs*: subject matter knowledge, pedagogical content knowledge and the perceived tasks as a teacher and (6) *micro-politics*: confidence regarding power and control of teachers in the school organization, active interaction with administrators and engagement in school administration.

Hong used this model to study beginning teachers' role perception and well-being in school. Among other things, it was found that negative perceptions such as unfulfilled commitment, lack of efficacy, unsupportive administrators and task beliefs emphasising teachers' heavy responsibilities appeared to be contributing factors to emotional burnout, and even drop out from the profession (Hong, 2010).

We think that this model (Hong, 2010, p. 1540) can be interpreted as a means for effectively identifying (interrelated) personal factors that may positively or negatively influence the development of beginning teachers' personal knowledge and beliefs. Thus, in our study, we apply Hong's model to identify amplifiers and filters influencing the student science teachers' development of pPCK.

Our Study

We can use the three theoretical frameworks described above to translate the global aim of our study into concrete research questions. Our goal is to provide in-depth, empirical input to design more sophisticated ways for educating and supporting (student) science teachers' pPCK development.

1. How can the development of student science teachers' pPCK be portrayed in terms of *developmental steps*, *PCK components* and *personal factors* using authentic data?
2. What general patterns can be identified to typify differences with respect to student science teachers' pPCK development?

Teacher Training Assignments

In light of the mechanisms discussed in the Introduction, the process of planning, enacting and reflecting on pedagogical constructions is the central source of information to monitor and analyse the student science teachers' pPCK development. Since the factors mentioned in the introduction are challenging, we incorporate explicit scaffolding of this process into the compact and intensive teacher training programme. This scaffolding will be made more explicit below.

Scaffolding Student Science Teachers' Lesson Design Activities

For more than 40 years, the pedagogical model of Van Gelder et al. (1973) has been dominant in educational literature and in teacher education programmes in The Netherlands and Flanders. The four dimensions of the model (learning goals, students' initial state, teaching and learning environment and assessment) are shaped by the answers to the following elementary questions (see Fig. 9.3):

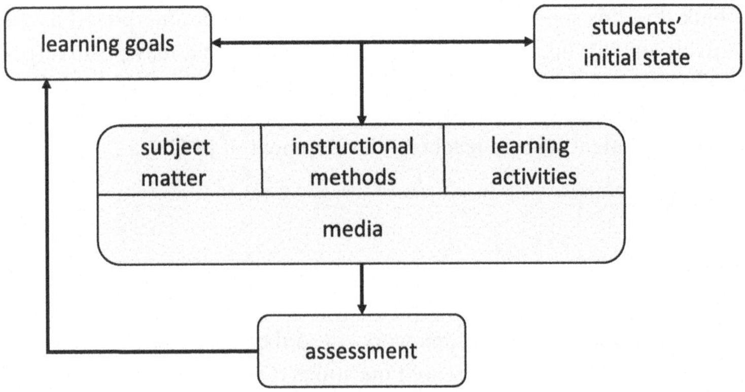

Fig. 9.3 Pedagogical model by Van Gelder et al. (1973) translated by authors, with permission from Wolters-Noordhoff, Groningen

- What is the intended learning outcome for the students? (Dimension 1: learning goals)
- How can we connect to the students' prior knowledge? (Dimension 2: students' initial state)
- How can we select and arrange the subject matter? What learning activities are triggered by suitable instructional methods? Which media are the most appropriate? (Dimension 3: teaching and learning environment)
- What is the result of the instructional activities? (Dimension 4: assessment)

The questions underlying the model naturally give rise to a lesson preparation method. This method is the basis of various lesson preparation forms that have been introduced into our science teacher education programmes to scaffold the student teachers' lesson designs. The answers to the above questions can be seen as part of the science teachers' pPCK with respect to the specific content and specific educational setting. The four components (M1 to M4) of the PCK model by Magnusson et al. (1999) can easily be recognised—that is, knowledge about: curriculum goals and objectives (M1), students' understanding (M2), instructional strategies (M3), assessment (M4), all with respect to certain content. Contrary to the Magnusson et al. model, the model by Van Gelder et al. is cyclic and the distinct dimensions are interconnected.

However, Van Gelder et al.'s model was conceived in the behaviourist and cognitivist traditions of the 1970s. In the meantime, scholars have developed a broader perspective on the content of various dimensions (e.g., Grossman, 1990; Magnusson et al., 1999; Shulman, 1986). For example, the second dimension has evolved also to incorporate the expected students' learning difficulties, preferences, alternative conceptions, etc. Moreover, formative assessment (assessment for learning) is an essential part of modern instruction, touching upon the components 2, 3 and 4. In our current science teacher education programme at TU Delft in The Netherlands, we are using an adapted version of the traditional lesson preparation forms, in order to reflect the modern constructivist situative perspective on learning and teaching (Greeno et al., 1996). Our preparation form contains topic and context-specific questions, based on the models by Van Gelder et al. (1973), Magnusson et al. (1999) and the Content Representation (CoRe) form by Loughran, Mulhall and Berry (2004), see Table 9.2.

The CoRe form was originally intended to capture the perceptions of a community of science teachers about the knowledge needed to teach a specific science topic. Thus, a CoRe serves to elicit a form of collective PCK (cPCK). Filling out a CoRe starts with identifying the 'big ideas' of the topic, followed by answering eight questions for each of these. The questions address the intended learning outcomes (3 questions), students' understanding (1 question), instructional strategies (3 questions) and assessment (1 question) (see Barendsen & Henze, 2017).

Table 9.2 Lesson preparation form

1. **What and why?** • What do you want your students to learn about this topic? Formulate the learning objectives in an operational way: "At the end of the lesson, students will be able to ..." • Why is this important for the students? (refer to your learning objectives)
2. **Possibilities and limitations** • What do you know about the students' thinking that influences your teaching [prior knowledge, level of cognitive development (Piaget)]? • What other factors (e.g., classroom, atmosphere in the class, group orientation, autonomy, differences between pupils, your development as a teacher) influence your teaching on this subject? • What difficulties for the students do you expect regarding to the subject of this lesson?
3. **Instructional strategy** • What instructional methods will you apply (what are the most important aspects of your lesson)? What are the particular reasons for choosing this approach? • How do you apply pedagogical link making to make this lesson meaningful for the students?
4. **Assessment** • How will you ascertain students' understanding or confusion around this topic?

Scaffolding Student Science Teachers' Lesson Enactment and Reflection

According to Clarke and Hollingsworth (2002), enactment and reflection are mediating processes for teachers' professional growth (Fig. 9.2). To support the student science teachers' ability to reflect on their teaching practices (c.f., Kegan, 1982, 1994), we have added a Lesson Evaluation Form and a Lesson Reflection Form, see Tables 9.3 and 9.4. The Lesson Reflection Form is based on the questions of a semi-structured (retrospective) interview that was used to capture science teachers' pPCK after conducting a lesson series on a specific science topic in a three-year longitudinal study by Henze et al. (2008). Again the four PCK components were addressed: learning goals (1 question), students' understanding (3 questions), instructional strategies (2 questions) and assessment (2 questions).

Table 9.3 Lesson evaluation form

• What were *meaningful moments* during the execution of your lesson?
• What was *successful* in this lesson, what can be improved?
• Justified conclusion: what is the 'take home message' (in terms of *practical rules*)?

Table 9.4 Lesson reflection form

1. **What and why?** *Retrospection:* • What learning objectives were addressed during the lesson? *Analysis:* • To what extent are you satisfied with the learning objectives (number, formulation, etc.) and with the way you conveyed their relevance? Please explain
2. **Possibilities and limitations** *Retrospection:* • Have you encountered any unexpected prior knowledge, perceptions or misconceptions? If so, give examples *Analysis:* • What did the students consider difficult and what was easy for them? How do you know? Discuss possible explanations
3. **Instructional strategy** *Retrospection:* • Have you executed your lesson as you prepared and planned? What was done and what was omitted? • Were the students able to carry out the learning activities? What was successful and what did not work? *Analysis:* • To what extent are you satisfied with the pedagogical link making during the lesson? Explain • To what extent are you satisfied with the way you engaged the students? Explain. • To what extent did the teaching and learning activities fit to the learning objectives? Explain • To what extent were the teaching and learning activities appropriate for the students and the classroom context? Explain
4. **Assessment** *Retrospection:* • Have your students reached the learning objectives? How do you know? *Analysis:* • In case students have not reached specific learning objectives: what are possible explanations? • To what extent are you satisfied with the way you monitored and assessed the students' understanding? Explain
Conclusion Which aspects in your lesson preparation form would you modify based on the above retrospections and analyses? Explain

Application of the Forms in Student Science Teachers' Activities

At the start of the one-year science teacher education programme, student teachers use the Lesson Preparation Form, Lesson Evaluation Form and Lesson Reflection Form with regard to single lessons. After this initial use, they switch to similar forms regarding entire lesson series (about 6 lessons) on a specific topic, that is, the Lesson Preparation Form *before* the lesson series, the Lesson Evaluation Form *after* each single lesson and the Lesson Reflection Form *after* completing the lesson series. When the student teachers have shown sufficient proficiency (in particular in evaluation and critical reflection), using the forms is made optional. They may choose only to apply it to series about a 'new' topic or pedagogical approach, for example. To monitor the student science teachers' pPCK progress during the lesson series, rubrics for formative (self and peer) assessment have been developed. These were not used in our study, however.

Methods

The study we carried out involved seven chemistry student teachers. The exploratory nature of the investigation meant we chose to conduct an in-depth study with an interpretivist approach as the underpinning paradigm. The study can be characterised as a multiple case study with a cross-case analysis. The participants were enrolled in a course on teaching methodology in the second half of their training programme. During this course, each student teacher developed two lesson series. The specific chemistry topics were determined by the school, the class and the curriculum at the schools of their current internships. Data were collected before, during and after the student teachers' teaching of these lesson series at their schools.

Data Collection

The student teachers planned, taught and reflected in the context of the two lesson series. Since our objective was to use authentic data, the student teachers' chemistry lesson preparations, evaluations and reflections (termed products) were the only data sources in our study. Each of these was investigated through their completed Lesson Preparation, Lesson Evaluation and Lesson Reflection forms (the forms in the Tables 9.2, 9.3 and 9.4, in their lesson series version).

Data Analysis

The student teachers' products were subjected to a qualitative analysis to investigate the process of pPCK development for chemistry teaching (see the Introduction). In the Lesson Evaluation and Lesson Reflection forms, we identified the stepwise development of the student teachers' professional knowledge, in terms of reflection and enactment in the External, Practice and Consequence Domains, respectively (Clarke & Hollingsworth, 2002) (see Fig. 9.1). To this end, we identified meaningful and coherent text segments, that is, segments containing reflection and/or enactment on a single matter or situation. These were the units of analysis in an analytic coding procedure (Gibbs, 2007) using Atlas-ti qualitative data analysis software. The codes were built up from the elementary steps *reflection* and *enactment*, together with the domain involved (*external*, *practice* and *consequence*)—'*reflection consequence*' is an example of such a code. It turned out to be useful to distinguish two forms of enactment: *concrete* action ('real enactment') and *plans* for future action, respectively.

Moreover, we identified the pPCK components involved (Magnusson et al., 1999), as well as possible moderating personal factors (Hong,2010), using the same units of analysis in the case of the evaluation and reflection forms. The codes for pPCK components were *M1* to *M4*. The codes for the personal factors consisted of one of the Hong factors (*value*, *commitment*, *self-efficacy*, *micro-politics* and *emotions*), wherever possible distinguishing positive and negative perceptions, for example '*self-efficacy* +'. We coded the materials related to one lesson series together, then divided the remaining coding work and compared the results afterwards, reaching consensus in cases of differences. The combinations of reflection steps, enactment steps and reflection-enactment cycles, as well as the manifestations of the four PCK components and the Hong personal factors were then quantified. Then we looked for patterns with respect to frequencies, combinations and progress in the course of time (cf. Cohen, Manion, & Morrison, 2013). In a subsequent inductive analysis phase, we analysed the text content more in-depth. The results were used to portray the content of the student teachers' pPCK and its development (Research Question 1).

In addition, the cross-case aspect was addressed by identifying differential features of pPCK development, with the purpose of typifying the student teachers' developmental processes (Research Question 2).

Results

We were able to conduct the complete qualitative procedure based on the data sources in our study for four student chemistry teachers, indicated by the pseudonyms Alice, Ben, Cindy and Debbie (i.e., four cases). The number of meaningful and coherent text segments varied considerably, ranging from 91–240 segments per participant.

Below, we give an impression of the results of the qualitative analysis across the four cases. We illustrate our findings with text segments taken from the student teachers' data (translated into English). The assigned codes are indicated below the segments.

Developmental Steps

Reflection on the Domain of Practice was seen more frequently in the data than reflection on the Domain of Consequence (i.e., student learning outcomes).

> Only while the students were doing their lab work I discovered that the lab assistant had interchanged the labels of the chemicals, because that would be fun, so that the students would have to find out which was which. Because I hadn't explained it that way (I just had

> prepared the lab assignment as described on the form) the students did not understand what to do and got confused. As a consequence, they did not like the lab session.
>
> (Debbie; M2, M3, reflection practice, micro-politics –)

Most enactment steps concerned plans rather than concrete actions.

> Because I wanted to work according to the prescribed procedure and the lab assistant always works slightly different, I feel forced to adjust the procedure again.
>
> (Ben; M3, reflection practice, concrete enactment practice, micropolitics –)
>
> I think it was especially hard for them to understand why an acid tastes sour and, in particular, they did not understand the relation between the COOH group and H+ . That they did not see that H+ could come off the acid group. In a 'normal' [not inquiry based] lesson I would have explained this immediately.
>
> (Alice; M2, M3, reflection consequence, plan enactment practice)

The number of enactment pathways (i.e., cycles) varies considerably among the students, ranging from 26% of the segments for Alice to 5% in the case of Debbie—her data contains lots of 'loose' reflection statements without enactment.

> Pupils obviously hadn't had a good time when doing the experiment and therefore were not interested in the discussion. They preferred to do something else.
>
> (Debbie; M2, M3, reflection practice)

The observed cycles mostly concern plans for action, except for Alice.

> [...] I ask N. if she understands why you have to add three times as much sodium hydroxide to citric acid than to hydrochloric acid. She doesn't know and I let R. explain it to the class once again. Now they understand.
>
> (Alice; M2, M3, reflection consequence, concrete enactment practice)

pPCK Components

In the student chemistry teachers' data, the pPCK components M2 (knowledge of students' understanding) and M3 (knowledge of instructional strategies) were most prominent. Comparing the results for the first and second lesson series, we observed a significant shift in the student teachers' data from instruction (M3)-oriented descriptions and reflections to more attention towards student learning difficulties and progress (M2). This shift is consistent with the more general finding that more advanced teacher development is accompanied by an increased awareness of the students' learning.

Personal Factors

Three personal factors were observed in the student chemistry teachers' data: efficacy, emotion and micro-politics.

> The various phases did not work out according to plan, due to the large number of absent students. I adjusted my planning on the spot, which did not cause any further problems. It did lead to a chaotic beginning of the lesson, however, and I started the lesson later than planned. I had no part in this, I hadn't been informed by the school management.
>
> (Cindy; M3, reflection practice, enactment practice, micro-politics –)

The observed personal factors appear in various frequencies and combinations. For example, Alice's texts contain by far the most positive self-efficacy indications. Texts of Alice and Ben showed positive emotions more frequently than those of Cindy and Debbie.

> One of the groups mentioned in their presentation: "Instructive and very nice to do. We have learned a lot and had a lot of fun." Isn't it just super if that is the case?
>
> (Alice; M2, M3, reflection practice, emotion +)
>
> After having read the first part together, I broke the students into groups. Last time dividing the students was hard. That is why I first explained the assignment clearly. I used a 'group maker' on the Internet to make student groups for the assignment: they got to work within 1 min. That went fabulous! The students noticed their groups and started working immediately. Meanwhile I had time to write things on the blackboard.
>
> (Ben; M3, reflection practice, enactment practice, efficacy +, emotion +)

Differential Features of Student Teachers' pPCK Development

We were able to typify the student chemistry teachers' pPCK development in terms of the three models above, using the following combinations of differential aspects: the reflection-enactment ratio and appearance of pathways/loops rather than single steps (for Clarke and Hollingsworth's model of professional growth), co-occurrence and chemistry content (for Magnusson's components), and the presence and relative frequencies of positive and negative perceptions of personal factors—especially emotion (for Hong's model).

The colours in Table 9.5 indicate the resulting typification of the student teachers' pPCK development, with green representing significant development, orange some development and purple little/no development.

Discussion

Our analysis using three analytical frameworks allowed us to effectively characterise individual differences within student chemistry teachers' pPCK development. The characterisation contained some surprising aspects that we had not been able to reveal before. One of these is the influence the amount of chemistry content has on student teachers' pedagogical reasoning (see pPCK components, Table 9.5).

Moreover, when examining the student teachers' developmental steps (see Table 9.5), we see a lack of subsequent pedagogical cycles, where only a reflec-

Table 9.5 Typification of student teachers' pPCK development

	Developmental steps	pPCK components	Personal factors
Alice	Enactment > plan in cycles and pathways	Rich description; combinations mostly on chemistry content	Positive factors predominate and increase
Ben	Mostly plans Some enactment	Combinations on chemistry content	Positive and negative factors
Cindy	Consequence reflection, but not contentwise Little enactment	Combinations not on chemistry content	Positive and negative factors; micropolitics Emotion almost absent
Debbie	Mostly loose steps No enactment in practice	Combinations hardly ever on chemistry content	Mostly negative factors No emotion

tion step was observed without any follow-up enactment step or just a conception of a plan. This observation suggests that three of the student chemistry teachers were tending to operate at the stage of planning actions based on their reflections rather than actually following up and enacting those plans (as with Alice). On the other hand, this finding can be seen as evidence that the mechanism we described for enhancing pPCK via ePCK (see Stepwise development of pPCK in the Theoretical Framework) can be supported by the applied assignments (i.e., Lesson Preparation Form, Lesson Evaluation Form and Lesson Reflection Form). How exactly the complexity or depth of the student teachers' reflections and the personal factors account (as amplifiers and filters, see Table 9.5) for the remaining individual differences in student teachers pPCK is an interesting subject for future research—so too is the relationship of pPCK development with the student teachers' (evolving) stages of cognitive development (see the Introduction). We also look forward to clarifying the role and position of extra-personal factors (i.e., aspects of the Learning Context in the RCM) as further research, which we look forward to. Finally, it would be interesting to relate the above findings to the four challenges connected to a one-year teacher training programme as mentioned in the introduction.

Despite many unanswered and/or new questions arising, we think our findings so far can inform new ways of tailored scaffolding of (student) science teachers' pPCK development. On a more theoretical level, our method allowed for a cross-sectional analysis combining three models connected to PCK development. The extreme cases in Table 9.5 show an interesting co-occurrence of differential features in terms of the respective models. This finding suggests a deeper relationship, which is worthwhile studying in more depth. Our typification of student science teachers' pPCK development was supported by their self-portrayals as a teacher, by means of metaphors elicited from them at the end of the teaching methodology course, independently of our study (cf. Saban, Kocbeker, & Saban, 2007). The student teachers' self-constructed metaphors were *silversmith*, *encyclopaedia*, *puppy* and *new-born deer* for Alice, Ben, Cindy and Debbie, respectively. These metaphors support the view that the student science teachers' personal knowledge and beliefs (including PCK) are related to perceptions of their professional role, as in Hong's (2002) study. As a consequence, one can imagine additional ways to stimulate PCK development by incorporating deeper levels of reflection that implicate the student science teachers' sense of mission in their work, and their perceptions of professional identity (c.f., CoRe Reflection; Korthagen & Vasalos, 2005).

A limitation of the method we used in our study is its sensitivity to incomplete data sources, which explains why the data concerning the remaining three student teachers had to be excluded from our analysis. Another limitation is the absence of method triangulation through our decision to use only authentic sources. The incorporation of two lesson series in the data collection (data triangulation) was meant to (partly) compensate for this shortcoming.

The Lorentz workshop in December 2016 followed up on the work of the PCK Summit held in Colorado Springs, 2012. Whereas the Colorado Springs Summit focused on the concept of PCK resulting in a consensus model, known as the Consensus Model (CM), at the Lorentz Center meeting we focused on the instruments used in PCK studies, the data that were collected with these instruments, and the procedures used to infer PCK from these data. Strengths and weaknesses of different instruments and procedures of PCK data collection and analysis were discussed leading to a refinement of the CM stemming from the First (1st) PCK Summit. For us personally, the fruitful discussions with fellow PCK investigators led to interesting reflections on the study discussed in this chapter, in particular about the value of student teachers' evaluations immediately following execution of a lesson. The evaluation forms (Table 9.2) turned out to be the richest information source for analysing the student science teachers' pPCK development. The apparent importance of this evaluation moment led us to extend the corresponding form, most importantly presenting the lesson as a learning opportunity for the student science teachers with a question regarding their most important learning experience, and a question referring to meaningful moments for the learning of their students, thus stimulating evaluation of the lesson in a more comprehensive way.

References

Barendsen, E., & Henze, I. (2017). Relating teacher PCK and teacher practice using classroom observation. *Research in Science Education*. https://doi.org/10.1007/s11165-017-9637-z.

Bishop, A. J., & Whitfield, R. C. (1972). *Situations in teaching*. New York: McGraw-Hill.

Borko, H., Roberts, S. A., & Shavelson, R. (2008). Teachers' decision making: From Alan J. Bishop to today. In P. Clarkson & N. Presmeg (Eds.), *Critical issues in mathematics education* (pp. 37–67). Berlin: Springer.

Clarke, D., & Hollingsworth, H. (2002). Elaborating a model of teacher professional growth. *Teaching and Teacher Education, 18*(8), 947–967.

Cohen, L., Manion, L., & Morrison, K. (2013). *Research methods in education*. London, New York: Routledge.
Garritz, A. (2010). Personal reflection: Pedagogical content knowledge and the affective domain of scholarship of teaching and learning. *International Journal for the Scholarship of Teaching and Learning, 4*(2), 26.
Gibbs, G. R. (2007). *Analysing qualitative data*. Sage.
Greeno, J. G., Collins, A. M., & Resnick, L. B. (1996). Cognition and learning. In D. C. Berliner & R. C. Calfee (Eds.), *Handbook of educational psychology* (pp. 15–46). New York, London: Teachers College Press.
Grossman, P. L. (1990). *The making of a teacher: Teacher knowledge and teacher education*. New York: Teachers College Press.
Hashweh, M. Z. (2005). Teacher pedagogical constructions: A reconfiguration of pedagogical content knowledge. *Teachers and Teaching, 11*(3), 273–292.
Henze, I., Van Driel, J. H., & Verloop, N. (2008). Development of experienced science teachers' pedagogical content knowledge of models of the solar system and the universe. *International Journal of Science Education, 30*(10), 1321–1342.
Hong, J. Y. (2010). Pre-service and beginning teachers' professional identity and its relation to dropping out of the profession. *Teaching and Teacher Education, 26*(8), 1530–1543.
Kegan, R. (1982). *The evolving self*. Harvard University Press.
Kegan, R. (1994). *In over our heads*. Cambridge, MA: Harvard University Press.
Korthagen, F., & Vasalos, A. (2005). Levels in reflection: Core reflection as a means to enhance professional growth. *Teachers and Teaching, 11*(1), 47–71.
Loughran, J., Mulhall, P., & Berry, A. (2004). In search of pedagogical content knowledge in science: Developing ways of articulating and documenting professional practice. *Journal of Research in Science Teaching, 41*(4), 370–391.
Magnusson, S., Krajcik, J., & Borko, H. (1999). Nature, sources, and development of pedagogical content knowledge for science teaching. In J. Gess-Newsome & N. G. Lederman (Eds.), *Examining pedagogical content knowledge* (pp. 95–132). Dordrecht: Kluwer.
McCaughtry, N. (2004). The emotional dimensions of a teacher's pedagogical content knowledge: Influences on content, curriculum, and pedagogy. *Journal of Teaching in Physical Education, 23*(1), 30–47.
Park, S., & Chen, Y.-C. (2012). Mapping out the integration of the components of pedagogical content knowledge (PCK): Examples from high school biology classrooms. *Journal of Research in Science Teaching, 49*(7), 922–941.
Park, S., & Oliver, J. S. (2008). Revisiting the conceptualisation of pedagogical content knowledge (PCK): PCK as a conceptual tool to understand teachers as professionals. *Research in Science Education, 38*(3), 261–284.
Rodgers, C. R., & Scott, K. H. (2008). The development of the personal self and professional identity in learning to teach. In M. Cochran-Smith, S. Feiman-Nemser, D. J. McIntyre, & K. E. Demers (Eds.), *Handbook of research on teacher education: Enduring questions and changing contexts* (3rd ed.). New York: Routledge.
Saban, A., Kocbeker, B. N., & Saban, A. (2007). Prospective teachers' conceptions of teaching and learning revealed through metaphor analysis. *Learning and Instruction, 17*(2), 123–139.
Schön, D. A. (1983). *The reflective practitioner: How professionals think in action*. New York: Basic Books.
Shulman, L. S. (1986). Those who understand: Knowledge growth in teaching. *Educational researcher, 15*(2), 4–14.
Shulman, L. S. (1987). Knowledge and teaching: Foundations of the new reform. *Harvard Educational Review, 57*(1), 1–23.
Van Driel, J. H. (2010). *Model-based development of science teachers' pedagogical content knowledge* (2010, February). Paper presented at the international seminar 'professional reflections', National Science Learning Centre, York.

Van Driel, J. H., & Henze, I. (2012). *Extended paper for PCK Summit, Colorado 2012.* Retrieved from http://pcksummit.bscs.org.

Van Gelder, L., Peters, J. J., Oudkerk Pool, T., & Sixma, J. (1973). *Didactische analyse*. Groningen: Wolters-Noordhoff.

Wehrmann, C. & Henze, I. (2016). A pedagogy to educate novices to professionalize. In M. A. Van der Sanden & M. De Vries (Eds.), *Science and technology education and communication: Seeking energy* (pp. 107–129). Sense Publishers.

Wongsopawiro, D. S., Zwart, R. C., & Van Driel, J. H. (2016). Identifying pathways of teachers' PCK development. *Teachers and Teaching*, 1–20.

Zembylas, M. (2007). Emotional ecology: The intersection of emotional knowledge and pedagogical content knowledge in teaching. *Teaching and Teacher Education, 23*(4), 355–367.

Chapter 10
Investigating Practising Science Teachers' pPCK and ePCK Development as a Result of Collaborative CoRe Design

Jared Carpendale and Anne Hume

Abstract This chapter reports on one case from a cross-case study exploring how collaborative Content Representation (CoRe) design can be used to develop science teachers' personal and enacted pedagogical content knowledge (pPCK and ePCK). These conceptualisations of PCK are components of the Refined Consensus Model (RCM) of PCK (see Chap. 2 of this book). The cross-case study focused on three cases involving science teachers with a limited physics background. Each case study teacher's initial pPCK and ePCK for teaching an *Electricity and Magnetism* topic to a class of 14-year-old New Zealand students were determined prior to the CoRe design intervention using data from interviews and classroom observations. These teachers then engaged in a collaborative CoRe design workshop with other science teachers and experienced physics teachers, where individuals shared their PCK with the whole group and together developed an agreed-upon collective PCK (cPCK) for teaching this topic. The case teachers were subsequently observed teaching a second class (similar age and ability students) and re-interviewed about their pPCK and ePCK development as a result of collaborative CoRe design. The findings from the reported case study reveal that the intervention had a discernible impact on the teacher's pPCK and ePCK, notably: deeper understanding of physics concepts; new ways to represent concepts to students; and greater awareness and consideration of what students may be thinking in their lessons.

Introduction

The study presented in this chapter was part of research conducted for a Doctor of Philosophy degree, which built on the previous work investigating ways to enhance and monitor science teachers' pedagogical content knowledge (PCK) development using collaborative Content Representation (CoRe) design. The conceptualisation

J. Carpendale (✉) · A. Hume
University of Waikato, Hamilton, New Zealand
e-mail: jared.carpendale@gmail.com

A. Hume
e-mail: annehume@waikato.ac.nz

A. Hume et al. (eds.), *Repositioning Pedagogical Content Knowledge in Teachers' Knowledge for Teaching Science*, https://doi.org/10.1007/978-981-13-5898-2_10

of PCK portrayed in the Refined Consensus Model (RCM) of PCK, in Chap. 2 of this book, informs this study into the development of science teachers' personal PCK (pPCK) and enacted PCK (ePCK) for teaching an *Electricity and Magnetism* topic to 14-year-old students in New Zealand after taking part in collaborative CoRe design. The rationale for using the RCM to inform this study was twofold: first, the model incorporated ideas from previous PCK conceptualisations and frameworks, along with opinions of well-respected members of the PCK research community; second, it provided a useful framework for viewing and researching the knowledge exchanges that occurred between a group of teachers working collaboratively, and how those exchanges may influence their individual knowledge and practise.

Two research questions from the Doctor of Philosophy project form the focus of this chapter. They are:

1. In the New Zealand context, what does the personal and enacted pedagogical content knowledge (pPCK and ePCK) of junior science teachers with a limited physics background look like for teaching *Electricity and Magnetism* to 14-year-old students?
2. What impact does collaborative CoRe design have on the pPCK and ePCK development of junior science teachers with a limited physics background for the topic of *Electricity and Magnetism* for 14-year-old students in New Zealand, when working collaboratively with experienced physics and junior science teachers?

Context

Data collection took place in a large boys' secondary school (approximately 2250 students, 13–18 years old) in New Zealand. The science department had 22 science teachers, nine of whom took part in this study. The junior science programme (first two years at secondary school) encompasses the disciplines of physics, chemistry, biology, and Earth science, while in the senior school (last two years at secondary school) these disciplines become separate programmes. All science teachers at the school are required to teach junior science, irrespective of their particular subject specialisation. Thus, it is very common for some teachers to be teaching topics in junior science where they have a limited background, especially in terms of content knowledge.

The researcher's (1st author) initial contact with the school, to introduce the study and identify potential participants, occurred in a meeting between the researcher and the school principal. During discussions about the intentions of the study and the roles of participants, the principal began singling out potential participants with respect to their attributes and how these aligned with various roles within the project. Nine teachers were identified and placed into three groups (three teachers in each). Whilst all nine teachers took part in the study, the Group One teachers were the primary focus and each teacher represented an individual case for investigating pPCK and ePCK development.

Details of the membership of each group are provided below, including key attributes and roles within the project. A summary of the teachers' personal background information is also provided in Table 10.1.

Teachers were assigned to the three groups:

Group One: Practising science teachers with a limited physics background. The principal identified these participants as teachers who would benefit from their PCK being strengthened for the *Electricity and Magnetism* topic. When approached to take part in the study, these teachers were enthusiastic about developing their PCK for this topic and willing to have their PCK development as the focus of the research.

Group Two: Experienced junior science teachers who do not have a strong background in physics. The principal regarded these teachers as effective teachers for junior science, but they were not physics specialist teachers. The principal and researcher felt the presence of these teachers would bring useful pedagogical insights into the CoRe design process, thus contributing to the PCK development of the focus teachers.

Group Three: Experienced physics teachers. The principal endorsed these teachers as effective junior science and physics teachers and felt these teachers would be able to tap into their extensive professional knowledge and experience to support and enhance the professional development of the whole group.

Table 10.1 Background information of participating teachers showing group memberships, names (pseudonyms), subject specialisations, years at the study school, and total years of teaching

Group	Name (pseudonym)	Subject specialisation(s)	Years at study school	Years of teaching (total career)
1	Tony	Biology	6	6
	David	Horticulture and Agriculture	20+	29
	Alan	Physical Education	3	10
2	Harry	Biology	20+	35
	Kate	Chemistry	8	8
	Lucas	Biology and Horticulture	7	10
3	Nick	Physics and Electronics	17	17
	William	Physics and Electronics	15+	40+
	Chris	Physics and Electronics	15	35

Study Design and Literature Review

An interpretivist-based methodology (Guba & Lincoln, 1989) was used in this study as it sought to develop theories and explanations that are contextually bound (Cohen, Manion, & Morrison, 2011; Treagust, Won, & Duit, 2014). The methodology at its heart had a multiple-case study approach (Yin, 2014), which utilised qualitative research methods. Each Group One teacher represented an individual case, which was developed separately. Cross-case comparisons followed and conclusions were drawn.

As stated earlier, the RCM presented in Chap. 2 of this book was used as a conceptual framework to inform this study (see Fig. 10.1). The strength of this model lies in the incorporation of multiple researchers' ideas and conceptualisations of PCK in science teacher education, its identification of the unique and personal nature of PCK (via pPCK), and its acknowledgement of the multiple sources and influences on a science teacher's personal professional knowledge for teaching particular content to particular students. By introducing enacted PCK (ePCK), as a form of PCK firmly placed within the classroom context, the model shows how science teachers access and utilise their pPCK when they are planning, teaching, and reflecting.

The study presented here focuses on the knowledge exchanges occurring between different layers of the RCM, represented in the model above by the double-headed arrows. Of special interest is the impact these exchanges have on the knowledge transitions that occur for individual science teachers as they transform collective PCK

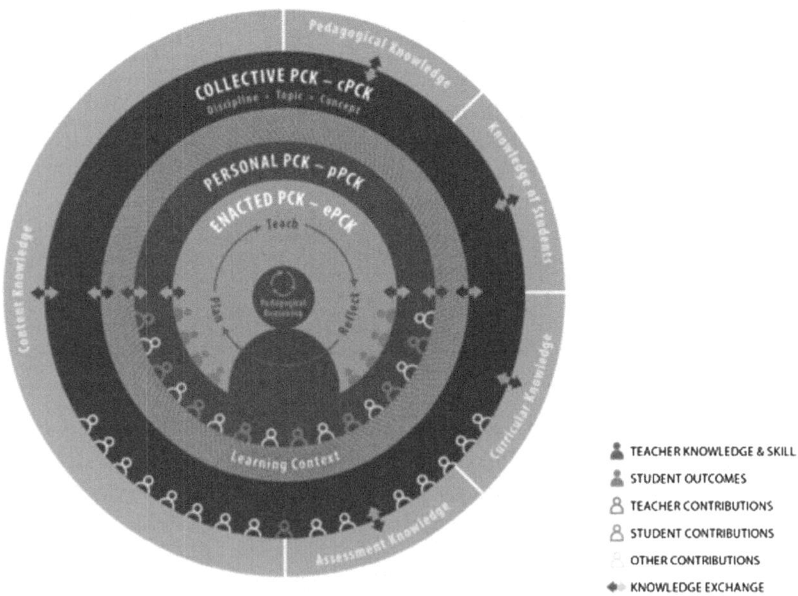

Fig. 10.1 Refined consensus model (RCM) of pedagogical content knowledge (PCK) (see Chap. 2)

(cPCK) to pPCK and ePCK. These transformations are influenced by the moderating effects of the learning context and the amplifying and/or filtering effects of teachers' attitudes and beliefs. In the RCM, ePCK encompasses a cycle of planning, teaching, and reflecting. However, while elements of teachers' planning and reflecting were revealed in this study during data gathering (e.g., discussions about the CoRe for planning after taking part in the workshops and asking teachers to reflect on lessons, and how that might influence future lessons in final interviews), the primary focus when exploring ePCK was on teaching science and the teachers' classroom actions. This focus was necessary to keep the project manageable for a doctoral study.

As a PCK form, cPCK is described in Chap. 2 as the knowledge shared by different science educators, which can be documented, shared, and understood by other teachers in a broader community. Formal documented and/or published cPCK provides a guide of canonical best-practice professional and pedagogical information for teaching particular science content to particular students, in a particular "learning context" as portrayed in the RCM (see Fig. 10.1). A less formal articulation of information and knowledge, such as that synthesised within a CoRe document by a group of science teachers working collaboratively, is potentially a useful practice-based conceptualisation of cPCK for professional learning and research purposes in science teacher education—the CoRe produced represents "localised" cPCK with respect to a particular learning context, which may/may not be in contrast with documented or canonical cPCK.

CoRes were originally devised in a template form (see Table 10.2 on the following page) in an effort to capture a holistic picture of the PCK possessed by a group of expert science teachers for a particular topic (Loughran, Berry, & Mulhall, 2006). These original CoRes proved to be valuable pedagogical tools for teacher educators because they unpack PCK in explicit ways that reveal the key ideas to be learned by students, their prior knowledge, learning difficulties and likely misconceptions, suitable instructional approaches and strategies, and appropriate assessment. Like any innovation in education, others took this original idea and gave it new uses. For example, some science teacher educators challenged their pre-service teachers or early-career teachers to create their own CoRes (e.g. Hume, 2010; Hume & Berry, 2011, 2013) using a CoRe template (see Table 10.2) to illustrate aspects of their emerging and developing pPCK for science teaching.

When filling in a CoRe template, either as an individual teacher (to represent their pPCK) or a group of teachers working collaboratively (to represent their localised cPCK), decisions must be made about what they believe are the big ideas of a science topic to be learned by students. Then a series of pedagogical prompts and questions, within the template, serve to interrogate and draw out teachers' pedagogical reasoning behind their choice of actions to help students develop an understanding of the big science ideas. When addressing these prompts and questions as they complete the CoRe, "teachers access canonical knowledge about their topic and organise it in a way that will be useful for planning instruction" (Gess-Newsome, 2015, p. 33). The resultant CoRe, which can be viewed as a manifestation of cPCK, can also create a platform for initiating and/or strengthening individual science teachers' pPCK development for each of the participating teachers. CoRe design has been shown to

Table 10.2 Template for a Content Representation (CoRe) (adapted from Loughran et al., 2006, p. 28)

Pedagogical questions/prompts	Big idea 1	Big idea 2	Big idea 3
What you intend the students to learn about this idea?			
Why is it important for the students to know this?			
What else you know about this idea (that you do not intend students to know yet)?			
Difficulties connected with teaching this idea			
Knowledge about student thinking which influences teaching about this idea			
Other factors that influence your teaching of this idea			
Teaching procedures (and particular reasons for using these to engage with this idea)			
Ways of ascertaining student understanding or confusion about the idea			

have positive effects on pPCK development of teachers (pre-service, early-career, and out-of-field), especially when done in collaboration with experienced mentor teachers or content experts (e.g., Hume, 2015; Hume & Berry, 2011, 2013; Hume, Eames, Williams, & Lockley, 2013; Nilsson & Loughran, 2012).

Building on the previous research that explored collaborative CoRe design for developing PCK in science, this study investigated the development of the Group One teachers' pPCK and ePCK after working collaboratively with six other science colleagues from their school to develop a CoRe. The resultant CoRe represents the cPCK of all nine participants who teach within the Learning Context of Year 10 *Electricity and Magnetism* at their school. To mitigate the identified issue of needing support from people outside of the school environment (Hume et al., 2013), pedagogical and content expertise was sourced within the school from current teaching staff (i.e., other participants detailed in Table 10.1).

This study involved three distinct phases:

Phase 1: Generating a baseline understanding of Group One teachers' pPCK and ePCK.
Group One teachers were interviewed to explore aspects of their initial pPCK about teaching *Electricity and Magnetism* to 14-year-old students and then observed teaching this topic to determine their initial ePCK.
Phase 2: CoRe design workshops for professional learning.
All nine teachers participated in two CoRe design workshops. The first workshop was an introduction to CoRe design, where participants shared experience and expertise while collaboratively creating a CoRe for teaching *The Nature of Science and Scientific Inquiry* topic to 14-year-old students. In the second workshop, all nine teachers worked collaboratively again to design another CoRe for the *Electricity and Magnetism* topic for 14-year-old students.

Phase 3: Evaluating the influence of CoRe design.
Guided by the collaborative *Electricity and Magnetism* CoRe, the Group One teachers planned and taught this topic again to a different class of similar ages and ability to the class in Phase 1 of the study. After CoRe design and prior to re-teaching this topic again, Group One teachers were interviewed about their self-perceived pPCK development and their experiences with collaborative CoRe design. They were then observed teaching this unit to determine their ePCK post-CoRe design. After teaching the topic again, they were interviewed one final time about changes to their professional knowledge and what they did differently in the classroom. After the CoRe design workshops, Group Two and Three teachers were interviewed about how collaborative CoRe design could enhance pPCK and their experiences with collaborative CoRe design.

Data Collection

Table 10.3 summarises the data that was collected during the study.

Detailed information is now provided about each of the data collection methods and tools.

Table 10.3 Data collected during this study

Phase	Data collected
1	Audio-recorded, semi-structured individual interviews with Group One teachers about teaching science and *Electricity and Magnetism* topic to 14-year-old students
	Video-recordings of Group One teachers' classroom lessons when teaching *Electricity and Magnetism* topic (Class 1)
2	Audio-recording and observations using field notes of teachers participating in *The Nature of Science and Scientific Inquiry* CoRe design workshop
	Audio-recording and observations using field notes of teachers participating in the *Electricity and Magnetism* CoRe design workshop
3	Audio-recorded, semi-structured individual interviews with Group One teachers exploring their perceptions of CoRe design and its effectiveness for enhancing PCK
	Audio-recorded, semi-structured focus group interviews with Group Two and Three teachers exploring their perceptions of CoRe design and to judge its effectiveness for enhancing PCK
	Video-recording of Group One teachers' classroom lessons when teaching *Electricity and Magnetism* (Class 2)
	Audio-recorded, semi-structured interviews with Group One teachers to explore how they think their pPCK and ePCK had developed as a result of collaborative CoRe design

Interviews

As each of the Group One teachers represented a separate case for this study, each semi-structured interview was individual. In contrast, when Group Two and Three teachers were interviewed, a semi-structured focus group format was utilised to increase time efficiency and to allow for rich data as multiple people can be interviewed at once, which can encourage participants to build on each other's ideas (Flick, 2014; Patton, 2014; Watts & Ebbutt, 1987). The semi-structured style was employed using a set of predetermined guiding questions, with the flexibility to explore responses further or seek clarification (Bogdan & Biklen, 2007; Kvale, 1996). The same guiding questions were used for all interviews, except for the final interviews as only Group One teachers were involved.

For the Group One interviews in Phase 1, the RCM, along with previous PCK models such as the work of Magnusson, Krajcik, and Borko (1999) and Gess-Newsome (2015), were used to develop interview questions about researchable entities of pPCK. Questions were asked about participants' knowledge of curricula, knowledge of students understanding in science, knowledge of instructional strategies, and knowledge of assessment strategies.

After taking part in collaborative CoRe design, all participants were questioned about their experiences with the process, including what they saw as being valuable and what limitations they faced. They were also asked about how the process could develop pPCK for science teaching and how their own pPCK may have been affected. For their final interview, after teaching their second class in the last phase of the study, Group One teachers were questioned about their self-perceived pPCK and ePCK development and any causal links with the content and/or process of collaborative CoRe design.

Observing Lessons

The complexity of capturing science teacher's professional knowledge meant that an approach using both interviews and observations was advisable. The approach used in this study reflects information and guidance found in the literature for capturing PCK and ensuring conclusions are trustworthy (e.g., Bryman, 2016; Henze & van Driel, 2015). However, during data collection, the principal researcher (1st author) was teaching full time, so personally attending and observing an adequate amount of lessons for a dependable and trustworthy analysis were not feasible. After discussion within the supervisory team, the pragmatic decision was made to video-recorded lessons.

Most of the *Electricity and Magnetism* lessons taught by the Group One teachers were video recorded and available for observing by the researcher later. Lessons were one-hour duration. Four lessons from each teacher were chosen for analysis pre-CoRe design and another four post-CoRe design. Those chosen for both pre-

CoRe and post-CoRe design included the introductory lesson, a practical lesson, and two others, which reflected discussions the teacher had during the workshop. These lessons were analysed using an observation protocol, which is discussed later in this chapter.

Facilitating the CoRe Design Workshops

Taking part in collaborative CoRe design can be a challenging experience (Hume & Berry, 2011, 2013), so it was decided that the first workshop would be a pilot/trial exercise for the participating teachers. To facilitate the development of the participants' capabilities in CoRe design, a *Nature of Science and Scientific Inquiry* CoRe was developed in the first workshop since elements of its design could be connected to an existing science topic at the study school. Thus, participants were likely to have varying levels of experience and understanding of teaching related science concepts. As further support, participants were asked to read an article about the nature of science and scientific inquiry (see Lederman, Antink, & Bartos, 2014a) before the workshop. This initial workshop, facilitated by the 2nd author, started with an introduction to the construct of PCK, the use of CoRes in capturing PCK, and the purpose of CoRe design in this study. Further discussions and works around the nature of science and scientific inquiry resulted in participants establishing the following key understandings: the nature of science refers to *knowledge* in science, while scientific inquiry refers to *practices* in science.

After the first hour of discussion, the teachers were asked to complete the views about scientific inquiry (VASI) questionnaire (see Lederman et al., 2014b) to explore their views on the nature of scientific inquiry. Responses were then discussed, after which the teachers were assigned to one of three Working Groups—each comprised three members, one from Group One, one from Group Two and one from Group Three, as shown in Table 10.4 below.

In these Working Groups, the teachers discussed what they understood to be key concepts and skills underpinning the nature of science and scientific inquiry, writing each on separate pieces of card. The facilitator promoted a whole group discussion where all the ideas were shared, collated, and themes identified. The key themes were recorded on the whiteboard and used by the whole group to develop big ideas (in the form of propositional statements) for the topic. These big ideas formed the basis upon which the teachers could begin working in their Working Groups to complete

Table 10.4 Working Group memberships for CoRe design workshops

Original study group	Working Group One	Working Group Two	Working Group Three
Group One	Alan	Tony	David
Group Two	Lucas	Harry	Kate
Group Three	Chris	Nick	William

a CoRe. As they worked, the facilitator moved amongst the groups and interacted with each group in turn as they addressed the pedagogical prompts within the CoRe template. In the final stage of the three-hour workshop, each Working Group shared and collated their outcomes as a single partially completed CoRe, which represented the groups' cPCK on the topic.

One week later, the teachers took part in another CoRe design workshop, this time for the *Electricity and Magnetism* topic (in the same Working Groups). Since they had recent experience working with CoRe design, introductory discussions were brief. The workshop facilitator recapped key PCK ideas and then addressed and/or revisited four CoRe-related points on the whiteboard:

1. Possible resources—what resources would be useful for this particular exercise?
2. Brainstorm concepts and skills—the first step in CoRe design for any topic.
3. Use of these concepts/skills to identify themes from which "big ideas" are developed.
4. Complete CoRe—for sharing of professional knowledge (stay in groups to create three separate CoRes, or work as one big group on one shared CoRe?)

This introductory discussion lasted 10 min, and then teachers commenced work on identifying key concepts and skills for *Electricity and Magnetism*. Resources, which the teachers identified and used to inform their decisions, included a copy of the New Zealand Curriculum (NZC) document (Ministry of Education, 2007) and NCEA assessment information for the Year 11 topic *Electricity and Magnetism* (NZQA, 2010). After 30 min, all identified concepts and skills were shared in the larger group where participants discussed the suitability of each particular science concept/skill for teaching to 14-year-old students in turn and collectively decided on its inclusion or not. Using these selected concepts/skills, the individual groups then worked to identify big ideas, which were to be written as propositional statements. Again, the information from the groups was shared, and seven big ideas were generated from this collective information. In the interest of time, the groups decided to select two or three big ideas per Working Group to address, and afterwards, the work of each Working Group was combined with that of others to produce a completed CoRe. This workshop was three hours in duration.

During both CoRe design workshops, all discussions were audio-recorded. Recording discussions was achieved by having digital recorders placed where each group was working—the recorders also captured whole group discussions. In addition, throughout the workshops, the first researcher also took detailed field notes about interactions that took place. At the end of each workshop, all CoRe materials were submitted to the researcher who collated the materials into one CoRe document. This single CoRe was then sent back to all participants for verification purposes.

Data Analysis

To determine the pPCK and ePCK of each Group One teacher, and any possible development, the data were analysed thematically via a deductive approach (using the RCM primarily to inform the analytical framework) and an inductive approach (by identifying any emergent themes from the data pertinent to the research objectives). To construct the analytical framework for analysing pPCK and ePCK, key parameters of the RCM needed to be identified. For example, these analytical parameters for analysing pPCK from interviews were the teachers' knowledge of curriculum; students' understanding and learning; topic-specific instructional strategies; and, assessment strategies.

To analyse the Group One teachers' ePCK from the video-recorded lessons, an observational protocol was developed that included a rubric identifying three components of ePCK along with 10 quality indicators. Each quality indicator was assessed as being either limited, basic, proficient, or advanced. The design of this rubric was based on: the previous PCK research (e.g., Alonzo, Kobarg, & Seidel, 2012; Gardner & Gess-Newsome, 2011; Lee, Brown, Luft, & Roehrig, 2007; Park, Jang, Chen, & Jung, 2011); the pedagogical prompts from the CoRe (i.e. Loughran et al., 2006); and, outcomes of discussions from the second PCK Summit about generating a "grand rubric" (reported in detail in Chap. 12 of this book).

The components and their quality indicators (and abbreviations) were:

Subject Matter Knowledge

- Appropriateness of the concepts (appropriateness)
- Scientific accuracy of the explanation of the concepts (accuracy)
- Links and/or connections made to other concepts (concept links)
- Links made (implicit or explicit) to the nature of science or scientific inquiry (NoS/SI links)

Knowledge of Student Understanding

- Recognition of possible prior knowledge, difficult concepts, or misconceptions (prior knowledge)
- Variations in student understanding and learning are identified which is used to guide instruction (variations in understanding)
- Questions are used to probe or extend student understanding (questions).

Knowledge of Instructional Strategies

- Appropriate sequence for teaching concepts (sequencing of concepts)
- Relevant examples and/or representations are used, which appear to be pedagogically effective at portraying the concept (example and representations)
- Strategies that allow for metacognition (metacognitive strategies).

A copy of the full rubric used for analysis can be found in Appendix 1 of this chapter.

Findings

This study yielded an extensive and rich database, which cannot be reported upon fully in this chapter. However, to give an indication of how collaborative CoRe design impacted on teachers' pPCK and ePCK in this study, the findings from one case study will be presented (i.e., one Group One teacher). Tony's case was selected for presentation since it signals, and at times confirms, the potential of the intervention for pPCK and ePCK enhancement.

Tony's Initial pPCK for Teaching Electricity and Magnetism to 14-Year-Old Students

During his initial interview, Tony saw science as an important school subject, reasoning it helps students to understand the world around them. However, when talking specifically about teaching the *Electricity and Magnetism* topic, his focus appeared to change as he commented, "I want them [*students*] to do well in the test". Tony explained that the test referred to a departmental assessment created by the head of junior science to assess "certain outcomes that we have to cover". Tony was unsure about the source of these outcomes, but suggested they could be from the NZC. Responding to further questions about the role of these outcomes in his teaching, he felt they were his top priority and he did not teach beyond their scope, emphasising "first and foremost, I need to make sure I cover those, because of the time limitations we have … I don't go beyond them, I try and cover them".

Tony felt the nature of science was about "finding out why things happen", and he tried to incorporate that notion into his lessons by promoting students to be inquisitive, stating "I like them to ask questions. I want them to be curious about what's going on. I want them to be interested in what's around them. I like to encourage then to, if you want to know something, then try and find out". He was unsure about the phrase "scientific inquiry"; however, with prompting from the researcher about scientific processes, he offered some insights about their inclusion in his lessons, notably around his use of questions and some contextual restraints. He explained, "if you want to do that, you need more questions at the start of the topic. It comes back to a time limitation to be honest, and then how do you measure how successful it is? How do you tell if they've learnt something?"

Regarding students' prior knowledge, Tony felt they needed some basic understandings before starting this unit. He was aware students had different learning needs and styles, commenting "they all learn differently". To accommodate for students' needs, he mentioned using videos initially for enjoyment purposes, but focused on ensuring that students had the opportunity to make circuits, so they could make cognitive linkages to concepts. He argued "I reckon a lot of students are tactile learners – they like to make things. Then once they can see it, hopefully they can make sense of it when it comes to circuit diagrams. Or doing stuff like that, it will make sense".

When he taught this unit, Tony's strategy was to "start with something fun", then "give them some notes about what is electricity, some definitions for voltage and current. I try and get them practical work as soon as possible ... So, try and get them onto those to make a circuit". He gauged students' understanding by their ability to "pass the test" and his approach to formative assessment during lessons featured strategies for preparing students for their test. When asked about how he might use information obtained from these formative assessment strategies in class, Tony explained "I don't record it. But I do identify students who I think need work. So, you know, you can figure out who's onto it and who's not".

Tony's Initial ePCK for Teaching Electricity and Magnetism to 14-Year-Old Students

The four video-recorded pre-CoRe design lessons selected to determine Tony's baseline ePCK included his: introductory lesson; third lesson, featuring explanations and discussions around series and parallel circuits; fourth lesson, where students made simple circuits; and sixth lesson, developing explanations about voltage, current, and resistance. In the lessons, most students were engaged in the learning tasks, particularly during practical activities.

Tony's teaching style was identified as teacher-centred with lessons frequently featuring a pre-made PowerPoint, unless the lesson involved practical work. He typically directed students to copy notes from the PowerPoint, drawing attention to underlined keywords that were to be tested. This PowerPoint was used with other classes, and on one occasion when the data projector failed, he expressed frustration and said "right, plan B... I have to write this, this [*is a nuisance*]".

Tony's lessons were analysed for his ePCK using the rubric discussed earlier, and Table 10.5 shows a summary of these results.

Linking Tony's Initial pPCK and ePCK

When comparing findings from Tony's interview and lesson observations, six key links were seen between his initial pPCK and ePCK, which are summarised below:

- The influence of assessment on Tony's teaching was a prevalent theme throughout his pPCK interview and classroom actions. During the interview, Tony spoke about wanting his students to pass tests, and whilst teaching, he made frequent references to taking notes from the PowerPoint and learning definitions as they would be in the test.
- Tony talked in his interview about wanting his students to be inquisitive in class and ask questions. However, the observations reveal when students did ask questions

Table 10.5 Summary of the results from analysing Tony's four video-recorded lessons using the rubric developed for this study

ePCK indicator	Lessons			
	1	2	3	4
Subject matter knowledge				
Appropriateness	Proficient	Proficient	Advanced	Advanced
Accuracy	Proficient	Proficient	Proficient	Proficient
Concept links	Basic	Basic	Basic	Proficient
NoS/SI links	Limited	Basic	Proficient	Basic
Knowledge of student understanding				
Prior knowledge	Basic	Basic	Proficient	Limited
Variations in understanding	Limited	Limited	Basic	Limited
Questions	Basic	Basic	Basic	Basic
Knowledge of instructional strategies				
Sequencing of concepts	Basic	Proficient	Basic	Proficient
Examples and representations	Proficient	Proficient	Basic	Proficient
Metacognitive strategies	Limited	Limited	Limited	Limited

in class, he rarely engaged with what they were asking and sometimes appeared to ignore them.

- Tony spoke about students learning science, so they could understand the world around them. He used examples in his practice, but these were largely ineffectual in the teaching of the desired concept.
- Tony's lack of understanding about the nature of science and scientific inquiry in the interview was also apparent in the lessons, as he made very few explicit or implicit references and/or links to these aspects in his teaching.
- In his interview, Tony talked about finding out about student understanding (in an informal way) and using that information to guide instruction. However, this strategy was rarely used in lessons with Tony following a tight schedule dominated by his PowerPoint notes.
- In the lessons, Tony did use practical work to help student understanding, as indicated during his interview. However, the practical work undertaken by the students appeared largely ineffectual in the teaching of the desired concept(s).

Tony's CoRe Design Contributions and Experiences

Tony collaborated with his teaching colleagues Nick and Harry for the CoRe design workshops. All three teachers engaged in relevant discussions around what the teaching and learning of the important *Electricity and Magnetism* concepts and skills entailed. In this discussion, Nick (experienced physics teacher) frequently took the lead, offering ideas that he believed to be pertinent to the pedagogy of *Electricity and Magnetism* for 14-year-old students. For example, early in the workshop, when analogies were first mentioned by Tony, Nick explained "teachers need to be careful and selective when using analogies for teaching some difficult-to-understand concepts in Electricity and Magnetism, because the result could be that students develop strong misconceptions".

During these discussions, Nick also offered detailed explanations to his colleagues about the importance of understanding the conservation of charge and energy to teach this topic. In one instance, he drew diagrams showing series and parallel circuits and explained that "it doesn't matter which way charges go, they lose energy". Tony mentioned that this concept can be confusing, to which Nick responded "yes, it is, you need to distinguish between everything … it is all about gaining and losing energy, and that energy in must equal energy out". This explanation and the diagrammatical representations Nick provided proved to be key points that Tony took away from the CoRe workshop. When interviewed about what impact the *Electricity and Magnetism* CoRe might have on his practice, Tony recalled this information and redrew the diagrams indicating that he would teach it this way in the future.

During the determination of the concepts and skills for determining the big ideas for the CoRe in his group discussion, Tony's first comment was students need to learn "definitions of current and voltage", to which Nick replied "it's not definitions. It's understanding of what it actually is". As the workshop progressed, Tony also commented that "you need to give them [*students*] enough to understand, but not too much to confuse them" in reference to some of the suggested ideas being pitched at a level that was too advanced. Tony's group subsequently identified the following six big ideas:

1. Voltage is the difference in energy between two points.
2. Magnets produce magnetic fields which exert a force on other magnets.
3. Rubbing materials together can lead to a separation of charge.
4. Current is the flow of charge.
5. Wires are full of charges and they all move, or none move.
6. Charges produce electric fields which exert a force on other charges.

Their big ideas were shared and compared with the other groups' big ideas, and through a process of negotiation and mediation, guided by the facilitator, seven key big ideas emerged that reflected the collective thinking of all nine participants.

The collective big ideas were:

1. Charges produce electric fields which exert a force on other charges.
2. Current is the flow of charge.
3. Voltage is the difference between the two points.
4. Ohm's law is the relationship between current, voltage, and resistance in a closed circuit.
5. Circuit diagrams are representations of electrical circuits.
6. Electrical circuits can be constructed to solve problems.
7. Magnetism is another effect of moving charge.

To ensure the CoRe was completed within the timeframe, the seven big ideas above were split amongst the groups. The first three listed above were addressed by Tony's Working Group, and their contribution is shown in Appendix 2.

In the post-CoRe workshop interview, Tony commented that he found working collaboratively in that style to be useful and would be interested working in that way in the future. He recognised that the CoRe design process "enables you to break a topic into smaller bits, so it seems less overwhelming", and during the design of the *Electricity and Magnetism* CoRe "it was useful to get some clarification and confirmation about teaching certain concepts". Tony also acknowledged how useful it was to have Nick in his group, as he could learn from his expertise, stating "having a guy who knows what he's doing, and then you can just get clarification and confirmation, or if you have a question you can ask straight away… If you want to know about physics, go and talk to a physics guy".

Tony's pPCK Development

Post-CoRe design, Tony felt that collaborative CoRe design can enhance a teacher's pPCK for science teaching as it has potential to "give people different ideas about doing stuff. There might be other ways of doing things that haven't been thought about". While he judged the first workshop did not enhance his pPCK about teaching *The Nature of Science and Scientific Inquiry*, he felt the second *Electricity and Magnetism* workshop did have an impact. In his post-CoRe design interviews, Tony identified aspects related to his subject matter knowledge and knowledge of topic-specific strategies as two areas of enhancement for his *Electricity and Magnetism* pPCK. He was particularly focused on a new way to teach students about energy and charge in both series and parallel circuits, which he had taken from the workshop:

> Nick talked to me about how to explain the concept of voltage… Why voltage is the same in a parallel circuit and different in a series circuit. He explained it in quite a good way. He drew a series and parallel circuit and explained how the voltage is shared. [Tony redrew Nick's diagrams]

He described how these explanations had helped his understanding and how he wanted to use them in the future with his students as it made the concept easier for them to visualise and understand. He also spoke of how he and David (another Group One teacher) had worked together with the completed CoRe to prepare for teaching their post-CoRe design class.

Tony indicated that his knowledge of students' understanding and learning had also improved, but did not elaborate or give examples. Similarly, he felt that the CoRe design process was focused on teaching concepts as opposed to assessment strategies, so he did not offer any information about how his knowledge of assessment strategies may have developed.

Tony's ePCK Development

The four video-recorded post-CoRe design lessons selected for analysis included his: introductory lesson; second lesson, featuring explanations about charged particles, voltage, and current; third lesson, where students explored differences between series and parallel circuits and the Ohm's law relationship; and fifth lesson, where students made simple circuits and took measurements. Again, most students were engaged in the lessons, particularly during practical work.

Tony's teaching style was again identified as predominantly teacher-centred with a focus on students taking notes. However, there were now some instances during his lessons where Tony engaged with students and challenged their thinking, particularly why they thought in a certain way. Again, Tony had a PowerPoint for this class, but the slides were different to those used previously.

To evaluate ePCK enhancement, post-CoRe observational data was compared to that obtained pre-CoRe. A summary of the post-CoRe rubric analysis is presented in Table 10.6, along with an indication in the last column of enhancement (or not) to each of the ePCK quality indicators compared to pre-CoRe results, where "–" represents no change, and "↑" represents development.

In both pre- and post-CoRe Y10 classes, the concepts that Tony taught were appropriate for students at that level. Links to the nature of science and/or scientific

Table 10.6 Summary of the results from analysing Tony's four video-recorded lessons (post-CoRe design) using the rubric developed for this study and an indication of enhancement

ePCK indicator	Lessons				Enhancement
	1	2	3	4	
Subject matter knowledge					
Appropriateness	Advanced	Advanced	Proficient	Advanced	–
Accuracy	Advanced	Advanced	Advanced	Advanced	↑
Concept links	Advanced	Advanced	Advanced	Proficient	↑
NoS/SI links	Basic	Basic	Basic	Proficient	–
Knowledge of student understanding					
Prior knowledge	Proficient	Basic	Proficient	Proficient	↑
Variations in understanding	Basic	Proficient	Advanced	Proficient	↑
Questions	Basic	Basic	Proficient	Advanced	↑
Knowledge of instructional strategies					
Sequencing of concepts	Advanced	Basic	Proficient	Proficient	↑
Examples and representations	Advanced	Proficient	Advanced	Proficient	↑
Metacognitive strategies	Basic	Basic	Basic	Proficient	↑

inquiry, by the way of implicit links, were similar into those in his pre-CoRe design class, so this aspect of his pPCK and ePCK appeared little changed.

The summary below outlines the developments that occurred in eight of 10 identified quality indicators of ePCK:

Subject Matter Knowledge

Accuracy

After the CoRe design workshop, Tony's explanations became more in depth, and he focused on the underlying principles as well as rules and definitions.

Concept links

Tony made some links in his pre-CoRe design class, but the explanation that linked the concepts needed further development. After taking part in the workshop, he made more links between concepts and offered students well-thought-out explanations about the linkage.

Knowledge of Student Understanding

Prior knowledge

While Tony sought some prior knowledge from students in his pre-CoRe design lessons, the information obtained was often used in a very limited way. However, after being involved with CoRe design he seemed much more aware of this information and attempted to use it more to inform his lessons.

Variations in understanding

Before CoRe design, it was apparent that Tony had certain content he wanted to get through during lessons, and he often did not deviate from that plan. However, after the workshop, he became more aware of students' needs and areas where they were having difficulty. As a result, he was able to change tack and vary his pedagogical approach at times to address learning issues that arose.

Questioning

Compared to his pre-CoRe design classes, Tony used many more questions with his students and his questions also had more variety. For example, extending beyond one-word factual questions to asking students to predict and explain phenomena.

Knowledge of Instructional Strategies

Sequencing of concepts

In his pre-CoRe design lessons, it appeared that the *intended* sequence of concepts was quite suitable and appropriate for that level of students most of the time. However, explanations to link changes in concepts were lacking, which resulted in students being unsure about what they were learning. In contrast, in his post-CoRe design lessons Tony offered insightful explanations to students about why they were changing concepts and how concepts were related.

Examples and representations

While Tony used these strategies in his pre-CoRe design lessons, they were often ineffectual for portraying the desired concept and his explanations linking the strategy to the concept were brief, incorrect, or missing. In contrast, his post-CoRe design examples and representations appeared much more effective at enabling student learning by being more targeted at building student understanding. He used Nick's analogies and diagrams that he encountered during the workshop.

Metacognitive strategies

Tony significantly improved his use of instructional strategies that provoked metacognition. In his pre-CoRe design class, there were no instances that indicated purposeful stimulation of students' metacognition. However, post-CoRe design he actively encouraged students to think about their own thinking and to express their ideas.

Discussion and Conclusion

This section interprets, discusses, and evaluates the findings from Tony's case study in relation to the research questions and pertinent literature. Each research question is restated and addressed in turn.

Research Question One

> In the New Zealand context, what does the personal and enacted pedagogical content knowledge (pPCK and ePCK) of junior science teachers with a limited physics background look like for teaching *Electricity and Magnetism* to 14-year-old students?

The findings presented above indicate that Tony's initial pPCK and ePCK were characterised by four features: one related to *what* he was teaching and three to *how* he was teaching. Regarding *what* concepts were being taught, Tony's decisions were dictated by the outcomes provided in the departmental guidelines; that is, he adhered to these outcomes in his planning and teaching. There is little evidence that he made autonomous decisions when selecting appropriate concepts to teach, which is an important attribute of a well-developed pPCK (Park & Oliver, 2008). As he worked through those outcomes, links to other concepts during lessons were often overlooked in his teaching, indicating a basic level of pPCK and ePCK (Gardner & Gess-Newsome, 2011).

When teaching in science, Alonzo et al. (2012) argue the need for teachers to appropriately sequence concepts, so students can identify the connections and develop their understanding of those concepts and their relationships with other concepts. As concepts transition to others, the sequencing often requires insightful explanations from teachers about how the concept(s) is changing to ensure students are developing their conceptual understanding appropriately. The findings revealed such sequencing of concepts in Tony's pre-CoRe design class were at a basic proficient level of pPCK and ePCK (Alonzo et al., 2012; Gardner & Gess-Newsome, 2011).

In terms of *how* Tony was teaching these concepts, the three identified features were: responsiveness pedagogically to student's understanding and learning; the use of representations and examples, and promoting metacognition; and, the influence of context.

Responsiveness is regarded as an essential attribute of well-developed PCK, that is, a teacher's ability to recognise students' learning and understanding, and then to vary his/hers next pedagogical move (Alonzo et al., 2012; Gardner & Gess-Newsome, 2011). In other words, teachers need to be pedagogically responsive to student needs during lessons and adapt their pedagogical approach as required. Since Tony was reliant on the provided outcomes, his lessons tended to be tightly organised around delivery of the required information to students. This approach meant he did not adapt his lessons to be pedagogically responsive when students required learning assistance, implying a limited to basic level of ePCK (Alonzo et al., 2012; Gardner & Gess-Newsome, 2011; Lee et al., 2007).

To help students develop their own conceptual understanding, teachers with a rich PCK for science teaching employ strategies where examples and representations are used to aid student understanding and metacognition is promoted (Alonzo et al., 2012; Gardner & Gess-Newsome, 2011; Lee et al., 2007). Students utilise these examples and representations to explain concepts and to relate new knowledge to their existing understanding and think about their thinking. In contrast, Tony's teaching was characterised by the transmission of information. There were times when he attempted to use examples and representations, but these were often ineffectual and instances that provoked metacognition were not seen in lessons, reflecting a limited to basic ePCK.

The RCM (see Fig. 10.1) places "learning context" as a key influence on teachers' pPCK and ePCK. This influence was clear for Tony, as contextual constraints within the learning environment (i.e. the school's focus on assessment and student achievement in national qualifications) underpinned his teaching decisions. The findings show that assessment requirements featured prominently in Tony's pPCK and ePCK.

Research Question Two

> What impact does collaborative CoRe design have on the pPCK and ePCK development of junior science teachers with a limited physics background for the topic of *Electricity and Magnetism* for 14-year-old students in New Zealand, when working collaboratively with experienced physics and junior science teachers?

When CoRe design is used as a collaborative process, it has been shown to enhance teacher's PCK, particularly for pre-service and early-career science teachers (e.g., Hume & Berry, 2011, 2013; Nilsson & Loughran, 2012). During the CoRe design workshop, there were many instances where knowledge was shared within the Working Group. Desimone (2009) and Daehler, Heller, and Wong (2015) predict this sharing of knowledge through collaborative efforts supports science teachers' professional learning, which was indicated in Tony's case. Findings showed his pPCK was enriched, which in turn enhanced his classroom practice, a key aspect of ePCK. This knowledge sharing underpins the knowledge exchange that occurred between cPCK, pPCK, and ePCK, as predicted and represented in the RCM of PCK by double-headed arrows, and evidenced in the knowledge transitions that Tony experienced.

Tony's case study reinforces the effectiveness of collaborative CoRe design as a means of developing pPCK and ePCK with significant enhancement to his subject matter knowledge, knowledge of instructional strategies, and knowledge of students' understanding and learning. During post-CoRe design interviews to explore his pPCK development, Tony explicitly identified his subject matter knowledge and knowledge of instructional strategies as areas of personal improvement. He particularly appreciated strengthening his understanding of voltage, charge, and energy concepts, and how to relay that information to students. Comparison of the two sets of classroom observational data, pre- and post-CoRe design, confirmed this enhancement. In addition, observational comparisons showed Tony's knowledge of students' understanding and learning had also improved, as he was more pedagogically aware of students learning needs, responsive to those needs, and used questions more effectively.

In conclusion, Tony's case signals that the use of collaborative CoRe design within a school learning community, to access and collate aspects of cPCK of teachers, promotes the pPCK and ePCK development of those science teachers with less content knowledge for that topic. One advantage of collaborative CoRe design in this setting is the ability of a school to capitalise on in-house expertise, rather than seeking it from outside sources, which may place undue pressure on a school's financial and organisational resources. This in-house use of collaborative CoRe design also addresses a limitation raised by Hume et al. (2013) about the logistical difficulty of organising various teachers (and content experts) from different locations to collaborate face-to-face.

The RCM has proved a useful and applicable conceptualisation of PCK for guiding this study. In particular, there are three features of this model that have facilitated this study. They are:

1. The conceptualisations of pPCK and ePCK, and showing how they interact. The separation of the professional knowledge that a teacher possesses and can talk about from the teacher's actions in the classroom aids comparisons and, at the same time, enables any synergy and/or dissonance between the two to be identified. To these ends, targeted research methods including quality indicators can be developed to investigate and capture these forms of PCK.

2. The introduction of cPCK. This form of PCK recognises the contributions of multiple people and encapsulates what a CoRe document represents when it has been developed as a collaborative process.
3. The emphasis on knowledge exchanges between different knowledge bases, including the different forms of PCK. These exchanges, represented by double-headed arrows in the diagrammatic form of the RCM, show how knowledge can be shared and how that process can influence/be influenced by classroom practice and the learning context. The discussion presented in this chapter reinforces the importance of this process, as knowledge that was shared within the cPCK realm was transferred into and enhanced the pPCK and ePCK of an individual science teacher via his knowledge transitions and/or transformations.

This study recommends that schools should consider the use of collaborative CoRe design, as portrayed in this study, as an effective professional development intervention for enhancing the cPCK, pPCK, and ePCK of its science teachers, particularly those without specialist science content knowledge.

Limitations

There are three main limitations, which should be taken into consideration when interpreting the findings and conclusion from this study. In this chapter, there is an account of only one teacher's experiences with collaborative CoRe design and his subsequent pPCK and ePCK development. While it is reported that collaborative CoRe design was a positive experience for him, enhancing his professional knowledge and practice in particular ways, this conclusion may not be drawn from these findings for others. However, it can be reported that both of the other Group One teachers (not included in this chapter) had positive experiences with CoRe design that enhanced their pPCK and ePCK, albeit in different ways and to different degrees.

In both the RCM of PCK and the previous Consensus Model (CM) of PCK (i.e., Gess-Newsome, 2015), student outcomes were included. However, in this study for pragmatic reasons, no data was obtained from students. In future studies, it would be important to make comparisons between students' science learning from the teacher pre- and post-CoRe design to see the effect changes in their teachers' pPCK and ePCK may have on their learning. This type of data would shed more light on the impact of collaborative CoRe design.

Similarly, the link between the pedagogical reasoning undertaken in pPCK and ePCK was not explored in this study. While Gess-Newsome (2015) encouraged the use of data collection methods such as stimulated recall interviews to investigate this aspect of teachers' PCK, these were not used in this study. Again, the researcher's commitment to teaching full time necessitated the decision not to explore this aspect of PCK. Researching science teachers' pedagogical reasoning in the act of teaching, after taking part in collaborative CoRe design, will also provide rich insights into the effects of collaborative CoRe design.

Appendix 1: Rubrics for Analysing ePCK

ePCK indicator	Limited	Basic	Proficient	Advanced
Subject matter knowledge				
Appropriateness of concept(s) in relation to NZC—physical world (level 5)	No alignment of concept(s) in lesson with NZC—physical world (level 5)	Little alignment of concept(s) in lesson with NZC—physical world (level 5)	Adequate alignment of concept(s) in lesson with NZC—physical world (level 5)	Close alignment of concept(s) in lesson with NZC—physical world (level 5)
Scientific accuracy of the explanation of the concept(s)	Explanation(s) were mostly inaccurate, which did not address the concept(s)	Explanation(s) were somewhat inaccurate, which loosely addresses the concept(s)	Explanation(s) were mostly accurate with only small inaccuracies seen, or they were too brief	Explanation(s) were accurate, which addresses the concept with no inaccuracies
Links and/or connections made to other concepts	No possible links and/or connections are made	Few of the possible links are made, but not connected with explanations	Some of the possible links and connections are made	Many of the possible links and connections are made
Links made (implicit or explicit) to the nature of science (NoS) and/or scientific inquiry (SI)	No links made to NoS and/or SI	Few of the possible links to NoS and/or SI are made	Some of the possible links to NoS and/or SI are made	Many of the possible links to NoS and/or SI are made
Knowledge of student understanding				
Teacher recognises and acknowledges possible student prior knowledge, difficult concepts, and misconceptions	No recognition or acknowledgement of possible student prior knowledge, difficult concepts, and/or misconceptions	Recognises some possible student prior knowledge, difficult concepts, and/or misconceptions	Recognises and acknowledges some possible student prior knowledge, difficult concepts, and/or misconceptions	Recognises and acknowledges most/all possible student prior knowledge, difficult concepts, and/or misconceptions

(continued)

(continued)

ePCK indicator	Limited	Basic	Proficient	Advanced
Teacher uses identified variations in student understanding and learning to guide instruction	No acknowledgement and/or use of variations in student understanding and learning to guide instruction	Acknowledgement of variations in student understanding or learning, but not used to guide instruction	Some acknowledgment of variations in student understanding or learning are used to guide instruction	Many instances where teacher acknowledged variations in student understanding or learning and used these to guide instruction
Teacher uses questioning to probe or extend student understanding	No questions are used to probe or extend student understanding	A few questions are used to probe or extend student understanding	An adequate range of questions are used to probe or extend student understanding	Many and varied questions are used to probe or extend student understanding
Knowledge of instructional strategies				
Appropriate sequence for teaching concepts	No overall flow between concepts and the sequence confuses students	Some flow between concepts and the sequence allows some concept building to occur	Suitable flow between concepts and the sequence allows satisfactory concept building to occur	Clear flow between concepts and sequence allows effective concept building
Relevant examples and/or representations are used in the lessons, which appear to be pedagogically effective at portraying the concept	No examples and/or representations used	Examples and/or representations used that do not appear to be pedagogically effective	Examples and/or representations used have some relevance, but appear pedagogically limited	Relevant examples and/or representations used that appear pedagogically effective
Use of strategies that allow for metacognition	No use of strategies that allow for metacognition	Limited use of strategies that allow for metacognition	Adequate use of strategies that allow for metacognition	Much use of strategies that allow for deep levels of metacognition

Appendix 2: Tony's Working Group's CoRe Contribution

	Big ideas (Tony's Working Group)		
Pedagogical prompts	Charges produce electric fields which exert a force on other charges	Current is the flow of charge	Voltage is the difference in electric potential energy between two points
What do you intend students to learn about this idea	• Rubbing different materials together can separate charges • Like charges repel and opposite charges attract	• Current flows from positive to negative • Charge is conserved • Current is the same in all parts of a series circuit • Current divides in a parallel circuit • Current (I) is measured in Amperes (A) • Ammeters are used in series so that all of the current flows through them	• Energy is conserved • The supply voltage is divided over the components in a series circuit • Voltage is the same for each branch of a parallel circuit • Voltage (V) is measured in Volts (V) • Voltmeters are used in parallel to measure the difference between two points
Why is it important for students to know this?	• It explains everyday phenomena—e.g. shocks on trampolines or lighting • Basis for current electricity	• These are foundational concepts for understanding the behaviour of all electrical circuits	
What else you know about this idea (that you do not intend students to know yet)	• Electromagnetic induction	• Conventional current versus electron flow	• Volts = joules per Coulomb
Difficulties and/or limitations connected with teaching this idea	• Humid conditions can wreck electrostatic experiments	• You can't see it • Analogies can lead to misconceptions • Conventional current versus electron flow	• You can't see it • Everyday use of the word—'power'

(continued)

(continued)

	Big ideas (Tony's Working Group)		
Pedagogical prompts	Charges produce electric fields which exert a force on other charges	Current is the flow of charge	Voltage is the difference in electric potential energy between two points
Knowledge about students' thinking which influences your teaching of this idea	• Students usually have some prior experience of static electricity	• Common misconception of single charge units moving as opposed to a wire full of charges that are all moving	• Students get hung up on wire colours
Other factors that influence your teaching of this idea	• Weather	• Students need to be able to build circuits	• Voltage is difficult to model
Teaching procedures (and particular reasons for using these to engage with this idea)	• Rods and clothes to demonstrate static charging—picking up paper and electroscopes • Van der Graaf Generator • YouTube videos	• Definitions • Measuring current in series and parallel circuits and establishing rules • Discussion of why the rules work • Can use model of students as charges moving single path/multiple paths	• Definitions • Measuring voltage in series and parallel circuits and establishing rules • Discussion of why the rules work
Specific ways of ascertaining students' understanding or confusion around this idea	• Can explain applications—e.g. why a person's hair stands up when touching Van der Graaf	• Can measure current and voltage in circuits • Can calculate current and voltage in series and parallel circuits	

References

Alonzo, A. C., Kobarg, M., & Seidel, T. (2012). Pedagogical content knowledge as reflected in teacher-student interactions: Analysis of two video cases. *Journal of Research in Science Teaching, 49*(10), 1211–1239.

Bogdan, R. C., & Biklen, S. K. (2007). *Qualitative research for education: An introduction to theory and methods* (5th ed.). New York, NY: Pearson.

Bryman, A. (2016). *Social research methods* (5th ed.). New York, NY: Oxford University Press.

Cohen, L., Manion, L., & Morrison, K. (2011). *Research methods in education* (7th ed.). New York, NY: Routledge.

Daehler, K. R., Heller, J. I., & Wong, N. (2015). Supporting growth of pedagogical content knowledge in science. In A. Berry, P. Friedrichsen, & J. Loughran (Eds.), *Re-examining pedagogical content knowledge in science education* (pp. 45–59). New York, NY: Routledge.

Desimone, L. M. (2009). Improving impact studies of teachers' professional development: Toward better conceptualizations and measures. *Educational Researcher, 38*(3), 181–199.

Flick, U. (2014). *An introduction to qualitative research* (5th ed.). London, UK: Sage.

Gardner, A. L., & Gess-Newsome, J. (2011, April). *A PCK rubric to measure teachers' knowledge of inquiry-based instruction using three data sources*. Paper presented at the Annual Meeting of the National Association for Research in Science Teaching, Orlando, FL.

Gess-Newsome, J. (2015). A model of teaching professional knowledge and skill including PCK: Results of the thinking from the PCK summit. In A. Berry, P. Friedrichsen, & J. Loughran (Eds.), *Re-examining pedagogical content knowledge in science education* (pp. 28–42). New York, NY: Routledge.

Guba, E. G., & Lincoln, Y. S. (1989). *Fourth generation evaluation*. Newbury Park, CA: Sage.

Henze, I., & van Driel, J. H. (2015). Toward a more comprehensive way to capture PCK in its complexity. In A. Berry, P. Friedrichsen, & J. Loughran (Eds.), *Re-examining pedagogical content knowledge in science education* (pp. 120–134). New York, NY: Routledge.

Hume, A. (2010). CoRes as tools for promoting pedagogical content knowledge (PCK) of novice chemistry teachers. *Chemistry Education in New Zealand, 119*, 13–19.

Hume, A. (2015). Getting to the core of the matter: Developing a primary school's science plan. *New Zealand Science Teacher, 134*, 18–22.

Hume, A., & Berry, A. (2011). Constructing CoRes—A strategy for building PCK in pre-service science teacher education. *Research in Science Education, 41*(3), 341–355. https://doi.org/10.1007/s11165-010-9168-3.

Hume, A., & Berry, A. (2013). Enhancing the practicum experience for pre-service chemistry teachers through collaborative CoRe design with mentor teachers. *Research in Science Education, 43*(5), 2107–2136.

Hume, A., Eames, C., Williams, J., & Lockley, J. (2013). The benefits of collaborative content representation (CoRe) design. *SET: Information for Teachers, 2*, 35–43.

Kvale, S. (1996). *Interviews: An introduction to qualitative research interviewing*. Thousand Oaks, CA: Sage.

Lederman, N. G., Antink, A., & Bartos, S. (2014a). Nature of science, scientific inquiry, and socio-scientific issues arising from genetics: A pathway to developing a scientifically literate citizenry. *Science & Education, 23*(2), 285–302.

Lederman, J. S., Lederman, N. G., Bartos, S. A., Bartels, S. L., Meyer, A. A., & Schwartz, R. S. (2014b). Meaningful assessment of learners' understandings about scientific inquiry—The views about scientific inquiry (VASI) questionnaire. *Journal of Research in Science Teaching, 51*(1), 65–83.

Lee, E., Brown, M. N., Luft, J. A., & Roehrig, G. H. (2007). Assessing beginning secondary science teachers' PCK: Pilot year results. *School Science and Mathematics, 107*(2), 52–60.

Loughran, J., Berry, A., & Mulhall, P. (2006). *Understanding and developing science teachers' pedagogical content knowledge*. Rotterdam, The Netherlands: Sense Publishers.

Magnusson, S., Krajcik, J., & Borko, H. (1999). Nature, sources, and development of pedagogical content knowledge for science teaching. In J. Gess-Newsome & N. G. Lederman (Eds.), *Examining pedagogical content knowledge* (Vol. 6, pp. 95–132). Dordrecht, The Netherlands: Kluwer Academic.

Ministry of Education. (2007). *The New Zealand curriculum*. Wellington, New Zealand: Learning Media Ltd.

Nilsson, P., & Loughran, J. (2012). Exploring the development of pre-service science elementary teachers' pedagogical content knowledge. *Journal of Science Teacher Education, 23*(7), 699–721.

NZQA. (2010). *Achievement standard physics 1.3: Demonstrate understanding of aspects of electricity and magnetism*. AS90937 version 3. Retrieved August 22, 2016, from http://www.nzqa.govt.nz/nqfdocs/ncea-resource/achievements/2015/as90937.pdf.

Park, S., Jang, J.-Y., Chen, Y.-C., & Jung, J. (2011). Is pedagogical content knowledge (PCK) necessary for reformed science teaching? Evidence from an empirical study. *Research in Science Education, 41*(2), 245–260.

Park, S., & Oliver, J. S. (2008). Revisiting the conceptualisation of pedagogical content knowledge (PCK): PCK as a conceptual tool to understand teachers as professionals. *Research in Science Education, 38*(3), 261–284.

Patton, M. Q. (2014). *Qualitative research and evaluation methods* (4th ed.). Thousand Oaks, CA: Sage.

Treagust, D. F., Won, M., & Duit, R. (2014). Paradigms in science education research. In N. G. Lederman & S. K. Abell (Eds.), *Handbook of research on science education* (Vol. 2, pp. 3–17). New York, NY: Routledge.

Watts, M., & Ebbutt, D. (1987). More than the sum of the parts: Research methods in group interviewing. *British Educational Research Journal, 13*(1), 25–34.

Yin, R. K. (2014). *Case study research: Design and methods* (5th ed.). Thousand Oaks, CA: Sage.

Chapter 11
A Grand Rubric for Measuring Science Teachers' Pedagogical Content Knowledge

Kennedy Kam Ho Chan, Marissa Rollnick and Julie Gess-Newsome

Abstract Rubrics are increasingly used to differentiate the quality of science teachers' pedagogical content knowledge (PCK), both qualitatively and quantitatively. Well-designed PCK rubrics can guide the judgement of PCK quality for valid assessment. This chapter considers the possibility of a "grand rubric" that allows measurement of different variants of PCK as depicted in the Refined Consensus Model (RCM). To achieve this goal, the chapter first reviews the characteristics of rubrics in current use in the science education field. It examines the critical considerations in the construction of a grand rubric through an analysis of an expert discussion group. Based on this analysis, the paper proposes a grand rubric and describes its layout and characteristics. The grand rubric is generic in nature and can be customised for use with different science content topics as well as for measurement of specific variants of PCK in the RCM, including individual science teachers' personal or enacted PCK (pPCK and ePCK) and the collective PCK (cPCK) of a group of science teachers.

Introduction

Assessing teacher knowledge has been a subject of interest for decades (Gitomer & Zisk, 2015). Teacher knowledge is important as multiple strands of evidence support the notion that what a teacher knows impacts the quality of classroom instruction and hence student learning (e.g., Baumert et al., 2010; Kersting, Givvin, Thompson, Santagata, & Stigler, 2012). Pedagogical content knowledge, or PCK, is an important province of knowledge within the professional base of teachers that is most germane

K. K. H. Chan (✉)
University of Hong Kong, Hong Kong, Hong Kong
e-mail: kennedyckh@hku.hk

M. Rollnick
University of Witwatersrand, Johannesburg, South Africa
e-mail: Marissa.Rollnick@wits.ac.za

J. Gess-Newsome
Oregon State University-Cascades, Bend, USA
e-mail: Julie.Gess-Newsome@osucascades.edu

A. Hume et al. (eds.), *Repositioning Pedagogical Content Knowledge in Teachers' Knowledge for Teaching Science*, https://doi.org/10.1007/978-981-13-5898-2_11

to teaching (Shulman, 1986). PCK includes the knowledge and skills needed for a teacher to teach a "particular topic in a particular way for a particular purpose to particular students for enhanced student outcomes" (Gess-Newsome, 2015, p. 36). In order to validate the efficacy of efforts to enhance teacher knowledge in teacher preparation programmes and professional development activities, as well as to certify teachers, measures of teacher knowledge and skill (including teachers' PCK) are needed.

Since the first (1st) PCK Summit in 2012, there has been an upsurge of interest in PCK research and its measurement particularly in the science education field (e.g., Kirschner, Taylor, Rollnick, Borowski, & Mavhunga, 2015; Park & Suh, 2015; Smith & Banilower, 2015). Given that data sources related to PCK are often qualitative in nature (such as interviews, completion of surveys and written prompts, teaching artefacts, and classroom observations), the use of scoring rubrics (hereafter, rubrics) has become popular. Rubrics are descriptive scoring schemes comprising scoring categories with specific pre-established performance criteria (Mertler, 2001). Well-designed PCK rubrics provide operational definitions of the key dimensions of PCK for measurement by demarcating the scope and range of the construct. As such, well-designed PCK rubrics can guide the analysis of performance and support the judgement of PCK quality.

At the second (2nd) PCK Summit in the Netherlands in 2016, the majority of the participants were keen on the idea of a "grand rubric" for measuring science teachers' PCK, so a subgroup of participants formed a discussion group to discuss this possibility (hereafter, the rubric group discussion). The driving force behind the discussion was the premise that a grand rubric would be valid, ubiquitously accepted, and support clear and unambiguous communication across researchers. If sufficiently generic in nature, a grand rubric could be customised to various science content topics and would allow for comparison of PCK scores across topics, for triangulation across data sources, and provide evidence of growth in PCK pre- and post-intervention (i.e., determine individual teachers' PCK development). Such a rubric would make a significant contribution to the establishment of international standards for articulating PCK for a number of commonly taught science topics.

In this chapter, we raise key considerations in the construction of a grand rubric for measuring science teachers' PCK that can be used to determine all variants of PCK as depicted in the Refined Consensus Model (RCM), including individual teachers' personal or enacted PCK (pPCK and ePCK) as well as the collective PCK (cPCK) of a group of science teachers. To achieve this goal, we reviewed the characteristics of PCK rubrics in current use, identified through a systematic literature review. As the authors of reviewed works seldom make explicit their underlying rationales or considerations in the process of rubric construction, we also analysed a recording of the rubric group discussion held at the 2nd PCK Summit to uncover critical considerations needed to create a grand rubric for measuring PCK. Using this information, we propose a generic grand rubric, which can be customised for use with different content and grain sizes as well as for measurement of specific variants of PCK for science teaching.

Research Questions

The following research questions guide this chapter:

1. What are the characteristics of rubrics used to differentiate the quality of science teachers' PCK in the existing literature?
2. What are critical considerations in the construction of a grand rubric for measuring science teachers' PCK?
3. What would a grand rubric for measuring science teachers' PCK look like?

Methods

This chapter employed a systematic review of published and unpublished literature (i.e., literature in the public domain as well as the research summary outlines provided by the 1st and 2nd PCK Summit participants) and qualitative data collected at the 2nd PCK Summit about the use of rubrics in PCK research. Bennett, Lubben, and Hogarth (2007) note the strengths of such reviews are found in the characteristics of the review process, such as: the use of explicit criteria for the selection of studies for review; exhaustive coverage of the studies published; and the involvement of at least two researchers in decision-making.

For the first research question, the three authors of this chapter selected studies involving the use of PCK rubrics through a systematic literature search. In the first round of the literature search, eleven peer-reviewed journals primarily in science education and three journals in the field of teacher education were searched using the keywords "pedagogical content knowledge" and "rubric". The journals searched included: *African Journal of Research in Mathematics; Science and Technology Education; EURASIA Journal of Mathematics, Science and Technology Education; Chemistry Education Research and Practice; International Journal of Science Education; International Journal of Science and Mathematics Education; Journal of Research in Science Teaching; Journal of Science Teacher Education; Research in Science Education; Research in Science and Technological Education; School Science and Mathematics; Science Education; Teachers and Teaching: Theory and Practices;* and *Teaching and Teacher Education*. The lead author screened each of the articles for inclusion using the following selection criteria:

- The empirical studies were written in English and focused on science teachers' domain, topic, and/or concept-specific PCK.
- The article contained sufficient description and details about the rubric.
- The rubric was primarily used to differentiate the quality of science teachers' PCK.
- The rubric was adequately informed by the PCK literature.

The above selection criteria resulted in studies that *self-identified* the use of a rubric to differentiate the quality of teachers' PCK. Studies that made use of a rating manual, coding manual, or simply described the scoring procedures were not included in the analysis, as the scoring scheme was not a rubric. The selection criteria also excluded articles that used rubrics to differentiate the quality of teachers' discipline level PCK such as PCK for argumentation or PCK for inquiry practices since PCK is related to the teaching of particular subject matter of different grain size (see Chap. 2). In the second round of the literature search, we further enriched our article sources by:

(1) searching the ERIC database (https://eric.ed.gov/) using the same keywords,
(2) identifying rubrics in the research summary outlines of the two PCK Summits, and
(3) inviting PCK Summit members to suggest additional rubrics.

All articles generated by all these three search processes were examined for inclusion using the criteria detailed above. The final list comprised 37 sources, including 10 outline papers from the second PCK Summit and 27 journal articles and chapters. The sources are listed in *Appendix 1* (obtainable from https://www.researchgate.net/project/Grand-Rubric-for-PCK).

To analyse characteristics of current PCK rubrics in use (Research Question 1), we first decided on eight rubric characteristics.[1] The lead author then analysed all papers according to these characteristics. The second author peer validated a subset of 13 of these papers. For the purposes of this chapter, 5 of these 8 characteristics were considered relevant: (1) structure and purpose of the rubric, (2) the variants of PCK investigated, (3) the PCK model and components, (4) the quality indicators for PCK and (5) data sources. Broad agreement was reached between the two authors and the differences were resolved by discussion. Based on the above analysis, we further clustered the rubrics into distinct groups and described their characteristics.

To identify critical considerations of a grand rubric for measuring science teachers' PCK (Research Question 2), we transcribed verbatim and analysed an audio recording of the rubric discussion group held at the 2nd PCK Summit (1 h and 10 min). The rubric discussion group comprised eleven PCK researchers[2] from eight countries, referred to hereafter as PCK experts. In the discussion, these PCK experts discussed the possibility of constructing a grand rubric for measuring science teachers' PCK. The discussion transcript and voice file were sent to all participants to check for transcript correctness and validation. The transcript was analysed inductively to identify, categorise, and explore the main themes that emerged on the issues involved in creating a grand rubric using standard qualitative research techniques

[1]The eight characteristics are: (1) primary research focus of the articles; (2) PCK model and components/categories; (3) PCK variant(s) explored; (4) the data sources; (5) the rubric development process; (6) the structures and purpose of the rubric; (7) the quality indicators for PCK; and (8) the scoring process.

[2]The eleven summit members involved in the rubric group discussion were Alicia Alonzo, Julie Gess-Newsome (USA), Amanda (Mandi), Berry (Australia), Jared Carpendale (New Zealand), Kennedy Chan (Hong Kong), Sophie Kirschner, Sven Liepertz (Germany), Elizabeth Mavhunga, Marissa Rollnick (South Africa), Pernilla Nilsson (Sweden), and Christopher (Chris) Wilson (UK, based in the USA).

(Patton, 2002). The analysis revealed several critical considerations for constructing a grand rubric.

Lastly, based on our analysis of existing rubrics and the rubric discussion from the PCK Summit, we conceptualised a potential structure for a grand rubric for measuring science teachers' PCK (Research Question 3).

We employed investigator triangulation (Denzin, 1989) to ensure the trustworthiness of the data. The three authors arrived at consensuses concerning the main themes as they relate to the analysis of the literature review, the themes from the 2nd PCK Summit, the implications derived for the grand rubric, the final grand rubric template and the sample rubrics via face-to-face meetings, email exchanges, and Skype meetings.

Findings and Discussion

Our findings and discussion are organised according to the three research questions. First, we present a detailed analysis of the rubrics found in the literature and their characteristics. We then discuss the main themes that emerged from the analysis of the rubric discussion group. Finally, we propose a grand rubric for measuring science teachers' PCK and describe its characteristics.

Characteristics of PCK Rubrics in Use

From the systematic review, 37 journal articles, chapters, and extended outlines produced for the 2nd PCK Summit met the search criteria. Further analysis revealed that several of the papers, though dealing with different science topics and methods, used rubrics that shared the same characteristics. These documents were grouped together, resulting in 26 distinct rubrics. The full list of papers and their grouping can be found in *Appendix I* (see https://www.researchgate.net/project/Grand-Rubric-for-PCK). Based on an analysis of the 26 distinct rubrics, we offer a summary of the five rubric characteristics that are most relevant to guiding the creation of a grand rubric for measuring science teachers' PCK. Other rubric characteristics are mentioned as appropriate.

(1) *Structure and purpose of the rubric.* When sorting the PCK rubrics according to their structure and purpose, all aimed to differentiate the quality of science teachers' PCK. Of these, 20 of the rubrics measured science teachers' PCK. The remaining 6 served more qualitative intentions.

While existing rubrics vary in intent, clearly there are a significant number of researchers that believe that PCK can be effectively *measured* using rubrics. We concur and believe that the grand rubric for measuring science teachers' PCK should be designed in a manner to allow measurement of science teachers' PCK against normative standards defined by researchers, experts, and/or best practice and empirically determined. In other words, we contend that PCK exists in a continuum from weak to strong and can be measured using a rubric.

PCK rubrics have different structures. Quality indicators may form the rows of the rubric, or alternatively, PCK components may be used as the rows—some rubrics delineate sub-dimensions of PCK components as the rows of the rubric. Most rubrics are analytic rubrics (rubrics specifying more than one key dimension) with PCK components constituting the rows of the rubrics. The number of performance levels commonly ranges from two to seven with the most common number being four.

(2) *Variants of PCK.* Berry, Depaepe, and van Driel (2016) describe PCK as static or dynamic. To these writers, static PCK is a fixed form of teacher knowledge, in contrast to dynamic PCK that interacts with other knowledge types and may develop in situ. This classification is in line with the consensus definition of PCK from the 1st PCK Summit that delineates two variants of PCK representing the opposite ends of an enactment spectrum: (1) "teachers' knowledge of, reasoning behind, and planning" and (2) "the act of teaching" (Gess-Newsome, 2015, p. 36). The former is related to investigating what teachers know or think (i.e., static PCK) or knowing '*that*', *without* investigating what teachers actually *do* inside the classroom (i.e., dynamic PCK), while the latter refers to know-*how* (i.e., skills and techniques) and knowing-to-*act* in the moment (Mason & Spence, 1999) that is inherently linked to, and situated in, the *act of teaching within a particular classroom*. In relation to the RCM, static PCK corresponds to collective PCK (cPCK), personal PCK (pPCK) or enacted PCK in the planning and reflection phases (ePCKp, ePCKr) while dynamic PCK pertains to ePCK in the interactive phase of teaching (ePCKi) - see Chap. 12.[3]

Of the 20 rubrics measuring PCK, seven targeted dynamic PCK (ePCKi) while the rest measured static PCK. Of the six rubrics with qualitative intentions, three targeted static PCK. It can thus be concluded that most rubrics, whether quantitative or qualitative, were more often used for static PCK.

(3) *The use of a model and components.* Another characteristic with implications for rubric development is the choice of a model to guide the work that locates PCK in relation to other categories of teacher knowledge. All but four of the rubrics made a commitment to a particular model. The most popular was the Magnusson's PCK model (Magnusson, Krajcik, & Borko, 1999) or an adaptation of it. Three used Shulman's initial conceptualisation (Shulman, 1986), two used the Consensus Model (CM) from the 1st PCK Summit (Gess-Newsome, 2015), and two used the Mavhunga model (Mavhunga & Rollnick, 2013).

In those rubric developments using models, almost all the rubrics are organised around, "knowledge of students understanding of science" and "knowledge of instructional strategies." These are two of the original components used in Shulman's oft-quoted original conceptualisation of PCK in 1986.

---

[3]For the purpose of clarity, in the following, the terms static PCK and dynamic PCK will be used below.

Other components that were emphasised as part of these characteristics for rubric development include content knowledge, use of representations, orientations towards teaching science, and pedagogical reasoning. Considerations of content also need to take into account grain size (i.e., whether the rubric considers PCK at the domain, topic, or concept level). Most rubrics in this survey targeted the topic level, for example force and motion, photosynthesis or chemical equilibrium.

The rubrics not committing to any existing models make interesting reading. Three of the rubrics (Alonzo & Kim, 2016; Gess-Newsome et al., 2017; Lee, Brown, Luft, & Roehrig, 2007) provide thorough reviews of the literature. While these three studies do not commit themselves to a single PCK model, they eventually emerge with empirically derived components similar to those in the Magnusson model.

(4) *Quality indicators for PCK.* A thematic analysis of the quality indicators suggested for the rubrics shows an emphasis on attributes flowing from a constructivist view of teaching and learning. The most common themes referred to conceptual approaches, sense-making, and teaching for meaning (in almost all rubrics), followed by an emphasis on awareness of student thinking and ideas, student-centred approaches, and links between student ideas and teaching strategies. Criteria not related to the above themes relate to accuracy, completeness, or nature of the content (in at least nine rubrics). Some rubrics make reference to big ideas, which also link to a conceptual view of content. Another recurring theme was the quality of pedagogical reasoning and the degree of integration between PCK components, although these two areas were not explicitly identified as a single dimension/row in the rubric.

(5) *Data sources.* Science teachers' PCK knowledge was most often determined using a single type of data source (e.g., open-ended written test). Seven distinct rubrics used paper-and-pencil responses and all of these measured static PCK (e.g., Davidowitz & Potgieter, 2016; Jin, Shin, Johnson, Kim, & Anderson, 2015). The remaining rubrics measuring static PCK used data sources such as interviews and videos, and one used a CoRe[4] (Loughran, Mulhall, & Berry, 2004). In the case of the video analysis (e.g., Alonzo & Kim, 2016), respondents were typically asked to analyse the science teaching of a teacher on video, thus calling for the respondents' knowledge rather than action. The rubrics measuring dynamic PCK (i.e., ePCKi) all used either lesson videos or observations of a teacher's teaching acts in the classroom.

Critical Considerations in Constructing a Grand Rubric for Measuring Science Teachers' PCK

The findings presented above related to the analysis of PCK rubrics reviewed in the current literature. We now turn to data from the rubric discussion group at the

[4]CoRe stands for Content Representation—an array to portray PCK structured by big ideas related to a topic with responses to key pedagogical prompts.

2nd PCK Summit to identify critical considerations in constructing a grand rubric for *measuring* science teachers' PCK. The discussion provided a window into the experts' thinking and rationales underpinning the *process* of designing the rubrics, including the critical considerations needed to create a rubric. Included below are quotations from the rubric discussion group, slightly edited to increase their clarity. Five main themes about the critical considerations in the construction of rubrics emerged from the analysis of the rubric group discussion transcript: (1) the role of content knowledge, (2) the integration of PCK components, (3) the placement of pedagogical reasoning, (4) the role of an underlying learning theory, and (5) the core components that represent the essence of PCK. These are discussed below:

(1) *Should content knowledge (CK) be assessed in the PCK rubric?* Early in the discussion, the experts reached a consensus about the centrality of content in the PCK construct. Although the content is considered important within the PCK construct, it was not clear to the experts *how* CK should be measured when assessing teachers' PCK. For example, Chris suggested that CK should be measured in a *separate* test:

Chris: I think we can measure content knowledge in ways that are well established. If in this rubric we're measuring (PCK), then we don't need to measure content knowledge in that in the same measure.

Alicia agreed with Chris that CK does not need to be assessed in the PCK rubric and added that:

Alicia: Content knowledge is part of how people understand students' ideas. So if you're thinking about student understanding and you have a misconception yourself, then you're going to have a weak understanding of student misconceptions. ... I wouldn't want to separate that out and say that content knowledge is a separate component [in the rubric].

Although it seems that Alicia agreed with Chris' rejection of a separate row in the rubric for CK, her reasoning was somewhat different; she believed that a teacher with inadequate CK would naturally be unable to identify student misconceptions (i.e. that CK is part of PCK). Pernilla echoed this idea and added:

Pernilla: I totally agree with this [i.e., your ability to tease out misconceptions depends on an accurate understanding of the relevant content]. ... I don't know if we lose something if we focus too much on content and not with *how* content is integrated *with pedagogy* in the classroom, i.e. PCK.

Pernilla was of the view that the rubric should focus on, 'how content is integrated *with pedagogy*' rather than content alone. This debate was well summarised by Julie's comment below:

Julie: It seems like this is the debate. Is there some kind of measure of CK that is separate, and is it a prerequisite to looking at teachers' PCK that could also include accuracy, or is there a component of PCK, which includes CK?

To summarise, the experts identified CK as a key attribute of PCK, but whether and how the assessment of CK should be included in the grand rubric for measuring science teachers' PCK remained unresolved.

(2) *How to measure the integration between PCK components?* Another consideration became obvious when the experts discussed how the PCK components should be included in the rubrics.

Elizabeth: Do we want to look at them [the PCK components] individually or are we looking at their interactions? … Maybe the listing of components is important, but going back to the criteria, are we looking at the criteria from the perspective of amalgamation or interaction? I don't know. I just find it really difficult to consider each component [separately].

Mandi: I think that's just the thing. We don't want to lose the problem. Once you start to disentangle it [the PCK construct] and those things [i.e., PCK components] become valued alone compared to the re-integration of those things as something that's also done.

Above, the experts were highlighting the importance of the interconnection between PCK components, in line with the thinking that PCK components interact in a complex and dynamic way that are synergistically applied in practice (Abell, 2008; Magnusson et al., 1999; Park & Chen, 2012). However, it remains unclear how the rubric should be structured to take into account the quality of the integration between the PCK components. With respect to this issue, Chris, coming from a measurement perspective, had this to say:

Chris: If we were to start with the premise of interconnectedness, our rubric would look very different. Our rubric might be something that's more akin to the way we might measure networks or social systems or the connectedness of ideas.

Collectively, the above discussion raises the issue of how to design a rubric that can take into account the assessment of the integration between the PCK components. How can the rubric take into account the assessment of each PCK component on the one hand and the integration between the components on the other?

(3) *How to measure quality of teachers' pedagogical reasoning?* As the discussion ensued, another distinct attribute of PCK became apparent. This acknowledgement was represented by the following quotes:

Julie: One thing that I don't see here is the idea of pedagogical reasoning.

Sven: It seems like selection [of instructional strategy] only becomes meaningful if the person can argue why he selects a certain strategy. This is especially true if you're not looking at what is happening in the classroom but more about his knowledge and how he works with his knowledge.

It appears that the experts subscribed to the views of Shulman (1987), who argued that a teacher's knowledge base only becomes useful when it is tied to judgement and teaching actions. For Shulman, the translation of knowledge to action involves a complex process called pedagogical reasoning where teachers reason about their judgment and decisions. Although the PCK experts affirmed the role of pedagogical reasoning within the PCK construct, how to define high-quality pedagogical reasoning remained less clear. Pernilla's statement illustrates this concern:

Pernilla: What I struggle with is, how do we find good quality reasoning in terms of how those different components interact and connect? If we can see that all the components actually interact and they are interconnected, [that is important]. I mean, [if actions] are reasoned and reflected [upon], is that high quality PCK? … I think it's more than only interaction between components.

Another issue that emerged was about *how* the assessment of pedagogical reasoning can be represented in the rubric, as evident in the following exchanges:

Mandi: [Pedagogical reasoning is to] explain and reflect on the "why." And I think it might be that pedagogical reasoning is more than a dimension of PCK. I might even say that pedagogical reasoning could be something which exists in all the different [components].

Elizabeth: I'm wondering whether, in each of these components in the rubric, if you'd have a particular criterion that elicits the reason for what you see. Even [for the PCK component], the next step is to decide whether those are appropriate [strategies]. You now have to find a reason and judge against your own understanding of what you see as well. You need to make the judgment whether this is an appropriate next step.

In this last comment, Elizabeth suggested embedding the measurement of pedagogical reasoning within each row of the rubric. In response, Alicia was quick to point out the drawbacks associated with this way of constructing the rubric:

Alicia: I think the thing you miss by putting pedagogical reasoning only at the top level of all the other [components] is you miss variation in the quality of the reasoning. So if you're just saying its present or absent as opposed to there's depth and quality of reasoning that might vary. I think [where you put pedagogical reasoning in the rubric] is a statement about how we think about it, whether we put it as something separate or at the top of everything.

In sum, it appears that the experts acknowledge the importance of pedagogical reasoning as part of the PCK construct. Their argument for placement of pedagogical reasoning in the rubric is an indication of the value placed on the construct; however, there is less clarity amongst the experts on how to construct a rubric that can measure the quality of pedagogical reasoning.

(4) *What is the learning theory underpinning the rubric?* While the experts discussed how to populate the different performance levels of the rubric, another issue emerged. The issue, implicit in the early part of the discussion, was pointed out explicitly by Julie:

Julie: Comments keep coming up [asking], do we have a learning theory behind this? It seems to me that at least part of our learning theory is that "teaching is not telling." What are we doing to promote student [learning], whether it is creating disequilibrium or helping students make meaning?

There were some dissenting voices as to whether it is important to specify a particular learning theory in the rubric. Chris, for example, described what he thinks below:

Chris: I just worry about taking the value approach with something like constructivism. I'm worried about it, especially since we're such an international group. Different countries value different approaches. I wonder what this [discussion] would look like in the Japanese context, German context, or an American.

Embedded in Chris' view above is the issue that teaching is a cultural activity (Stigler & Hiebert, 2009), which may cast doubts on whether the same learning theory is equally valued in different contexts when assessment of PCK is concerned. Proposing a single learning theory also goes against the very nature of PCK which is context-specific (Park & Oliver, 2008). With respect to this issue, Alicia took another perspective.

Alicia: Fundamentally, I think there's a connection. For PCK there has to be a connection between what you're doing and the students. So even if there's not an explicit learning theory like constructivism, it would be difficult to describe high PCK in a manner of "I'm just going to stand up and talk and not care about who's in front of me," right?

The above discussion suggests that while a learning theory may help guide the delineation of quality in different levels of performance in the rubric, there has yet to be a consensus on whether it is really needed and, if so, which learning theory should be drawn upon in the construction of a rubric.

(5) *What are the core components of PCK?* The experts were aware that the CM that emerged from the 1st PCK Summit (Gess-Newsome, 2015) did not adequately unpack the composition of PCK. With this need to expand upon PCK composition, the experts' discussion also revolved around the core attributes that should be included in the rubric. Julie commented:

Julie: So my observation is, as people talk about PCK, they have stayed or gone back to the Magnusson model. But if you look at the research that's been done, almost nobody, as Kennedy points out, has done much with the assessment or curriculum [components]. Everybody included student understanding and instructional strategies. … As I think about what is essential to PCK, it seems that student understanding and instructional strategies are a large part of that.

As most of the existing research on PCK studies drew on Magnusson's PCK model, it appears that this model provided a good starting point for the experts to think about a consensual view on the core components of PCK for science teaching. As the discussion continued, the experts in the discussion group identified several more core components within PCK. The discussion group members finally reached an agreement on the following key PCK components:

1. *Selection and connection of big ideas*: The big ideas selected are relevant to the students and are pedagogically appropriate. There is coherence among big ideas.
2. *Selection of instructional strategies and representations*: The instructional strategies and representations selected are appropriate for the students and content. A student-centred learning approach that promotes meaningful learning is used.
3. *Recognition of variations in student understanding*: There are opportunities for students to reveal their thinking, a climate for students to expose their thinking, and activities that engage students´ interests and student misconceptions/prior knowledge are included in the teaching.
4. *Selection of next appropriate steps*: The teacher is adjusting instructional moves based on student learning of concepts. The teacher uses productive representations to advance student thinking.
5. *Pedagogical reasoning*: There is an interaction between the components above and the teacher possesses the ability to justify his/her teaching.

From a measurement perspective, a clear delineation of the exact composition of PCK is a prerequisite for valid measurement of PCK. As the RCM does not delineate the composition of PCK, the discussion demarcates the scope and range of the PCK, as a specialised integrated form of professional knowledge and skills, for measurement.

To summarise this section, the discussion of the PCK experts raised important considerations in the possible construction of a grand rubric for measuring science teachers' PCK. These include: what the critical PCK components should be; the placement of content knowledge in the grand rubric; the possible need of a learning theory in populating a rubric; as well as how to measure the interaction of PCK components and pedagogical reasoning.

The Grand Rubric for Measuring Science Teachers' PCK

Across conversations in the two PCK Summits, the review of the literature using PCK rubrics, and the transcripts from the rubric discussion group, there are a number of implications for the key characteristics of a grand rubric for measuring science teachers' PCK. Most importantly, there was a commitment to *measuring* PCK and a recognition that a rubric offered an effective means of doing so. Characteristics of the grand rubric are presented below along with justifications from the literature review and rubric discussion group for their inclusion.

Overall Characteristics of the Grand Rubric for Measuring Science Teachers' PCK

- Although rubrics appear to be a useful way to measure PCK, current science PCK research uses over 20 distinct rubrics for measuring teachers' PCK, making communication across researchers almost impossible. The construction of a grand rubric allows for more effective communication and aggregation of results across studies.
- The rubric needs to be flexible enough to measure the different variants of PCK in the RCM: the cPCK of a group of teachers (what a group of teachers know); the pPCK of a teacher (what a teacher knows); and ePCK (what a teacher does); and pedagogical reasoning (the reasons for his/her judgment and actions). Such flexibility allows more versatility in PCK research and education.
- To be universally adoptable and adaptable, the grand rubric must be sufficiently generic to allow its customisation for use with different content and grain sizes—discipline, topic, or concept levels. The final rubric would be customised to each study, though the basic structure remains the same in order to compare data across studies on science teachers.
- The rubric needs to be designed to be used with multiple data types, allowing triangulation of data from different sources. Prior research suggests that the quality of science teachers' PCK may be different when different data sources were used to determine the teachers' PCK (Gardner & Gess-Newsome, 2011).

Structure and Components of the Grand Rubric for Measuring Science Teachers' PCK with Rationales

- For our model, we are proposing five components generated from the 2nd PCK Summit. We believe that this structure, based on expert opinions, establishes the content validity (i.e. relevance and representativeness of PCK) of the rubric. These five components are not explicitly articulated in the RCM (see Chap. 2).
- The rubric is composed of five rows, each corresponding to one of the five components that resulted from the 2nd PCK Summit These components are named below, with insights for each into the types of evaluation criteria that might be used to measure the component. The arrows between the lowest level and highest level boxes in Fig. 11.1 indicate a need to determine the number of the column in the rubric and to establish quality indicators for each evaluation criteria identified.

1. Knowledge and Skills Related to Curricular Saliency: appropriate selection, connection, and coherence of big ideas; accuracy of content;
2. Knowledge and Skills Related to Conceptual Teaching Strategies: selecting and using appropriate instructional strategies; using multiple representations;

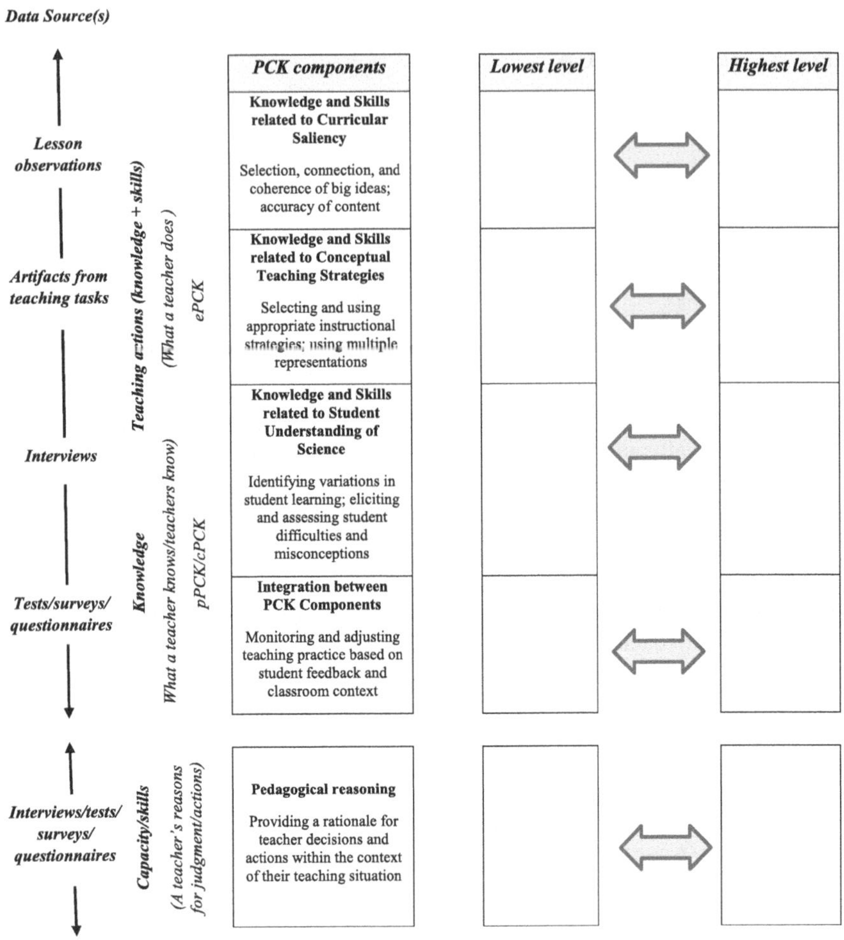

Fig. 11.1 Grand rubric template for measuring science teachers' PCK

3. Knowledge and Skills Related to Student Understanding of Science: identifying and acknowledging variations in student learning and eliciting and assessing student difficulties and misconceptions;
4. Integration Between PCK Components: monitoring and adjusting teaching practice based on student feedback and learning of the big ideas as well as the classroom context;
5. Pedagogical Reasoning: providing a rationale for teacher decision-making and actions within the context of their teaching situation.

- The first three components represent ideas that are consistent with the rubric discussion group and are similar to those found in the literature review (see Chap. 1).

- Integration between PCK components (row 4) is becoming a focus of investigation in recent PCK studies (see Chap. 1) and is considered important by the rubric discussion group. Such a measurement establishes that PCK is more than the sum of its parts.
- Pedagogical reasoning (row 5) acknowledges the importance of decision-making behind a science teacher's actions and is considered essential in the RCM (see Chap. 2).
- Since PCK represents a distinct category of knowledge distinct from CK in the RCM (see Chap. 2), we chose not to use a separate role for CK in the rubric to reflect this consensual view. However, it is quite clear that CK can influence the quality of the proposed PCK components.
- Quality indicators should be based on evidence from the research literature. For instance, when considering Knowledge and Skills related to Student Understanding of Science, teachers' ability to identify students' most common wrong answer is related to student learning is supported by research evidence (Sadler, Sonnert, Coyle, Cook-Smith, & Miller, 2013). Similarly, accuracy and hierarchical organisation of content knowledge is recognised as a feature of knowledge held by experts (Bransford, Brown, & Cocking, 1999).
- The rubric delineates a spectrum of performance levels (lowest level to highest level). Each categorical end of the spectrum represents the level of performance achievable by a subset of science teachers.
- While we did not gain consensus on the importance of an underlying learning theory, it was agreed that an underlying learning theory will influence the construction of quality indicators. We acknowledge that the indicators provided here are based on a broad constructivist framework. By citing research to support the selection of indicators, the theoretical features of the rubric will become more evident and should be acknowledged explicitly.

Data Sources

- The first column highlights the types of data sources that lend themselves to measurement.
- The first four components may be used to measure science teacher knowledge and/or actions. Teacher knowledge can be measured through the teachers' articulation of their pedagogical decisions in their planning (interviews, tests/surveys/questionnaires/lesson plans), or through examination of reflections. Alternatively, teacher knowledge may be *inferred* from their teaching actions (lesson observations) or teaching artefacts. Science teachers' PCK may be manifested in teaching actions through the use of pedagogical moves. The interaction between PCK components can also be measured directly or indirectly from teacher's statements, actions, or artefacts.
- The last row (i.e., pedagogical reasoning) relates to the science teachers' capacity to provide rationales for justifying their teaching actions. Pedagogical reasoning

cannot be accessed using *only* observation data. Stimulated recall interviews may be conducted with the teacher to access the teacher's pedagogical reasoning. In addition, a teacher may have a very good understanding of what they should do and why (i.e., knowledge and reasoning), but limited ability to implement that knowledge/skill in the classroom due to poor pedagogical skills (i.e., classroom management), contextual considerations (i.e., mandated curriculum), or motivational factors. The inclusion of teachers' pedagogical reasoning takes into account the teacher's sensitivity and responsiveness to the context.

How to Use the Grand Rubric Template

Generic guidelines for creating a rubric already exist. For example, the construction of a rubric often involves an iterative process comprising one or more of the following steps: articulating observable attributes; identifying characteristics for each attribute; identifying performance levels and corresponding criteria; and, revising the rubrics based on the empirical data (Mertler, 2001). We hope that by using the grand rubric template (Fig. 11.1) described in this chapter, science education researchers and practitioners can create a PCK rubric for the specific context in which it is needed.

The grand rubric can act as a generic template since it is designed to be customised to each setting. The science content topic under consideration and its grain size will need to be explicitly noted, as well as the age group of the students. To create or use a scoring guide, the researchers themselves will need to have strong PCK on the science topic and use evidence from empirical and canonical research and best practice. The terms that describe the level of performance will need to be proposed by the developer to better articulate the level of performance (e.g., limited, basic, proficient, and exemplary) and empirically defend via the data. Concrete descriptions/descriptors of each performance level, as well as detailed exemplars, would need to be included in the scoring guide, making them available to other researchers drawing on the research. A scoring guide will need to identify issues, such as the appropriateness of various instructional strategies or representations, the range of potential student misconceptions and those that are most common, and the evidence that will be used to judge the soundness of a rationale for specific actions. Published scoring guides will assist in articulating PCK for a given science topic and allow for their use across multiple settings. Scorer training would need to include the application of the scoring guide to actual data and inter-rater agreement, which might also result in refinement of the scoring guide.

Data collection tools should be carefully designed to elicit the ideas included in the rubric. For instance, purposeful questions about the selection of big ideas or specific misconceptions of concern might need to be asked directly, rather than assuming that such topics will arise spontaneously in a data source.

Finally, considerations for the validation and use of the rubric will need careful attention. What scoring strategy will be used? What is the meaning of a score related to a single component row? Is there an overall PCK score associated with the rubric?

How are component scores derived when there are different numbers of evaluation criteria? Does the rubric discriminate between individuals that we judge to have high and low PCK? Will a factor analysis provide evidence for the proposed PCK components?

Conclusion

This chapter explored considerations in the construction of a grand rubric for measuring science teachers' PCK. We examined the existing literature where PCK rubrics were developed and used, as well as critical considerations surfacing at the second PCK Summit. The proposed characteristics of the grand rubric for measuring science teachers' PCK provide it with several advantages: it can be used with multiple data sets and with different content and grain sizes, it is built on the RCM, and it draws on best practices found in the PCK literature. This grand rubric contributes to the field as it is an important tool for the measurement of science teachers' PCK. We hope that this chapter provides science education researchers and/or practitioners with guidance in the important work of creating purpose-built rubrics and associated data collection tools and scoring guides by customising the grand rubric template for use in their own contexts.

References

Abell, S. K. (2008). Twenty years later: Does pedagogical content knowledge remain a useful idea? *International Journal of Science Education, 30*(10), 1405–1416.

Alonzo, A. C., & Kim, J. (2016). Declarative and dynamic pedagogical content knowledge as elicited through two video-based interview methods. *Journal of Research in Science Teaching, 53*(8), 1259–1286.

Baumert, J., Kunter, M., Blum, W., Brunner, M., Voss, T., Jordan, A., … Tsai, Y.-M. (2010). Teachers' mathematical knowledge, cognitive activation in the classroom, and student progress. *American Educational Research Journal, 47*(1), 133–180.

Bennett, J., Lubben, F., & Hogarth, S. (2007). Bringing science to life: A synthesis of the research evidence on the effects of context-based and STS approaches to science teaching. *Science Education, 91*, 347–370.

Berry, A., Depaepe, F., & van Driel, J. H. (2016). Pedagogical content knowledge in teacher education. In J. Loughran & M. L. Hamilton (Eds.), *International handbook of teacher education* (Vol. 1, pp. 347–386). Singapore: Springer Singapore.

Bransford, J. D., Brown, A. L., & Cocking, R. R. (1999). *How people learn: Brain, mind, experience, and school*. Washington: National Academy Press.

Davidowitz, B., & Potgieter, M. (2016). Use of the Rasch measurement model to explore the relationship between content knowledge and topic-specific pedagogical content knowledge for organic chemistry. *International Journal of Science Education, 38*(9), 1483–1503.

Denzin, N. K. (1989). *The research act: A theoretical introduction to sociological methods* (3rd ed.). Englewood Cliffs, NJ: Prentice Hall.

Gardner, A. L., & Gess-Newsome, J. (2011). *A PCK rubric to measure teachers' knowledge of inquiry-based instruction using three data sources*. Paper presented at the annual meeting of the National Association for Research in Science Teaching (NARST), Orlando, Florida.

Gess-Newsome, J. (2015). A model of teacher professional knowledge and skill including PCK: Results of the thinking from the PCK Summit. In A. Berry, P. J. Friedrichsen, & J. Loughran (Eds.), *Re-examining pedagogical content knowledge in science education* (pp. 28–42). New York: Routledge.

Gess-Newsome, J., Taylor, J. A., Carlson, J., Gardner, A. L., Wilson, C. D., & Stuhlsatz, M. A. M. (2017). Teacher pedagogical content knowledge, practice, and student achievement. *International Journal of Science Education*, 1–20.

Gitomer, D. H., & Zisk, R. C. (2015). Knowing what teachers know. *Review of Research in Education, 39*(1), 1–53.

Jin, H., Shin, H., Johnson, M. E., Kim, J., & Anderson, C. W. (2015). Developing learning progression-based teacher knowledge measures. *Journal of Research in Science Teaching, 52*(9), 1269–1295.

Kersting, N. B., Givvin, K. B., Thompson, B. J., Santagata, R., & Stigler, J. W. (2012). Measuring usable knowledge: Teachers' analyses of mathematics classroom videos predict teaching quality and student learning. *American Educational Research Journal, 49*(3), 568–589.

Kirschner, S., Taylor, J. A., Rollnick, M., Borowski, A., & Mavhunga, E. (2015). Gathering evidence for the validity of PCK measures: Connecting ideas to analytic approaches. In A. Berry, P. J. Friedrichsen, & J. Loughran (Eds.), *Re-examining pedagogical content knowledge in science education* (pp. 229–241). New York: Routledge.

Lee, E., Brown, M. N., Luft, J. A., & Roehrig, G. H. (2007). Assessing beginning secondary science teachers' PCK: Pilot year results. *School Science and Mathematics, 107*(2), 52–60.

Loughran, J., Mulhall, P., & Berry, A. (2004). In search of pedagogical content knowledge in science: Developing ways of articulating and documenting professional practice. *Journal of Research in Science Teaching, 41*(4), 370–391.

Magnusson, S., Krajcik, J., & Borko, H. (1999). Nature, sources, and development of pedagogical content knowledge for science teaching. In J. Gess-Newsome & N. G. Lederman (Eds.), *Examining pedagogical content knowledge: The construct and its implications for science education* (pp. 95–132). Dordrecht: Kluwer Academic.

Mason, J., & Spence, M. (1999). Beyond mere knowledge of mathematics: The importance of knowing-to act in the moment. *Educational Studies in Mathematics, 38*(1–3), 135–161.

Mavhunga, E., & Rollnick, M. (2013). Improving PCK of chemical equilibrium in pre-service teachers. *African Journal of Research in Mathematics, Science and Technology Education, 17*(1–2), 113–125.

Mertler, C. A. (2001). Designing scoring rubrics for your classroom. *Practical Assessment, Research & Evaluation, 1–7*.

Park, S., & Chen, Y.-C. (2012). Mapping out the integration of the components of pedagogical content knowledge (PCK): Examples from high school biology classrooms. *Journal of Research in Science Teaching, 49*(7), 922–941.

Park, S., & Oliver, J. S. (2008). Revisiting the conceptualisation of pedagogical content knowledge (PCK): PCK as a conceptual tool to understand teachers as professionals. *Research in Science Education, 38*(3), 261–284.

Park, S., & Suh, J. (2015). From portraying toward assessing PCK. In A. Berry, P. J. Friedrichsen, & J. Loughran (Eds.), *Re-examining pedagogical content knowledge in science education* (pp. 104–119). New York: Routledge.

Patton, M. Q. (2002). *Qualitative research & evaluation methods* (3rd ed.). Thousand Oaks, California: Sage Publications.

Sadler, P. M., Sonnert, G., Coyle, H. P., Cook-Smith, N., & Miller, J. L. (2013). The Influence of teachers' knowledge on student learning in middle school physical science classrooms. *American Educational Research Journal, 50*(5), 1020–1049.

Shulman, L. S. (1986). Those who understand: Knowledge growth in teaching. *Educational Researcher, 15*(2), 4–14.
Shulman, L. S. (1987). Knowledge and teaching: Foundations of the new reform. *Harvard Educational Review, 57*(1), 1–22.
Smith, S. P., & Banilower, E. R. (2015). Assessing PCK: A new application of the uncertainty principle. In A. Berry, P. J. Friedrichsen, & J. Loughran (Eds.), *Re-examining pedagogical content knowledge in science education* (pp. 104–119). New York: Routledge.
Stigler, J. W., & Hiebert, J. (2009). *The teaching gap: Best ideas from the world's teachers for improving education in the classroom*. New York: Free Press.

Chapter 12
Unpacking the Complexity of Science Teachers' PCK in Action: Enacted and Personal PCK

Alicia C. Alonzo, Amanda Berry, and Pernilla Nilsson

Abstract This chapter focuses on enacted PCK (ePCK), i.e. the specific knowledge and skills that science teachers use in their practice, as it plays out in specific classroom contexts while teaching particular content to their students. In unpacking this aspect of the Refined Consensus Model (RCM) of PCK, we consider both the nature of ePCK and its interactions with other realms of PCK, primarily personal PCK (pPCK). Recognising the complexity of classroom practice—in terms of both the uniqueness of each classroom situation and the necessarily spontaneous nature of classroom interactions—we propose a mechanism through which pPCK is transformed into ePCK, and vice versa, throughout the plan-teach-reflect cycle. We then illustrate these ideas using several empirical examples of efforts to capture and analyse science teachers' ePCK (and associated pPCK). We conclude with discussion of some of the opportunities, challenges and implications of using the RCM, along with our unpacking of ePCK and its relationship to pPCK, as a means of understanding the knowledge that science teachers utilise in the midst of planning, teaching and reflecting.

The original version of this chapter was revised: Reference citation and list were updated. The correction to this chapter is available at https://doi.org/10.1007/978-981-13-5898-2_15

A. C. Alonzo (✉)
Michigan State University, East Lansing, USA
e-mail: alonzo@msu.edu

A. Berry
Monash University, Melbourne, Australia
e-mail: amanda.berry@monash.edu

P. Nilsson
Halmstad University, Halmstad, Sweden
e-mail: Pernilla.Nilsson@hh.se

A. Hume et al. (eds.), *Repositioning Pedagogical Content Knowledge in Teachers' Knowledge for Teaching Science*, https://doi.org/10.1007/978-981-13-5898-2_12

Introduction

The Refined Consensus Model (RCM) of PCK (Carlson, Daehler et al., 2019) builds on a model of teacher professional knowledge and skill developed from the First (1st) PCK Summit (Gess-Newsome, 2015). As compared to the earlier model and in the context of science education, the RCM has a stronger emphasis on making explicit the different variables, layers and complexities associated with PCK and highlighting in a clearer way the relationship between PCK and teaching practice. The RCM identifies three distinct "realms" of PCK: collective PCK (cPCK), representing the specialised professional knowledge held by multiple science educators in a field; personal PCK (pPCK), representing the personalised professional knowledge and skills held by an individual science teacher; and enacted PCK (ePCK), the unique subset of knowledge and skills that a science teacher draws on and that play out while planning, teaching and reflecting on a lesson. Within the model, these realms are represented as concentric rings, with cPCK in the outer ring, pPCK in the middle ring and ePCK in the centre (see Chap. 2, Fig. 2.3). The design of the model is intended to emphasise the practitioner perspective through the central placement of ePCK.

To date, research on science teachers' PCK has mostly focused on cPCK, e.g. assessing whether teachers know "canonical" PCK, and pPCK, e.g. getting teachers to articulate what they know about teaching a particular science topic in a particular context. However, there has been relatively little research focused on ePCK, i.e. how PCK is utilised in teachers' actual practice. Therefore, in this chapter, we focus on ePCK,

> the specific knowledge and skills utilised by an individual science teacher in a particular setting, with a particular student or group of students, with a goal for those students to learn a particular concept, collection of concepts, or a particular aspect of the discipline. (see Chap. 2)

We unpack this aspect of the RCM, providing our interpretation of ePCK in order to focus attention on the knowledge that science teachers make use of in action. Consistent with the interpretations in Chap. 2, we note that ePCK plays out not only when enacting instruction (i.e. when interacting directly with students), but also when planning for and reflecting on instruction. Thus, we consider ePCK to exist in three forms: $ePCK_P$ (for planning), $ePCK_T$ (for teaching) and $ePCK_R$ (for reflecting). Below we argue that because ePCK focuses on specific and, thus, unique classroom situations, it must involve more than static, declarative knowledge or scripts and procedures. Further, we explore how ePCK, as constantly evolving in response to these unique classroom situations, not only relies upon but also drives modifications to science teachers' pPCK.

Thus, in the sections below, we start with a brief overview of pPCK. We then unpack our interpretation of ePCK as a form of knowledge in action. Next, we explain how we view ePCK and pPCK as mutually influential, proposing a mechanism through which these two realms of knowledge interact and evolve through the plan-teach-reflect cycle, as pPCK is transformed into ePCK, and vice versa. In order to illustrate these ideas, we then present several examples of efforts to capture and analyse science teachers' ePCK and pPCK. Finally, we discuss some of the opportunities, challenges and implications of using the RCM and, in particular, our unpacking of ePCK and its relationship to pPCK, as a means of understanding the knowledge that science teachers use while planning, enacting and reflecting on instruction.

The Nature of pPCK

Personal PCK (pPCK) refers to the knowledge resources that an individual science teacher brings to the classroom enabling her/him to think and perform as a teacher in order to promote student learning about specific science subject matter. In understanding pPCK as a form of *personal knowledge*, we draw on Eraut (2000) who defines personal knowledge as

> the personal, available for use, version of a public concept or idea…[that] incorporates codified knowledge in its personalised form, together with procedural knowledge and process knowledge, experiential knowledge and impressions in episodic memory. Skills are part of this knowledge, thus allowing representations of competence, capability or expertise in which the use of the skills and propositional knowledge are closely integrated. (p.114)

Hence, pPCK is a specialised form of personal knowledge that includes different knowledge resources related to the teaching and learning of specific science topics. Consistent with Eraut (2000), who considers skills to be part of knowledge, in this chapter we refer to knowledge and skills collectively as knowledge. pPCK includes both explicit (i.e., articulable) knowledge and tacit knowledge (e.g., experiential knowledge, impressions in episodic memory) and is therefore unique for each science teacher. pPCK differs from cPCK in that cPCK represents publicly held (i.e., shared) codified knowledge.

The Nature of ePCK

Consistent with its connection to practice in the RCM, we consider ePCK to be "tacit knowledge in action" (Eraut, 2000, p. 123), i.e., knowledge that science teachers draw on in the moment of action, where the action may include planning, teaching or reflecting on teaching. This interpretation has two important implications. First, ePCK exists only in action (i.e., as tacit, unarticulated knowledge). Second, ePCK is flexible and generated in the moment of action. Since action occurs in the moment, the underlying ePCK is also adaptive, created and used in that moment. Thus, we contrast ePCK with pPCK and cPCK, which are more declarative and relatively more stable (or static) forms of knowledge.[1,2]

Science teaching is responsive to students and context, so each classroom situation is (at least) slightly different from others that a teacher has experienced (or knows about). Thus, the ePCK utilised in each classroom situation is unique, and it is unlikely (and even impossible) for a science teacher to already possess the exact ePCK required to plan, enact and reflect on instruction about a particular topic for a particular group of students in a particular setting. Thus, ePCK must be constructed anew for each teaching episode. Of course, ePCK for a given classroom situation might be almost identical to that for another similar situation, but differences in terms of the context and/or students will necessitate (even very small) tweaks, resulting in unique ePCK for that setting. Therefore, new ePCK is constantly being generated for each science teacher during every act of planning, enacting and reflecting on instruction.

Therefore, we view ePCK as the knowledge in action generated during, and made visible in, science teachers' planning ($ePCK_P$), enactment ($ePCK_T$) and reflection ($ePCK_R$) on instruction in a particular classroom situation. As such, ePCK is the unarticulated knowledge that underlies action in each of these activities. $ePCK_T$ is perhaps easiest to imagine, as the knowledge that underlies science teachers' in-the-moment instructional decisions. Teachers respond to students—e.g., with feedback, with explanations or demonstrations, and questions—in the midst of science instruction, without articulating (even to themselves) the reasoning behind those decisions. Similarly, when planning, teachers may propose particular instructional activities, with the intuitive sense that they will be appropriate for a given upcoming classroom situation ($ePCK_P$). Reflections may start with a teacher's sense that a given activity did not "go well" or that a particular student was confused about part of the lesson ($ePCK_R$). Such reflections, tied to specific instances and/or specific students, do not already exist as part of a teacher's ePCK—and a teacher may not have associated declarative knowledge to express the basis for his/her concerns. As discussed in the section below, these intuitive actions (planning, teaching and reflecting) are all influenced by a science teacher's pPCK; however, in the moment, they exist as ePCK.

[1]This is not to say that pPCK and cPCK do not evolve over time (indeed, as detailed below, we argue that pPCK changes through the construction of ePCK). However, both pPCK and cPCK are static in the sense that it is (theoretically) possible to articulate this knowledge and, thus, to measure it, whereas ePCK is inarticulable and fleeting, existing only in the moment (before potentially being transformed into pPCK). In other words, we fully expect that all three realms of teachers' PCK will change over time, but that change in ePCK will occur at a much shorter timescale.

[2]In this contrast, i.e., a focus on knowledge that is not declarative and not static, we connect with literature that refers to "dynamic PCK" (e.g. Alonzo & Kim, 2016; Schmelzing et al., 2013) as opposed to "declarative PCK".

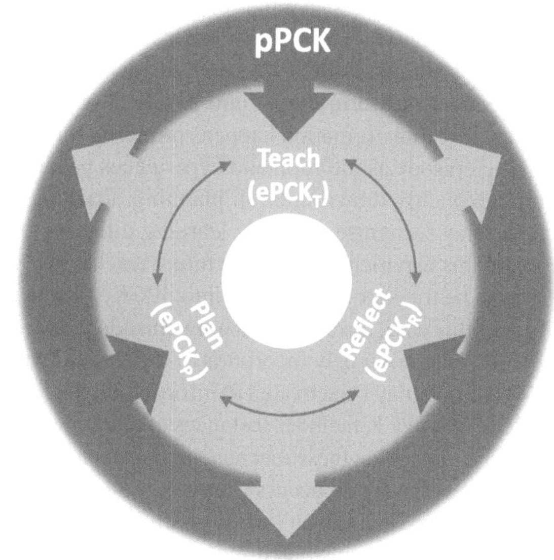

Fig. 12.1 Relationships among ePCK stages and between ePCK and pPCK

Relationships Between ePCK and pPCK

In this section, we describe how—through the constant generation of ePCK and the interaction between ePCK and pPCK—teaching experience can lead to changes in both science teachers' ePCK and pPCK. We depict this process in Fig. 12.1, which is an expansion of the ePCK and pPCK parts of the RCM (see Chap. 2), depicting in more detail both the different forms of ePCK and the specific points at which pPCK influences ePCK and vice versa. To illustrate the fuzziness that we see between ePCK and pPCK (particularly in their tacit forms), we have blurred the line representing the interface between ePCK and pPCK.[3] As shown in Fig. 12.1, in the RCM, double-sided arrows on the interface between ePCK and pPCK indicate a bidirectional flow between these two realms of PCK, representing how pPCK influences ePCK and vice versa.

First, pPCK provides the basis for ePCK at each step of the plan-teach-reflect cycle. In other words, ePCK is generated in the moment, but not out of thin air. All of a science teacher's knowledge, from past teaching and learning experiences, including classroom situations that are similar to the current one, serve as resources. The three dark blue arrows pointed inwards in Fig. 12.1 represent this sourcing of extant knowledge. Second, ePCK is transformed into pPCK, i.e., part of the store of knowledge available for future planning, teaching and reflecting. Consistent with the composition of pPCK as including both explicit and tacit knowledge, ePCK may be

[3] Although not discussed here, we expect that similar ambiguities exist at the pPCK–cPCK interface; thus, the outside of the pPCK ring (i.e. the boundary between pPCK and cPCK) is likewise blurred in Fig. 12.1.

transformed into pPCK in either of these forms. The three light blue arrows pointed outwards in Fig. 12.1, following each stage of the plan-teach-reflect cycle, represent the transformation of ePCK into both explicit and tacit forms of pPCK. A conscious process may transform ePCK into pPCK in a form that can be articulated by the teacher. This transformation happens primarily through reflection in, or on, a science teaching episode as intuition and experiences become part of future knowledge that can be explicitly drawn upon in planning, teaching and reflection. For example, a teacher may recognise a student learning difficulty during class and later explicitly draw on this experience to inform future teaching. In a subconscious process, ePCK may also be transformed directly into pPCK without the teacher's conscious awareness.[4] In this case, a science teaching episode (e.g., recognising a student difficulty) becomes subconsciously incorporated into memory that forms part of a tacit knowledge base that may be activated to inform future action (tacit pPCK). Transformation of ePCK into pPCK includes instances of planning and reflecting as well as teaching.

Before unpacking these mechanisms for each stage of the plan-teach-reflect cycle, we note that this cycle occurs on two timescales: a "macro" one focused on a unit of instruction (e.g. a lesson) and a "micro" one focused in-the-moment during a unit of instruction (i.e. many such moments in a lesson). At the lesson level, a teacher plans the lesson, teaches the lesson and then reflects on learning and instruction during the lesson. The teaching of the lesson includes all of the instructional moves that the science teacher makes (whether planned or unplanned). When reflections at the "macro" level are made explicit, ePCK is transformed into pPCK as articulable knowledge.

As illustrated in Fig. 12.2, we can also "zoom in" to investigate how the teaching of the science lesson (as a series of instructional moves) arises. At this level, we see a reflect-plan-teach cycle associated with each instructional move in the "macro" cycle. Here, instruction ("teach" in the macrocycle) comprises a series of instructional moves ("teach" in the microcycle). In contrast, the planning and reflection that occur as part of the microcycle happen during "teach" in the macrocycle (i.e. distinct from the planning and reflection that occur before and after a lesson, respectively). In a microcycle, a particular instance (e.g. an interaction with a student) prompts reflection (i.e., noticing and identifying the significance of a student's question or contribution to a class discussion), a plan for how to respond and the instruction (i.e., the response, such as a follow-up question to the student or a revision to the instructional plan). As this entire cycle takes place in one instance, in the moment between the student's contribution and the teacher's response, the ePCK generated is likely to remain tacit and, thus, unless included in reflection as part of the "macro" cycle, more likely to be transformed into pPCK in tacit form.

As described above, since each student and each classroom context is a little bit different, most teaching situations will present science teachers with some similarity

---

[4] While repeated encounters with similar situations may eventually lead to tacit knowledge becoming explicit, the opposite may also be true, i.e. explicit knowledge may become tacit, for instance, through the routinisation of certain instructional moves over time, as is the case with highly expert teachers. Thus, ePCK that is transformed into pPCK in tacit form may eventually become explicit pPCK, and ePCK that is transformed into pPCK in explicit form may eventually become tacit pPCK.

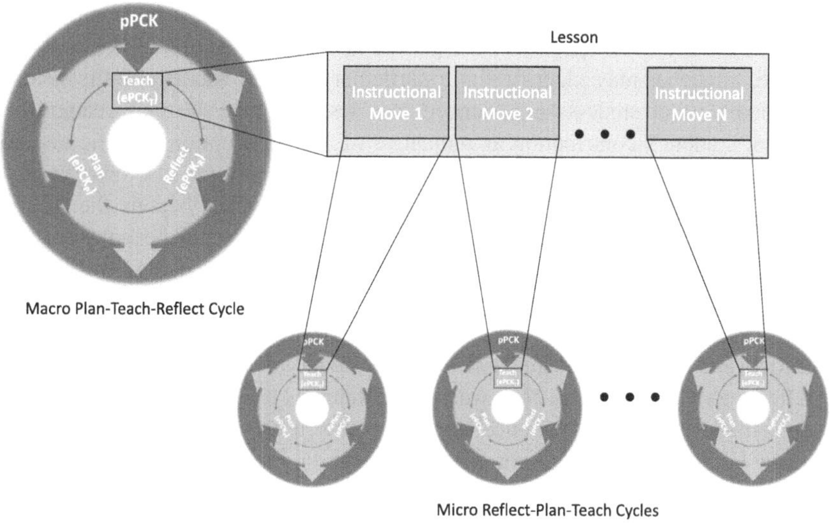

Fig. 12.2 Macro- and microplan-teach-reflect cycles

to past teaching situations and/or teachers' prior knowledge, but also some uniqueness—such that existing pPCK is relevant and useful, but ePCK must be generated for a particular situation. Thus, when planning instruction, science teachers draw on their existing pPCK, using knowledge of common ways students interact with the content and instructional strategies that can be used to address that content in order to identify a particular set and sequence of learning activities. As teachers tailor instruction to a particular classroom context and group of students, they may propose learning sequences and/or instructional moves without explicitly articulating the underlying reasoning (e.g., knowledge of common student learning difficulties, knowledge of the conditions under which a particular instructional strategy is most beneficial)—or even being aware of it themselves. Through this process, science teachers generate their $ePCK_P$.

Science teaching is complex and uncertain, requiring continuous in-the-moment responses to students' learning needs and features of the classroom context. While teachers' pPCK may include a range of instructional strategies associated with particular classroom conditions, teachers are unlikely to find themselves in those precise conditions in any given teaching situation. Therefore, to support student learning, they must generate responses appropriate for the moment. Through this process, science teachers generate their $ePCK_T$.

During and after instruction, teachers may reflect on their planned instruction ($ePCK_P$), their in-the-moment adaptations ($ePCK_T$) and/or the foundational knowledge (pPCK) underlying both. When reflecting on the outcome of enacting a strategy in the unique situation of a particular set of interacting factors in a particular classroom context, science teachers generate $ePCK_R$. While drawing on pPCK (e.g., knowledge of common student difficulties or common student expressions of content understanding) that is applicable across classroom situations, teachers engage in in-the-moment reflections specific to the particular incident under consideration. For example, a teacher may identify that a particular moment was key to the success (or difficulty) that students experienced in a lesson, or he/she may recognise a particular student's contribution as indicative of a preconception that she had not encountered before. When this $ePCK_R$ is articulated and/or stored as knowledge that, while contextualised in the teacher's classroom, exists for use beyond the specific students and classroom conditions under which it was generated, it becomes part of a science teachers' pPCK. In this way, insights gained from the specific situation may contribute to new knowledge that can be applied in other situations.

Thus, whether coming up with instructional strategies appropriate for a given classroom situation ($ePCK_P$), recognising new evidence of student thinking ($ePCK_T$), or reflecting on the outcome of instructional strategies or a response to evidence of student thinking ($ePCK_R$), science teachers build on existing pPCK and generate new ePCK. When articulated, the new ePCK can be incorporated into a teacher's pPCK. In this way, the interplay between these different realms of PCK operates in both directions: ePCK informs and is informed by pPCK.

To illustrate how these different forms of ePCK play out in a science teaching episode, consider the following example. Recalling how her students have struggled to understand natural selection (pPCK), a biology teacher designs an activity to address common learning difficulties ($ePCK_P$). While teaching the lesson, a student expresses an understanding of natural selection that the teacher was not expecting. On the spot, she decides to use Darwin's Galapagos finches to respond to the student ($ePCK_T$). After school, the teacher thinks about how the student may have come up with his idea ($ePCK_R$). She remembers the student idea and her explanation so that she can anticipate this response when she teaches natural selection again (pPCK). Considering just the teacher's instructional response in the lesson, we can zoom in further to see how ePCK plays out at the level of the microcycle described above. While some evidence of student thinking may be presented in ways that match perfectly with teachers' prior knowledge (i.e., pPCK), most classroom situations require teachers to recognise/notice something they have never encountered before—whether a particular student's way of expressing a known pattern of student thinking or evidence of truly novel student thinking. Thus, when the student expresses her understanding of natural selection, the teacher must immediately make sense of the student idea (i.e., what it indicates about student understanding, what the student does and does not understand; generating $ePCK_R$). Still acting in the moment, the teacher must then make a decision about how to respond, (i.e., plan an instructional move; generating $ePCK_P$) and enact the planned response (generating $ePCK_T$). In these in-the-moment instances, ePCK is likely to be transformed into pPCK only tacitly, but this decision is also available for reflection in the macrocycle and, thus, could contribute to the development of more explicit pPCK.

Illuminating the Complexity of Science Teachers' PCK in Action: Empirical Examples

In the sections above, we laid out a conceptualisation of ePCK and its relationship to pPCK in order to unpack how these realms of PCK are brought to bear in the moment of planning, teaching and reflecting. In this section, we provide examples from empirical work on PCK that help to both illustrate our conceptualisation and illuminate the complexity of the knowledge in action that we seek to understand by articulating ePCK and pPCK. We start with an example of the processes by which pPCK is transformed into ePCK and then ePCK is transformed into pPCK, both through pedagogical reasoning. This example helps to make concrete specific features of ePCK and pPCK described above and provides further elaboration of the pedagogical reasoning inherent in the transformation from ePCK into pPCK.

Since ePCK is tacit knowledge, the best efforts to capture ePCK may still only result in approximations of this realm of science teachers' PCK. The next two examples in this section represent different approaches to making such approximations, both seeking to understand the ePCK that is utilised in the moment of instruction (i.e., in microcycles of plan-teach-reflect). These examples serve to illustrate the complexity of capturing ePCK, pointing out where reasonable approximations can and cannot be made.

All three examples highlight tools and approaches that have been developed to capture and/or support science teachers' PCK in action. While standardised instruments can be used to evaluate whether teachers have acquired particular cPCK, the contextualised nature of pPCK and ePCK requires different kinds of tools and approaches. Below, we describe the use of some of these tools and approaches and the extent to which they can be used to gain insights into science teachers' ePCK and/or pPCK and the interaction between them both.

Pedagogical Reasoning: Transformations Between ePCK and pPCK in Macro- and Microcycles of Plan-Teach-Reflect

For the purpose of stimulating science teachers' reflections and developing their PCK, Content Representations (CoRes) have been shown to be a useful pedagogical tool (Hume & Berry, 2011; Loughran, Berry, & Mulhall, 2006; Nilsson & Loughran, 2012). Further, in her review on PCK, Kind (2009) argued that the CoRe tool offers the most useful technique devised to date in science education research for eliciting and capturing PCK directly from teachers. Constructing a CoRe requires the teacher(s) to reflect upon how to teach a specific topic in order to promote students' learning. It prompts the teacher(s) to articulate what is called "big ideas" and address queries that include: what students should learn about each big idea; why it is important for students to know these ideas; students' possible difficulties with learning the ideas; and how these ideas fit in with the knowledge the teacher holds about that content. In this way, working with the CoRe as a reflective tool has the potential for transforming science teachers' tacit pPCK into explicit pPCK but also, when implemented into teaching practice, informing teachers' ePCK for planning ($ePCK_P$), teaching ($ePCK_T$) and reflecting ($ePCK_R$). CoRes may also be used to represent the collective views of a group of science teachers for teaching a specific topic, so that a CoRe also represents a form of cPCK for that teacher group.

In Nilsson and Karlsson´s (2018) research, the CoRe was introduced to student science teachers as a tool to stimulate their thinking about links between the content, teaching and student learning as they individually planned and tailored science instruction to a particular secondary classroom context and group of students. As such, each student teacher's individual CoRe was used to stimulate the transformation of pPCK into ePCK (for planning, teaching and reflecting). During the planning process, the student teachers were also encouraged to use resources such as curriculum materials and educational research, thus supporting the process of transforming cPCK into pPCK. The student teachers then taught a science lesson based on their constructed CoRes. Following their teaching, the student teachers viewed their video-recorded lessons and were encouraged to reflect upon their teaching performance to identify unexpected moments (expressed as critical incidents) in relation to their CoRes. Each student teacher chose two science teaching episodes, each about 4–8 min in length, representing: (1) a critical incident where she/he had succeeded in accordance with the big ideas in the CoRe and (2) a critical incident where she/he had experienced difficulties in fulfilling ambitions as expressed in the CoRe. The student teachers made annotations in the videos pinpointing these two critical incidents and providing reasoning as to why they felt they had succeeded or not in achieving their aims as expressed in the CoRes. In this way, the student teachers' video-recorded lessons were used to scaffold and structure their articulation of their in-the-moment pedagogical reasoning, transforming their $ePCK_T$ and their $ePCK_R$ into pPCK.

The outcomes of this research indicate that CoRe design prior to teaching episodes raises student science teachers' awareness of teaching issues around certain science content and engages them in reflection and decision-making that they enact in classrooms. As such, the research supports the notion that reasoning about specific instances of practice can help student teachers develop different aspects of their pPCK (e.g. knowledge of content and knowledge of students' understanding) as well as their ePCK (i.e., knowledge that teachers draw on in the moment of action, where the action may include planning, teaching or reflecting on teaching). The use of the CoRe as a tool for planning the science lesson illustrates the macrocycle of the unit of instruction. At the same time, the use of video annotations highlighting critical incidents illustrates the microcycle. Such a way of organising student teachers' reflective work during their practicum implies a transformation from pPCK to ePCK to more sophisticated form of pPCK through the process of pedagogical reasoning, from both a macro- and a microlevel perspective. As such, the CoRe, together with the video annotation tool, proved to be successful in scaffolding, structuring and even transforming student teachers' reflections, and consequently contributed to their pPCK development.

Approximating ePCK in Microcycles of Plan-Teach-Reflect

The tacit nature of ePCK presents a clear challenge for researchers seeking to capture this realm of PCK. Even when connected to a particular instance of science instruction, artefacts such as lesson plans or annotated videos capture pPCK (expressed when teachers' reasoning is made explicit as part of macroprocesses of planning or reflecting), rather than ePCK. Because ePCK is transformed into pPCK as it is made explicit, we argue that it is impossible to capture the true nature of ePCK. An alternative approach is to try to infer ePCK through evidence of the planning, teaching and reflecting that occurs in association with a single instructional move in science teaching (i.e., a microplan-teach-reflect cycle). In this section, we describe two examples of this approach.

Cognitive science research suggests that, even a short time after a given activity, people are unable to recall exactly what they were thinking when engaged in that activity (e.g., Ericsson and Simon, 1993; Leighton, 2004). Therefore, there is reason to believe that inferring the ePCK associated with a given instructional move would require teachers to "think aloud" (Ericsson and Simon, 1993) while teaching (i.e., to articulate pedagogical reasoning associated with the planning, enacting and reflection on that instructional move).[5] "Thinking aloud" would allow inferences of $ePCK_T$ to be made directly from the observed instructional move, but also provide opportunities (a) to elicit pPCK associated with planning and reflecting (as a proxy for $ePCK_P$ and $ePCK_R$) and (b) to elicit pPCK associated with teaching (to check inferences about $ePCK_T$ made directly from teaching actions). Unfortunately, this ideal is clearly not feasible in real classroom settings. Thus, researchers turn to work with science teachers outside of the classroom context to try to recapture or to simulate aspects of the plan-teach-reflect cycle that happen in-the-moment during instruction. We describe a method of each type in the sections below.

Documenting Evidence of ePCK and Associated pPCK

Pedagogical and Professional-experience Repertoires (PaP-eRs) (Loughran, Milroy, Berry, Mulhall, & Gunstone, 2001) offer one means of representing science teachers' in-the-moment instructional decisions and actions. PaP-eRs are short (1–2 pages) vignettes intended to represent the thoughts and actions of a knowledgeable science teacher in teaching a specific aspect of the content to students in a particular context. PaP-eRs include information about the classroom context, the teacher's thinking

[5]While acknowledging that video stimulated recall is often used to elicit teachers' recollections of in-the-moment reasoning (e.g., Akerson, Flick, & Lederman, 2000; Nilsson, 2008), following Ericsson and Simon (1993), it seems that such efforts may be accessing existing pPCK (i.e., the way a teacher has made sense of a given classroom event after the fact), rather than pPCK that is being transformed directly from ePCK during the stimulated recall (i.e., pPCK that could serve as a direct proxy for ePCK).

about the content, examples of students' responses, and what it is about the content that shapes the approach to teaching and learning and why. PaP-eRs are constructed by researchers in consultation with teachers from data gained while observing a particular science teacher's classroom and/or through interviewing a teacher about an instance of practice where he/she came to understand the content differently as a consequence of teaching it. Through making explicit these components of classroom practice and associated teacher reasoning, PaP-eRs capture aspects of a teacher's $ePCK_P$, $ePCK_T$ and $ePCK_R$, within the microcycles of instructional moves occurring in the lesson, and since PaP-eRs are constructed post-lesson, their $ePCK_R$ in the macrocycle of instruction, as teachers think back on their planned instruction and its subsequent student outcomes.

For example, Bertram and Loughran (2012) used CoRes in combination with PaP-eRs to investigate the development of experienced secondary science teachers' PCK over a two-year period. In this study, participating teachers ($n = 6$) individually created CoRes for a science topic they planned to teach, then reflected on the process of making the CoRe and how that process influenced their thinking about teaching and learning, and how it influenced their understanding of PCK. As Bertram and Loughran (2012) noted:

> in creating the CoRe, it forced these teachers to explicitly think about and connect with their tacit knowledge about teaching and learning. Thus, the process of working through developing a CoRe encouraged these participants to find ways of articulating that which they knew and how they developed their knowledge of practice. (p.1036)

Following their teaching of the topic, participants were then asked to develop a PaP-eR (in collaboration with the researchers) illustrating a particular classroom teaching episode in science based on their CoRe. As one participant noted:

> "So, what I feel is - that this [PaP-eR] is articulating, documenting, making explicit - that kind of process which … on reflection, is a process … that I have going on in my head all the time, in relation to teaching…". (p.1040)

Bertram and Loughran's study showed the use of the CoRe and PaP-eR tools enhanced science teachers' knowledge of practice (i.e., transformation of ePCK to pPCK) through making explicit and sharing their knowledge about teaching and helping to highlight the ways in which content and purpose are closely linked in teaching. In particular, all participants claimed that developing their PaP-eRs encouraged their self-reflection and self-evaluation of their specific contexts and teaching practices (pPCK and $ePCK_R$) and helped to pinpoint areas in which they could improve (e.g., connecting with particular students and their learning needs).

Stimulating Generation of ePCK Outside of the Classroom

Simulating aspects of the plan-teach-reflect cycle that happen in-the-moment during science instruction, outside of the classroom involves a trade-off between the authenticity of a real classroom situation (such as represented in the PaP-eRs) and the ability to capture approximations of ePCK that would be unfeasible in real classroom situations. While not engaging teachers with their own students in their own teaching contexts, this method often incorporates elements of real teaching situations, such as authentic prompts (e.g., video of students expressing their ideas) and authentic response formats (e.g., interacting with a live actor). To date, these methods have not captured all three types of ePCK, focusing either on teachers' articulating in-the-moment decision-making ($ePCK_P$ and $ePCK_R$) or researchers making inferences on the basis of teachers' in-the-moment actions ($ePCK_T$).

In order to simulate a science teacher's encountering of unexpected student thinking in a classroom situation, Alonzo and Kim (2016) presented teachers with videos of students expressing ideas about force and motion. The videos, all drawn from real physics classrooms similar to those of the participating teachers, highlighted unusual student thinking—i.e., "unexpected or novel student ideas or questions" (p. 1268). Teachers were asked first to describe the student thinking in the video and then to explain how they might respond to the student. The intent was to capture teachers' in-the-moment reasoning if a student were to offer the same statement or question in their own classrooms, by asking teachers to make explicit (i.e., transform into pPCK) the $ePCK_R$ and $ePCK_P$, respectively, that might underlie a classroom instructional response.

In contrast, two German research groups have devised methods to simulate science teaching situations and teachers' actual responses to students (i.e., opportunities to infer $ePCK_T$), but do not require teachers to describe their planning or reflecting processes and, thus, do not capture $ePCK_P$ or $ePCK_R$. In the domain of mathematics education, Lindmeier and colleagues (Knievel, Lindmeier, & Heinze, 2015; Lindmeier, 2011) used videos of classroom situations highlighting student thinking; however, rather than describing potential instructional moves to an interviewer, teachers were asked to speak (to a computer) as if directly to the student. With this method, researchers capture teachers' instructional moves in response to the video and, thus, infer their underlying $ePCK_T$. As the video-recorded student cannot react to the teacher's instruction, this method (like the one used by Alonzo and Kim) involves a single instructional move.

The method used by Kulgemeyer and Schecker (2013) entails multiple instructional moves. In this method, teachers are given time to prepare an explanation of a particular physics problem and then are asked to provide that explanation to a "student" (a specially trained live actor). The student asks questions or provides other responses to the teacher's explanation, using a predetermined script. With this method, researchers can capture instructional moves that the science teacher makes throughout the explanation interaction and, thus, infer evidence for $ePCK_T$ across multiple plan-teach-respond cycles.

In the above described examples, Alonzo and Kim captured $ePCK_P$ and $ePCK_R$, while Lindemeir, Kulgemeyer and colleagues captured $ePCK_T$. In order to capture all three forms of ePCK, one might imagine a hybrid situation, in which science teachers are presented with evidence of student thinking and are then asked to (a) articulate not only a proposal for how to respond to the student thinking, but also the reflection and planning underlying the proposed instructional response (i.e., transform $ePCK_R$

and $ePCK_P$ into pPCK) and (b) enact that response (i.e., provide evidence from which $ePCK_T$ might be inferred).

One advantage of all of these approaches is that they permit comparison across teachers. While it is impossible to observe multiple science teachers in the exact same "real" classroom situation, the same video can be shown over and over again, and actors can be trained to behave similarly when interacting with many different teachers. At the same time, this advantage is a limitation, in that ePCK—like the pPCK on which is it based—is specific to a teacher's own teaching context. Simulations outside of the classroom strip that context away from the enactment. Thus, it is likely that multiple approaches, in combination, will be required to fully approximate a teacher's ePCK. Methods such as PaP-eRs provide authentic contextualisation, whereas simulations outside of the classroom may capture closer approximations of ePCK.

Conclusion

To date, research on PCK in the science education field has largely focused on relatively static forms of propositional knowledge and, thus, has deepened our understanding of the composition and structure of teachers' cPCK and pPCK, i.e., the outer rings of the RCM (see Fig. 2.3, Chap. 2). Like other chapters in Part III, ours illustrates how the RCM can be used to classify different realms of PCK and, therefore, more clearly articulate the focus of a given research or teacher education effort. As shown in Figs. 12.1 and 12.2, we found it useful to identify the different types of enactment and, thus, the different types of ePCK that are entailed in enacting macro- and microplan-teach-reflect cycles. In doing so, we highlight the growing body of research that draws attention to the centre of the RCM, exploring science teachers' ePCK (i.e., PCK in action) and the relationships that exist between ePCK and pPCK. We argue that this work is essential if we are to understand not just what science teachers know, but how that knowledge is transformed into learning experiences for students.

We bring to the RCM a strong interest in and commitment to the aspects of teachers' work that take place "in action". While the RCM acknowledges this realm of PCK (i.e., ePCK), it has not yet been fully elaborated. Thus, in this chapter, we have sought to unpack ePCK and its relationship to pPCK. By considering ePCK to be tacit knowledge in action, we emphasise that teachers' knowledge is often not made explicit, especially in the midst of interacting with students. Our perspective on the relationship between ePCK and pPCK allows us to explain how pedagogical reasoning facilitates the gradual growth of pPCK in response to the experience of teaching particular content to particular students in particular contexts. This perspective also helps us to articulate why it is so difficult to capture exactly what enables a given moment of instruction. So much of what happens in the moment is tacit. While teachers make a number of instructional moves throughout a lesson—many of them unplanned and, thus, generated in the moment—it is extremely rare for the knowledge resources (e.g., knowledge, decision-making) underlying a given move to be made explicit as part of instruction. We cannot directly observe the ePCK involved in teachers' planning, teaching or reflecting and, thus, do not know exactly what motivates a given instructional move.

We put forth this interpretation of ePCK and its relationship to pPCK with the goal of enabling other researchers to utilise this critical area of the RCM. As others heed the call to focus more attention on PCK in action (e.g., Henze & van Driel, 2015), we see the constructs of ePCK and pPCK as especially valuable for clarity in communicating the aims and challenges of our research and in devising ways to capture particular aspects of PCK in action.

References

Akerson, V. L., Flick, L. B., & Lederman, N. G. (2000). The influence of primary children's ideas in science on teaching practice. *Journal of Research in Science Teaching, 37,* 363–385.

Alonzo, A. C., & Kim, J. (2016). Declarative and dynamic pedagogical content knowledge as elicited through two video-based interview methods. *Journal of Research in Science Teaching, 53,* 1259–1286.

Bertram, A., & Loughran, J. (2012). Science teachers' views on CoRes and PaP-eRs as a framework for articulating and developing pedagogical content knowledge. *Research in Science Education, 42,* 1027–1047.

Carlson, J., Daehler, K. R., Alonzo, A. C., Barensden, E., Berry, A., Borowski, A., ... & Wilson, C. D. (2019). The Refined Consensus Model of pedagogical content knowledge in science education. In A. Hume, R. Cooper, & A. Borowski (Eds.), Repositioning pedagogical content knowledge in teachers' knowledge for teaching science (pp. 77-92). Singapore: Springer.

Eraut, M. (2000). Non-formal learning and tacit knowledge in professional work. *British Journal of Educational Psychology, 70,* 113–136.

Ericsson, K. A., & Simon, H. A. (1993). *Protocol analysis: Verbal reports as data.* Cambridge, MA: MIT Press.

Gess-Newsome, J. (2015). A model of teacher professional knowledge and skill including PCK: Results of the thinking from the PCK summit. In A. Berry, P. Friedrichsen, & J. Loughran (Eds.), *Re-examining pedagogical content knowledge in science education* (pp. 28–42). New York: Routledge.

Henze, I., & Van Driel, J. H. (2015). Toward a more comprehensive way to capture PCK in its complexity. In A. Berry, P. Friedrichsen, & J. Loughran (Eds.), *Re-examining pedagogical content knowledge in science education* (pp. 120–134). New York, London: Routledge.

Hume, A., & Berry, A. (2011). Constructing CoRes—a strategy for building PCK in pre-service science teacher education. *Research in Science Education, 41*(3), 341–355.

Kind, V. (2009). Pedagogical content knowledge in science education: perspectives and potential for progress. *Studies in science education, 45*(2), 169–204.

Knievel, I., Lindemeier, A. M., & Heinze, A. (2015). Beyond knowledge: Measuring primary teachers' subject-specific competences in and for teaching mathematics with items based on video vignettes. *International Journal of Science & Mathematics Education, 13*(2), 309–329.

Kulgemeyer, C., & Schecker, H. (2013). Students explaining science—assessment of science communication competence. *Research in Science Education, 43,* 2235–2256.

Leighton, J. P. (2004). Avoiding misconception, misuse, and missed opportunities: The collection of verbal reports in educational achievement testing. *Educational Measurement: Issues and Practice, 23*(4), 6–15.

Lindmeier, A. (2011). *Modeling and measuring knowledge and competencies of teachers: A threefold domain-specific structure model for mathematics*. Münster, Germany: Waxmann.

Loughran, J., Milroy, P., Berry, A., Mulhall, P., & Gunstone, R. (2001). Science cases in action: Documenting science teachers' pedagogical content knowledge through PaP-eRs. *Research in Science Education, 31*, 289–307.

Loughran, J., Berry, A., & Mulhall, P. (2006). *Understanding and developing science teachers' pedagogical content knowledge*. Dordrecht: Sense Publishers.

Nilsson, P. (2008). Teaching for understanding—the complex nature of PCK in pre-service teacher education. *International Journal of Science Education, 30*(10), 1281–1299.

Nilsson, P., & Karlsson. (2018). *Capturing student teachers' pedagogical content knowledge (PCK) using CoRes and digital technology*. Manuscript submitted for publication.

Nilsson, P., & Loughran, J. (2012). Exploring the development of pre-service elementary teachers' pedagogical content knowledge. *Journal of Science Teacher Education, 23*(7), 699–721.

Schmelzing, S., van Driel, J. H., Jüttner, M., Brandenbusch, S., Sandmann, A., & Neuhaus, B. J. (2013). Development, evaluation, and validation of a paper-and-pencil test for measuring two components of biology teachers' pedagogical content knowledge concerning the "cardiovascular system". *International Journal of Science and Mathematics Education, 11*, 1369–1390.

Part IV
So What and What Next?

Andreas Borowski

Overview

In many discussions at international conferences, great interest has been shown in the model of teacher professional knowledge and skill including PCK - Consensus Model (CM) of PCK as it is more commonly known) (Gess-Newsome, 2015) - which was developed at the first PCK Summit in 2012. However, there was often feedback from discussants that it is difficult to imagine concrete research projects using this model. For this reason, most of the chapters in this book discuss the application or adaptation of the RCM of PCK in science, presented in Chap. 2, in various existing studies. Note most of these studies were originally planned and conducted without the RCM, but these have now been retrospectively related to the new model. Thus, the chapters in this book show that the RCM is not something completely new; rather, it is a combination of ideas already existing in the community that have now been merged.

This last part of the book wants to look forward, but also backward. So it is about what we as a community can do with the RCM now, but also, what have we learned from the Summit meetings so far and how can we continue to make use of this learning.

Chapter 13 deals with the possibilities of further research in the field of science PCK and beyond using the RCM and raises questions about planning new research projects. Which areas of the model have perhaps not yet been studied so intensively? Which new research areas can be established? How could research methods be used to investigate the complexity of RCM? The chapter seeks to point out and promote the benefits of making the model the basis for future research. For example, if many research projects refer to the model, the results can be linked more closely.

Chapter 14 discusses the research on PCK in general, especially that emanating from the two PCK Summits in Colorado Springs, USA, and Leiden, Netherlands. The chapter describes the productivity of working together at Summits, but also the responsibility involved. The role of the Summits in moving the field of PCK research forward is also discussed. Above all, it describes how much fun it can be to join this field of research via events like Summits and advance joint research on PCK!

Chapter 13
Perspectives on the Future of PCK Research in Science Education and Beyond

Christopher D. Wilson, Andreas Borowski and Jan van Driel

Abstract This book demonstrates that PCK is studied with different intentions, different methodologies, and in different contexts. Nevertheless, the two PCK Summits of researchers in science education have made significant progress in conceptualising PCK, representing it first in a consensus model, and now refining that model in response to the successes and failures from studies that have applied the original model to current research. The Refined Consensus Model (RCM) of PCK for science teaching consists of a set of related components, allowing researchers to locate their work in specific components of PCK, connect their work across components, and to connect their work with broader issues and policies. Clearly, the RCM and research on PCK are closely intertwined. The intention of this chapter is to use the refined model to help outline some possible strands for future PCK research, particularly with the new PCK researcher in mind. Using the RCM as a framework, we describe possible studies on the structure of PCK, the development of PCK, the measurement of PCK, and the broader impacts of the refined model itself in science education and potentially other domains.

Introduction

In the National Academies report *Scientific Research in Education*, Shavelson and Towne (NRC, 2002) highlight the current importance of educational research, stating:

> In today's rapidly changing economic and technological environment, schooling cannot be improved by relying on folk wisdom about how students learn and how schools should be

C. D. Wilson (✉)
Biological Sciences Curriculum Study, Colorado Springs, CO, USA
e-mail: cwilson@bscs.org

A. Borowski
University of Potsdam, Potsdam, Germany
e-mail: andreas.borowski@uni-potsdam.de

J. van Driel
University of Melbourne, Melbourne, Australia
e-mail: j.vandriel@unimelb.edu.au

A. Hume et al. (eds.), *Repositioning Pedagogical Content Knowledge in Teachers' Knowledge for Teaching Science*, https://doi.org/10.1007/978-981-13-5898-2_13

> organised. No one would think of designing a rocket to the moon or wiping out a widespread disease by relying on untested hunches; likewise, one cannot expect to improve education without research. (NRC, 2002, p. 12)

While there is widespread agreement that the educational process can be improved through research (the American Educational Research Association (AERA) alone has more than 25,000 members), educational research has, at times, received significant criticism. Some critiques focus on insufficient progress, such as the lack of (a) systematic accumulation of knowledge; (b) evidence-based refinement and replacement of theories; and (c) policy-relevant consensus understandings, each of which have led to significant improvements in other areas of scientific research such as medicine and agriculture (Klahr, 2010, Slavin, 2008). Suggested actions to address this problem include calls to develop consensus and operational definitions of constructs, approaches, and theories; to better define the intended goals of instruction; and to conduct replicable empirical studies that examine the relationship between instructional interventions and desired outcomes (BMBF, 2017; Klahr, 2010; NRC, 2012).

Enter the PCK Summits—two workshops held in Colorado, USA, and Leiden, the Netherlands, in 2012 and 2016, respectively, involving expert researchers and emerging researchers in science education (and mathematics education at the First (1st) Summit). The organisers of these meetings recognised that research on PCK was important, and even very common, especially in the domain of science education. However, its impact on practice and policy could be greatly improved by developing a consensus understanding and operational definition of the PCK construct. Further, by bringing together different perspectives on PCK research—from areas such as teacher education, learning theory, policy, educational measurement, as well as representatives from multiple research traditions in different countries—the two workshops sought to take an evidence-based approach to understanding what is currently known about PCK, and what we still need to learn. This chapter describes the latter. Specifically, it is a humble attempt to interpret and summarise the complex and challenging (but always constructive) discussions at the Second (2nd) PCK Summit, with a focus on how they might inform future research on PCK. And to expand beyond current work, we encourage new educational researchers and researchers new to PCK, both within and outside of science education, to consider the directions suggested in this chapter and to use the RCM to both inspire and frame their work.

(A Brief) Summary of Implications for Research Suggested in the Preceding Chapters

While rarely their focus, all the chapters in this book either describe the limitations of existing research on PCK, recommend future directions for research on PCK, and/or discuss the role of the RCM in shaping future lines of research in science education. Most importantly in this regard is of course Chap. 2, which reframes research on PCK around the RCM. This chapter describes how the RCM repositions PCK as three distinct realms of knowledge and skill within science teachers' overall professional knowledge base that are centred round the practice of teaching. We expand on the implications of this reframing for research below. Similarly, there are implications for research from the RCM's emphasis on pedagogical reasoning and knowledge exchanges between these realms of PCK, other professional knowledge forms and contextual and affective influences.

Other chapters explore the relationships between several existing models of PCK and the RCM. Park in Chap. 4, for example, highlights several issues related to research on PCK, including the need for longitudinal studies on PCK development; how PCK develops through interactions with professional communities; the need to examine contextual factors associated with PCK; as well as providing recommendations for research across the full plan–teach–reflect pedagogical cycle in science teaching. Sorge, Stender and Neumann in Chap. 6 connect models of the development of science teachers' professional competence with the RCM. This chapter points to the need for future research on the complex relationships between the components of PCK in the refined model, especially as they relate to the development of PCK. While their data are largely quantitative, the authors point to the need for complementing such work with qualitative studies. Schneider in Chap. 7 explores the potential of learning studies that allow researchers to reveal and characterise science teacher PCK, an approach grounded in the plan–teach–reflect cycles that represent the work of science teachers in the RCM. Schneider recommends the use of learning studies in future PCK research aimed to describe learning progressions for science teachers.

Henze and Barendsen in Chap. 9 apply non-intrusive techniques to the measurement of PCK development in student science teachers for formative purposes, allowing instructors to adjust instruction without additional external tools or tasks. They also explore patterns in PCK development, and point to the need for more research on models of these trajectories. Carpendale and Hume in Chap. 10 examine the impact of the collaborative Content Representation (CoRe) design on science teacher PCK and encourage more research that (a) includes multiple teachers, allowing generalisability; (b) examines student outcomes to better explore interventions designed to increase science teacher PCK; and (c) examines teacher reasoning in the act of science teaching.

Chan, Rollnick and Gess-Newsome in Chap. 11 explore the development of a grand science rubric for measuring PCK. This work extends the value of a consensus model of PCK by suggesting a consensus approach to measurement, allowing researchers to look across studies with common metrics. Alonzo, Berry and Nilsson in Chap. 12 examine how to reveal and characterise PCK in action, that is, the interaction of science teachers' personal and enacted PCK (pPCK and ePCK) that is used to inform instructional decisions in the classroom. Their work points to the need for research not just on teachers' professional knowledge, but on how that knowledge is enacted in teaching, which has implications for measuring PCK in situ. Further, their chapter raises questions about the model that require further research.

A Framework for Future PCK Research

In looking across the findings and suggested future directions in the chapters described above, and the recommendations from the 1st PCK Summit (Berry, Friedrichsen, & Loughran, 2015), to other research in this field and/or other domains that might not have been represented at the Summits, as well as current research and policy contexts, several themes emerge. Below, we describe some of these major themes and suggest some research questions that address these themes and their potential contribution to the field. While certain research questions suggest certain research methods, we attempt to remain somewhat agnostic and open towards methodology. Building a complete and informative body of understanding requires both quantitative and qualitative inquiry, as well as synthesis studies, design research, experimental studies, replication studies, and theoretical reflection.

1. **Research on the Structure of PCK**

It is important to recognise that the evidence base for the RCM is incomplete. The review of research conducted during the 1st PCK Summit found agreement around several trends, such that PCK exists on a continuum from weak to strong and that teachers with strong PCK are better able to improve student learning. However, the evidence for most broad claims is not strong, and certainly in science education includes very few studies employing rigorous research designs or providing high confidence in causal claims (see Chap. 1). The RCM invites us to revisit these trends, together with a systematic examination of the individual components of the model, the relationships between components of the model, and perhaps most importantly, how these components correlate with and predict student learning.

The RCM describes three components or realms of PCK in science teaching—collective PCK (**cPCK**), personal PCK (**pPCK**), and enacted PCK (**ePCK**)—with knowledge exchange occurring between them via processes of pedagogical reasoning that ultimately result in learning opportunities for students. Numerous research questions exist within and between each of the three PCK realms of ePCK providing perhaps the richest and most accessible research opportunities, particularly for beginning researchers.

ePCK

In our thinking about future research directions in PCK research, we now turn attention to the centre circle of the RCM model (see Chap. 2) representing ePCK, which is the realm or component of PCK comprising the knowledge and skills science teachers use in the act of teaching. Within ePCK, research on the knowledge and resources that inform science teachers' instructional decisions, particularly for certain content or for certain students, is greatly needed. For example, research on how teachers select, evaluate, adapt, and integrate instructional materials has a strong history in science education, but must now explore how science teachers interact with the overabundance of digital resources of vastly varying quality that is currently available to them. Similarly, research should examine tools and processes that can support science teachers in integrating a range of instructional resources while still providing coherent and effective learning experiences for all students, that is, support their pedagogical reasoning. Alternatively, researchers might examine the plan–teach–reflect pedagogical cycle, such as questions around how the cycle varies across science teachers with different backgrounds or experience, or across content areas or topics. Equally, one might study how science teachers reveal and respond to student thinking, or how intentional they are in their practice and pedagogical decision-making. In this context, it may be relevant to connect research on ePCK with research on teacher professional noticing. The construct of teacher professional noticing (observing, understanding and responding to student thinking during instruction) as a key element of teaching expertise has gained increased attention over the last decade (e.g., Mason, 2002; Scheiner, 2016; Sherin, Russ, & Colestock, 2011). Research on teacher professional noticing, in particular, in the domain of mathematics education, has focused on teachers' in-the-moment reasoning and decision-making, typically using video recording of classroom situations as input. Both conceptually and methodologically, professional noticing and enacted PCK are related, and research in these domains may benefit from each other in an attempt to improve our understanding of how and why teachers make certain pedagogical decisions when they interact with students about particular subject matter.

Some of the most important questions related to ePCK connect it with **student learning**. While the 2nd Summit's focus on PCK is grounded in the belief that a science teacher's PCK is important in providing students with meaningful opportunities to learn, the basis for this belief is perhaps where evidence from current research is the weakest. There are many reasons for this weakness, among them being the problem that research studies reporting on this relationship were rarely designed to reveal it. Instead such studies were designed and statistically powered to look at impacts of professional development (PD) interventions on teacher learning, teacher practice, and student learning separately. In other studies, measures of student learning (e.g., standardised tests across a certain domain) were not well aligned with the measures of teachers' PCK (e.g., focus on a specific topic or application of knowledge). As such, our science education field needs studies that are designed to reveal the extent to which ePCK, as visible in planning, classroom practice, and reflection, results in or contributes to student learning (e.g., Liepertz & Borowski, 2017). Such research must also recognise that what one values in student learning of science is an important consideration (e.g., factual recall, motivation, scientific literacy, and reasoning), and examine the relationship for specific outcomes, as well as for different students.

pPCK and its Connections to Other Components

Moving outward in the RCM to the increasingly larger concentric circles, we identify many potential research questions that connect ePCK with **personal PCK (pPCK)**, and on pPCK itself. As described in Chap. 2, pPCK is both informed by and informs ePCK, which suggests research on how pPCK is translated into instruction, and conversely how experiences in the classroom shape and revise a science teacher's professional knowledge. For example, one might study how pPCK varies across science teachers, and how that variation influences what happens in their classrooms. Equally, we know relatively little about how feedback from students and their own science teaching experiences in the classroom impacts science teachers' pPCK and inform their understanding of effective practice. Research into that feedback can be extended to studying how it shapes a science teacher's own teaching, and how it supports them in collaborating with colleagues.

cPCK
Finally, moving to the outer circles of the model, other important research questions emerge. Science teacher educators and professional development providers can benefit greatly from research that advances our understanding of how science teachers' pPCK emerges through interaction with the collective knowledge of the field (**cPCK**). How do science teachers access this knowledge for example, or what filters mediate the impact of that collective knowledge on their individual practice? And further beyond, to the outermost circle beyond PCK representing other forms of professional knowledge contributing to PCK, some basic questions about PCK remain largely unanswered. What is the nature of the relationship between content knowledge and PCK? What types of content knowledge do science teachers need? How does general knowledge of effective teaching strategies contribute to effective teaching in different disciplines of specific topics? How can assessment be used to help science teachers develop a more student-centred mindset? Despite the fact that these questions have been studied in PCK research over the past 25 years (see for an overview: Van Driel, Berry, & Meirink, 2014), they have not been resolved. The structure of PCK for science teaching, as represented in the RCM (Fig. 13.1), provides a rich vein for research that can inform areas such as teacher education, professional development, and instructional materials development.

2. **Research on the Development of PCK**

In considering the development of teacher professional knowledge, the 1st PCK Summit concluded that PCK can be strengthened through teaching experience, professional development, or other interventions, but that teaching experience does not

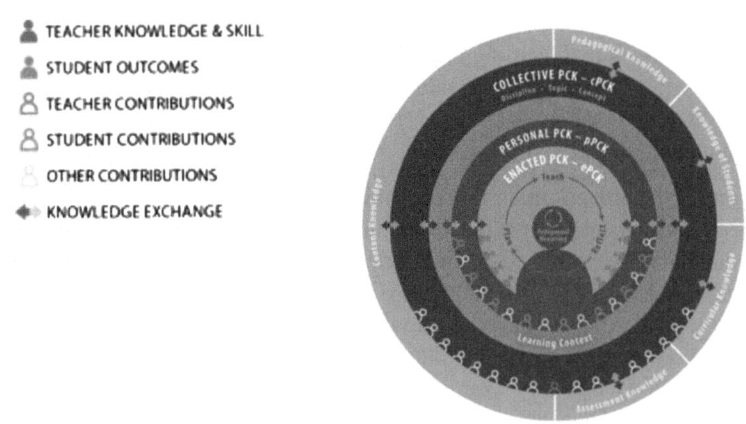

Fig. 13.1 Refined Consensus Model of PCK for teaching science (see Chap. 2)

necessarily result in increased PCK. The chapters in this book build on these findings in the context of science education, and delve deeply into the development of PCK, highlighting the need to future research in this area. Here, we consider three areas of inquiry related to the development of PCK.

(a) Longitudinal Studies
A number of researchers in this book point at the need for more research on the trajectories or pathways that science teachers travel as they move along a continuum from novice to expert. Such research holds great potential to inform science teacher education and the development of programmes for professional learning and to guide professional development (PD) leaders in making instructional decisions. Learning progressions research is one potential avenue for research here, and the potential and challenges of applying this approach were described in the report from the 1st PCK Summit (Friedrichsen & Berry, 2015; Schneider, Chap. 7).

(b) Intervention Studies
Many questions exist around how to most effectively develop teacher PCK (Darling-Hammond, Hyler, & Gardner, 2017). The focus here is often on teacher education and professional development programmes, where research seeks to identify which programmes are more effective than others (for what students and for what outcomes), and how the features of those programmes help support the development of PCK. Ideally, such studies compare the impact of different interventions on intended outcomes, while controlling for other variables. Such comparative experimental studies are largely absent from the research base, with correlational studies and teacher self-reports being the dominant research methodologies (NASEM, 2015; Yoon et al., 2007). Causal research on PCK interventions might compare a professional learning programme focused on teacher content knowledge with one with a more pedagogical focus (e.g., Roth et al., 2017), or compare the impact of educational curriculum materials with and without professional development to reveal important information on how to support the use of educative materials (e.g., Gess-Newsome et al., 2017). Whenever possible, intervention studies should examine multiple teacher outcomes (pPCK, cPCK, and ePCK, classroom practice, as well as science content knowledge). Finally, impacts on student outcomes are particularly important to explore since little is still known about the impact of teacher learning experiences on student learning (NASEM, 2015; Wayne, Yoon, Zhu, Cronen, & Garet, 2008; Wilson, 2013). In a systematic review of research on professional development of science teachers Van Driel, Meirink, Van Veen, and Zwart (2012) identified only six studies (out of 146 examined) that applied "measures to assess student learning outcomes" (p. 152). Future studies should pay attention to both the type of student learning that is valued and measured, as well as mediation of student learning by teacher outcomes.

(c) Contextual Studies
As we develop more robust understandings of how science teachers' PCK develops over time and what interventions are effective in supporting that develop-

ment, complementary research needs to examine the factors that influence that growth. Such research questions might be explored within a longitudinal or intervention study, for example, through analysis of the mediation and moderation of teacher outcomes by grade level, topic, teacher experience, or district and school characteristics. Alternatively, contextual factors might be the primary focus of studies, which examine how PCK develops in specific systems, cultures, or contexts.

3. **Research on the Measurement of PCK**

Multiple measures of science teacher PCK currently exist in the literature. Since PCK is, as Shulman described, "uniquely the province of teachers, their own special form of professional understanding" (1987, p. 8), it is most authentically measured in the professional work of teachers. Consequently, many current assessments of PCK measure its manifestations in different parts of the plan–teach–enact pedagogical cycle. Other measures take the form of more formal knowledge assessments, such as asking science teachers to provide written responses to classroom situations, or engaging them in watching and analysing classroom video recordings. Researchers continuing work on these measures should clearly situate their approach in the RCM and describe which components of PCK they do and do not measure. Concurrent validity across measures is also an important area of research, such as exploring the extent to which different measures of PCK locate science teachers in the same place on a continuum from novice to expert, or to what extent different measures predict student achievement.

A primary concern in educational measurement is of course the validity of measures, which is a critical area for future research on both new and existing measures. Work in this area should recognise that assessments are not inherently valid, but that assessments are only valid for particular uses or decisions. As such, those intended uses need to be clearly defined and various forms of validity evidence must be gathered to support validity arguments. A validity framework is helpful here, which can guide research on PCK measurement (Kirschner, Taylor, Rollnick, Borowski, & Mavhunga, 2015). Various such frameworks exist in the literature, and some are quite accessible to those researchers less embedded in educational measurement. Pellegrino, DiBello, and Goldman (2016), for example, describe a validity framework with three components—a *cognitive* component that evaluates the extent to which an assessment measures the intended cognitive processes; an *instructional* component that evaluates the extent to which an assessment aligns with the goals of instruction and can inform instruction, and an *inferential* component that explores the extent to which an assessment provides accurate information on the performance of respondents. Applied to PCK research, the cognitive component might look at the extent to which an assessment measures a specific component of the refined PCK model, the instructional component might examine the alignment between a PCK measure and the goals of a specific PD programme, and the inferential component might explore how information from an assessment can provide a science teacher educator with useful information on teacher learning. While the Pellegrino et al.'s framework was designed for student assessments, transferring such work to the measurement of science teacher learning and the development of PCK can support and guide instrument development research, and improve the quality of measures in research on PCK and its development.

Finally, future research on the measurement of PCK should also explore:

(a) The sensitivity of assessments—that is, the extent to which measures are able to detect change over time, the differential impact of interventions, or the differences between novices and experts.
(b) The formative value of assessments—including the extent to which measures can effectively inform instructional decisions.
(c) The efficiency of assessments—how different measures can provide the same information with less time or intrusion into science teaching and learning.
(d) The authenticity of assessments—that is, the extent to which measures are situated in the professional work of science teaching, and the value of that on measurement.

4. **Research on the Impact of the Refined Consensus Model of PCK**

One goal of both of the PCK Summits was to build a consensus understanding of PCK that could shape future research, reduce ambiguity, and provide a common framework and language that would facilitate more effective accumulation of knowledge in the science education field. We therefore hope future researchers will explore the impact of the RCM, as well as other findings from the 2nd PCK Summit. Researchers might examine the uptake of the model, and the extent to which it becomes pervasive in studies of PCK in science education and other domains, or in practice and policy. Similarly, studies could examine if knowledge accumulation is supported by the findings from the 2nd PCK Summit, by supporting research synthesis or meta-analytic work. The work of Gastaldo, Castro, Homen-de Mello, and Leal (2017) is one such example.

Conclusion

The four themes described here represent one attempt to frame the future needs, actions, and impact of some avenues for research on PCK in science education and beyond. The themes or topics within this chapter are certainly not exhaustive, and indeed, even major areas are not included. Connecting PCK research with policy was not addressed, for example. Discussion of policy filled a whole chapter (Sickel,

Banilower, Carlson, & Van Driel, 2015) in the previous book emerging from the 1st PCK Summit and could probably be the topic of several in this volume, given the current international agenda on teacher quality, various reforms that are reimagining student outcomes, as well as contemporary discussions on the role of schooling. Similarly, a discussion of PCK research and equity is absent here and deserves more thorough treatment than we had room for in this chapter. Both policy and equity are important areas that future research should address.

The RCM holds great promise for guiding future research, but as with any model, it should be continually evaluated and refined as our knowledge of PCK develops through the very research it informs. While some may have responded with frustration that the model evolved so quickly from its first iteration (Gess-Newsome, 2015), especially in science education, this evolution was driven by its successes and failures as it was released into its natural environment. All scientific models change over time and adaptations that make them more effective should be encouraged.

Finally, we encourage new researchers in science education, and in domains other than science, to explore this terrain deliberately and rigorously, but also with an open and optimistic mind about how their work can help build knowledge on the structure, development, and implications of PCK for teaching and learning. Returning to Shavelson and Towne's commentary in *Scientific Research in Education* (NRC, 2002), the contribution of any research study does not live or die on finding significant effects, but rather:

> … many scientific studies in education and other fields will not pan out. Research is like oil exploration—there are, on average, many dry holes for every successful well. This is not because initial decisions on where to dig were necessarily misguided. Competent oil explorers, like competent scientists, presumably used the best information available to conduct their work. Dry holes are found because there is considerable uncertainty in exploration of any kind. Sometimes exploration companies gain sufficient knowledge from a series of dry holes in an area to close it down. And in many cases, failure to find wells can shed light on why apparently productive holes turned out to be dry; in other words, the process of failing to make a grand discovery can itself be very instructive. Other times they doggedly pursue an area because the science suggests there is still a reasonable chance of success. Scientific progress advances in much the same way. (NRC, 2002, p. 25)

We look forward to seeing the impact the RCM has on future research into PCK in science education and other domains. All signs point to fewer dry holes as well as deeper exploration and productivity.

References

Berry, A., Friedrichsen, P., & Loughran, J. (2015). *Re-examining pedagogical content knowledge in science education*. New York, London: Routledge.

BMBF (Bundesministerium für Forschung und Bildung) [Federal Ministry for Education and Research]. (2017) Rahmenprogramm empirische Bildungsforschung des Bundesministeriums für Bildung und Forschung [Framework Programme empirical educational research of the Federal Ministry for Education and Research] http://www.empirische-bildungsforschung-bmbf.de/media/content/Rahmenprogramm%20empirische%20Bildungsforschung%20(BITV).pdf.

Darling-Hammond, L., Hyler, M. E., & Gardner, M. (2017). *Effective teacher professional development*. Palo Alto, CA: Learning Policy Institute.
Friedrichsen, P., & Berry, A. (2015). Science teacher PCK learning progressions: Promises and challenges. In A. Berry, P. Friedrichsen, & J. Loughran (Eds.), *Re-examining pedagogical content knowledge in science education* (pp. 214–228). New York, London: Routledge.
Gastaldo, B. C., Castro, P. M., Homen-de Mello, P., & Leal, S. H. (2017). *Five years of PCK summit. What has endured? A descending hierarchical classification of the science PCK published papers*. X Congreso Internacional sobre Investigación en la Didáctica de las Ciencias, Sevilla.
Gess-Newsome, J. (2015). A model of teacher professional knowledge and skill including PCK: Results of the thinking from the PCK summit. In A. Berry, P. Friedrichsen, & J. Loughran (Eds.), *Re-examining pedagogical content knowledge in science education* (pp. 28–42). New York, London: Routledge.
Gess-Newsome, J., Taylor, J. A., Carlson, J., Gardner, A. L., Wilson, C. D., & Stuhlsatz, M. A. (2017). Teacher pedagogical content knowledge, practice, and student achievement. *International Journal of Science Education*. https://doi.org/10.1080/09500693.2016.1265158.
Kirschner, S., Taylor, J., Rollnick, M., Borowski, A., & Mavhunga, E. (2015). Gathering evidence for the validity of PCK measures: Connecting ideas to analytic approaches. In A. Berry, P. Friedrichsen, & J. Loughran (Eds.), *Re-examining pedagogical content knowledge in science education* (pp. 229–241). New York, London: Routledge.
Klahr, D. (2010). Coming up for air: But is it oxygen or phlogiston? A response to Taber's review of constructivist instruction: Success or failure. *Education Review, 13*(13), 1–6.
Liepertz, S., & Borowski, A. (2017). Testing the consensus model: The relationship between physics teachers' professional knowledge, interconnectedness of content structure and student achievement. *International Journal of Science Education* (Submitted).
Mason, J. (2002). *Researching your own practice: The discipline of noticing*. London: Routledge Falmer.
National Academies of Sciences, Engineering, and Medicine (NASEM). (2015). *Science teachers learning: Enhancing opportunities, creating supportive contexts*. Committee on Strengthening Science Education through a Teacher Learning Continuum. Board on Science Education and Teacher Advisory Council, Division of Behavioral and Social Science and Education. Washington, DC: The National Academies Press.
National Research Council. (2002). *Scientific research in education*. Committee on Scientific Principles for Education Research (R. J. Shavelson & L. Towne, Eds.). Center for Education. Division of Behavioral and Social Sciences and Education. Washington, DC: National Academy Press.
National Research Council. (2012). *A framework for K-12 science education: Practices, crosscutting concepts, and core ideas*. Committee on a Conceptual Framework for New K-12 Science Education Standards. Board on Science Education, Division of Behavioural and Social Sciences and Education. Washington, DC: The National Academies Press.
Pellegrino, J. W., DiBello, L. V., & Goldman, S. R. (2016). A framework for conceptualizing and evaluating the validity of instructionally relevant assessments. *Educational Psychologist, 51*(1), 59–81. https://doi.org/10.1080/00461520.2016.1145550.
Roth, K., Bintz, J., Wickler, N., Hvidsten, C., Taylor, J., Beardsley, P., ... Wilson, C. D. (2017). Design principles for effective video-based professional development. *International Journal of STEM Education*, 4–31.
Scheiner, T. (2016). Teacher noticing: Enlightening or blinding? *ZDM Mathematics Education, 48*(1–2), 227–238.
Sherin, M. G., Russ, R. S., & Colestock, A. (2011). Accessing mathematics teachers' in-the-moment noticing. In M. G. Sherin, V. R. Jacobs, & R. Philipp (Eds.), *Mathematics teacher noticing: Seeing through teachers' eyes* (pp. 79–94). New York, NY: Routledge.
Shulman, L. (1987). Knowledge and teaching: Foundations of the new reform. *Harvard educational review, 57*(1), 1–23.
Sickel, A. J., Banilower, E., Carlson, J., & Van Driel, J. H. (2015). Examining PCK research in the context of current policy initiatives. In A. Berry, P. Friedrichsen, & J. Loughran (Eds.),

Re-examining pedagogical content knowledge in science education (pp. 199–213). New York, London: Routledge.

Slavin, R. E. (2008). What works? Issues in synthesizing educational program evaluations. *Educational Researcher, 37,* 5–14.

Van Driel, J. H., Berry, A., & Meirink, J. A. (2014). Research on science teacher knowledge. In N. Lederman & S. K. Abell (Eds.), *Handbook of research on science education* (2nd ed., pp. 848–870). London: Taylor & Francis.

Van Driel, J. H., Meirink, J. A., Van Veen, K., & Zwart, R. C. (2012). Current trends and missing links in studies on teacher professional development in science education: A review of design features and quality of research. *Studies in Science Education, 48,* 129–160.

Wayne, A. J., Yoon, K. S., Zhu, P., Cronen, S., & Garet, M. S. (2008). Experimenting with teacher professional development: Motives and methods. *Educational Researcher, 37,* 469–479. https://doi.org/10.3102/0013189x08327154.

Wilson, S. M. (2013). Professional development for science teachers. *Science, 340,* 310–313.

Yoon, K. S., Duncan, T., Lee, S. W.-Y., Scarloss, B., & Shapley, K. (2007). *Reviewing the evidence on how teacher PD affects student achievement*. Washington, DC: U.S. Department of Education, Institute of Education Sciences, National Center for Education Evaluation and Regional Assistance, Regional Educational Laboratory Southwest. Retrieved from http://ies.ed.gov/ncee/edlabs.

Chapter 14
Developing Research on PCK as a Community

Rebecca Cooper and Jan van Driel

Abstract After its introduction, a group of scholars, led by Lee Shulman, performed several studies on PCK in a range of disciplines. Since the 1990s, PCK studies have become a prominent strand in science education research. Initially, most of these studies were done in the USA, but once PCK was picked up by science education researchers in other continents, a proliferation of conceptions and models of PCK, and instruments to study it, became apparent. This chapter describes the ways in which scholars in science education have communicated with each other, through books, articles, presentations at conferences and, significantly, the PCK Summits to continue the conversation around PCK. The chapter will focus on the process of developing a consensus model of PCK among the scholars that participated in the two PCK Summits, how they communicated with each other during and after the Summits, and with the broader community of researchers with an interest in PCK. The chapter includes personal reflective narratives to exemplify key features of the PCK Summit processes and outcomes and looks to offer insights into the impact and possible next steps post the Second (2nd) PCK Summit.

Introduction

After Lee Shulman introduced pedagogical content knowledge (PCK) in his 1986 presidential lecture for American Educational Research Association (AERA), he led a group of scholars who performed studies on PCK in disciplines ranging from language and social studies to mathematics and science. Since the 1990s, PCK studies have become a prominent strand of research, especially in the domains of mathematics and science education. Initially, most of these studies were conducted in the USA,

R. Cooper (✉)
Faculty of Education, Monash University, Melbourne, Australia
e-mail: rebecca.cooper@monash.edu

J. van Driel
Melbourne Graduate School of Education, The University of Melbourne, Melbourne, Australia
e-mail: j.vandriel@unimelb.edu.au

A. Hume et al. (eds.), *Repositioning Pedagogical Content Knowledge in Teachers' Knowledge for Teaching Science*, https://doi.org/10.1007/978-981-13-5898-2_14

but once PCK was picked up by science education researchers in other continents, a variety of conceptions and models of PCK, and instruments to study it, was developed and subsequently disseminated through books, articles, joint projects, and presentations at seminars and conferences. This chapter focuses on the processes of sharing research methods and outcomes among the mostly science education scholars who participated in two international meetings known as the PCK Summits, including personal reflections by the authors as Summit participants. The chapter describes how attendees communicated with each other during and after the Summits, and with the broader community of researchers with an interest in PCK, to develop a shared language and arrive at a consensus about models and methods.

The processes of sharing research methods and outcomes should be seen in the broader context of academic research in the twenty-first century, which in all disciplines is a global enterprise. The numbers of international conferences, books, and journals have increased exponentially in the last decades. Publications and presentations are vital to share research and discuss how new research outcomes contribute to the field. Decisions about acceptance of papers are often based on considering what the community of researchers in the domain of the paper can learn from it or what the paper adds to the existing body of knowledge. At the same time, research output measured by publications and presentations has become very important for the careers of individual researchers and in the assessment of research institutes. The increase of scale comes with several challenges. Conferences with an attendance over 10,000 delegates have had to organise their schedules to accommodate as many presentations as possible, for instance, by limiting the time per presentation (5 talks per slot of 90 min is not unusual) and increasing the number of parallel sessions. Obviously, such measures compromise the opportunity for discussions about research. Not only the time to discuss is limited, but by organising a conference schedule around specific themes or strands (in divisions or Special Interest Groups) that run parallel in time, the chances to meet researchers from adjacent research areas are minimised. In addition, research journals increasingly tend to specialise on specific strands of research. As a result, groups of researchers tend to communicate within specific channels (e.g., Special Interest Groups and specialised journals) and are thus not always aware of what happens in other communities. This narrowing of communication lines is particularly problematic for multidisciplinary research, which thrives on connections between groups of specialists in different areas. To counter these problems, interactions between researchers are organised in a variety of ways, such as summer schools around a specific (multidisciplinary) theme or exchanges of staff and Ph.D. students between institutes, often facilitated by scholarships and sabbaticals.

The PCK Research Community: 1986–2010

The origins of research on PCK have been well documented. For instance, in an interview with Lee Shulman in 2007, he reflected on what was called the Teacher

Knowledge Project in the early 1980s, which came out of a growing concern about the role of content in teaching. The initial question for that project was:

> 'How does somebody that really knows something, teach it to somebody who doesn't?' Simple as you can get. So, [...] somebody who really knows evolutionary theory, what do they do if they have to teach it to somebody that not only that doesn't know, but if he or she did, wouldn't believe it? Thinking like that meant that we had to do these studies subject by subject and it just happened that at Stanford we prepare teachers, all secondary, in Science, Math, English and Social Studies and I just happened to have wonderful doctoral students in each of those areas. (Shulman, quoted in Berry, Loughran, & Van Driel, 2008, p. 1274)

In the Teacher Knowledge Project, about a dozen research projects were conducted across areas as diverse as English, mathematics, history, science, and social studies. In later years, research on PCK spreads around the globe; however, most research in PCK since the 1990s has focused on the domains of mathematics and science. A recent review of the literature on PCK in the context of pre-service education (Berry, Depaepe, & Van Driel, 2016) found 66 empirical studies, the large majority of which were located in mathematics (34) and science (24). The remaining eight studies were conducted in the domains of physical education (3), language (2), history (1), geography (1), and drama education (1). The numbers of studies in elementary and secondary pre-service teacher education were more or less the same. The review revealed that researchers in mathematics and science education have developed different conceptual models of PCK and associated methods to study PCK. More problematic, these researchers typically publish and present in different journals and conferences, with a focus on either mathematics or science education and very rarely cross-reference each others' work.

Research on PCK in science education, until 2000, was mostly done in the USA and presented at conferences such as National Association for Research on Science Teaching (NARST) and Association for the Education of Teachers in Science (AETS). Some of this research was brought together in a book *Examining Pedagogical Content Knowledge*, edited by Gess-Newsome and Lederman (1999), commonly referred to as "the purple book". This volume included a section called "The literature", with chapters on conceptual models of teacher knowledge, as well as sections with reports of empirical studies on PCK and their impact on the development of teacher education programs. The PCK model presented in a chapter by Magnusson, Krajcik, and Borko (1999) became very influential (over 1900 cites in Google Scholar to date) and has informed many PCK studies in science education in the last 20 years (Friedrichsen, Van Driel, & Abell, 2011). In the 2000s, research on PCK in science education proliferated across the globe with concentrations in a number of places, in particular Monash University, Australia (Loughran and colleagues); University of Missouri at Columbia (Abell and colleagues); BSCS, Colorado (Carlson and colleagues); University of Leiden, the Netherlands (Van Driel and colleagues); University of the Witwatersrand, South Africa (Rollnick and colleagues); UNAM-Mexico (Garritz and colleagues); and University of Duisburg-Essen, Germany (Fischer and colleagues). These researchers met at conferences, especially NARST and European Science Education Research Association (ESERA), and jointly organised symposia. Also, researchers from these groups visited each other's institutes, leading to joint projects and publications. In 2008, a special issue of the *International Journal of Science Education* appeared, edited by Berry, Loughran and Van Driel. This issue contained eight contributions from the aforementioned research groups, most of which were based on presentations during symposia at ESERA 2005 and NARST 2006.

The First PCK Summit: Colorado Springs, USA, October 2012

The 1st PCK Summit was an initiative from a group of US scholars led by Gess-Newsome, Carlson, and Gardner. The main purpose was to bring a group of around 25 PCK researchers together for a number of days to share and discuss their work, with the aim of forming "a professional learning community to explore the potential of a consensus model of PCK to guide science education research in this area through multiple research approaches" and to identify "specific next steps that would move the field forward" (Carlson, Stokes, Helms, Gess-Newsome, & Gardner, 2015, p. 15).

The organisers decided to invite a combination of senior and junior researchers, mostly from the groups mentioned in the previous section, rather than individuals. In addition, a small number of researchers from the domain of mathematics education were invited. The participants were asked to submit an abstract summarising the PCK research in their group several months prior to the Summit. These summaries were published on a website that was specifically created for the Summit. Somewhat later, each group was asked to write an elaborated version of their summary following a particular template. To prepare for the Summit, participants were asked to read the extended summaries of all groups. The website, http://pcksummit.bscs.org/, which was initially only accessible to participants, has been made public after the Summit and gives access to the Summit agenda (including papers and presentations), online study modules, and a discussion forum.

The Summit began with a presentation via Skype from Lee Shulman, who reflected on the context in which PCK had been introduced and the history of PCK research and gave his opinion about the relevance of PCK research today. Next, short group presentations followed from the participants. Rather than talking about their own (past) research, each group was asked to address a certain theme. The presentations basically served as a framework for a discussion with all participants about the theme. These discussions were led by two convenors (Taylor and Settlage), both of whom were science education researchers, however, not PCK specialists. During the second half of the Summit, most time was spent in subgroups of 4–5 participants who explored certain issues in depth, such as the relevance of PCK research for policy and practice. Finally, all groups were asked to produce a conceptual model of PCK. During the final session, these models were compared and discussed, working towards an outcome, that is, a consensus model of PCK.

In the following personal narrative, Cooper (one of the authors of this chapter) shares some of her experiences as one of the early career researchers at the 1st PCK Summit.

Intermezzo 1: The PCK Journey of an Early Career Researcher

The 1st PCK Summit presented an exciting and challenging opportunity for me [Cooper] as an early career researcher. At the time that I was invited, I had just completed writing up the first draft of my Ph.D. research and submitted it to my supervisor. Initially, my research was looking to investigate the development of PCK in science teacher educators, but as the research progressed it became clear to me that my thinking about PCK did not align with what I was inferring from my collected data. To me, PCK was about the knowledge that teachers develop over time through experience related to teaching particular content in particular ways to enhance student learning (Loughran, Berry, & Mulhall, 2012). However, the data I collected from the participants in my research was not focused on science content but more on pedagogy and sharing expertise for teaching using science as a context for this practice. Thus, I felt that it was more appropriate to infer pedagogical knowledge (PK) from my data, and so I changed frameworks and worked with Morine-Dershimer and Kent's model for pedagogical knowledge (Morine-Dershimer & Kent, 1999). Looking back, this was the right decision because I was better able to represent the participants' experiences through the facets of this model and to analyse them in ways that offered reasonable and justifiable insights in my thesis.

One of the outcomes of the 1st PCK Summit was an agreed-upon definition for PCK; i.e., "PCK is the knowledge of, reasoning behind, and enactment of the teaching of particular topics in a particular way with particular students for particular reasons for enhanced student outcomes" (Carlson et al., 2015). As I was part of the creative process leading to its development, this definition resonated with my understanding of PCK and also leant further support to my decision to shift to PK for my thesis. This experience and understanding gave me the confidence to describe my own research with clarity and to be sure of how and why it was not PCK that I was studying. In addition to the agreed-upon definition, a consensus model of teacher professional knowledge and skill, including PCK (CM) and influences on classroom practice and student outcomes, was developed. In order to arrive at these shared understandings, there were many in-depth discussions borne out of group activities and inspired by presentations by other participants, as outlined in the previous section. To be an active part of the discussions, I needed to be brave and articulate my thoughts. I needed to think about how my research aligned with what was being discussed at the Summit and how it might influence my own thinking and that of other participants. One of the discussions centred either on what is and what is not PCK, or when knowledge is and is not PCK. Participating in this discussion challenged me to elaborate on my stance for my research and why I was investigating PK and not PCK. Having to participate this way in the Summit helped me recognise that to be an academic and a part of the PCK research community, then I needed to contribute my arguments, be willing to justify my stance and be open to the critique and comments of others. Further, I was going to need to find productive ways to work with this feedback and turn it into productive deliberations that would further my thinking and thus my research agenda.

The 1st PCK Summit not only introduced me to other researchers but also to their research in a more detailed way. While I had read the work of many great researchers, several of whom were at the Summit, it was not until I met them and had time to explore and discuss their research that I realised what it truly meant to be an academic pursuing a research agenda. The researchers that I met were very willing to share their expertise and to share the evolution of their work and the progress of their thinking around PCK, which I found so helpful and inspirational. It left me thinking how could I work on my research, shaping it so that it continues my agenda but also becomes a significant contribution to the PCK research community? It made me realise that becoming an academic involves becoming a part of the bigger picture and thus a member of a research community that you contribute to through networking, reviewing, researching, and collaborating. To contribute in all these forums, I needed to be clear about what my research goals were and how they formed part of that bigger picture of PCK research.

After the First PCK Summit

It has been the experience of the authors of this chapter that maintaining the momentum when everyone returns home after a research meeting like the PCK Summit is vital. It is easy to leave after such an experience and become immersed in work at home. Technology can help as outlined earlier, but there is value in continuing to meet. The participants of the 1st Summit regularly arranged to meet while attending the major international science education conferences. These meetings were often held in the afternoon and followed with dinner where a participant would chair discussion, following a brief agenda, to continue the focused conversation. Often, rapid progress was made and participants left with a clear understanding of what needed to be followed up on or done, by when, and how to keep track of the shifts in thinking and discussion in order to continue to move forward.

Sometime after the 1st Summit, the participants agreed to produce a book, titled, *Re-examining Pedagogical Content Knowledge in Science Education* (Berry, Friedrichsen, & Loughran, 2015), but affectionately known as "the blue book". As part of the Summit workshops, participants sorted themselves into small groups that focused on topics that had been raised during the Summit (such as assessment of PCK, or the role of PCK research in policy initiatives). The chapters in Part III of the "blue book", called *Emerging themes*, were written by teams comprising two to five co-authors from different institutes, emerging from the small groups formed at the Summit. Teams often used Skype to stay in touch with each other to discuss their writing and to continue conversations and progress their thinking. Some writing teams also used email exchanges to continue the writing process. Others employed Google Docs, which enabled them to write collaboratively and save time by not having to maintain versions of documents and wait for email replies.

As follow up, presenting as groups at conferences (NARST 2013, ESERA 2013) assisted in maintaining the momentum and helped us, the Summit participants, to articulate our thinking so that it could be shared with the broader educational research community. Further, it meant that we could incorporate the feedback provided by those who attended our conference presentations into our future work. In fact, this chapter was inspired by feedback we, the authors of this chapter, received at a conference presentation.

The discussions during these conferences ultimately led us to question whether another Summit was needed, and if so, why, and for what purpose? How would it build on what had already been done so we wouldn't keep doing the same thing?

The Second PCK Summit: Leiden, the Netherlands, December 2016

Preparing for the Second (2nd) PCK Summit

Preparations for the 2nd Summit were made by a team consisting of Van Driel, Berry, Kirschner, Borowski, and Carlson (who had been one of the organisers of the first summit). After much discussion, it was agreed that the focus for the 2nd Summit should be on sharing data and instruments. The idea was to build an understanding of each other's research and to consider how scholars infer PCK from their data. In addition, the organising group made it a priority to bring in new Summit participants, both senior and beginning, in an effort to broaden the thinking of the group and to continue sharing the experience of the PCK Summit with more members of the PCK research community. In total, 25 participants were invited to this Summit.

The Second (2nd) PCK Summit

The 2nd PCK Summit was designed to provide international researchers working on PCK in general science, biology, chemistry, and physics education the opportunity to share (1) how their data from PCK studies were collected, (2) the different kinds of instruments used to collect these data, and (3) the procedures used to infer PCK from these data. The aims of this Summit were to: develop a shared set of criteria to identify PCK for each kind of instrument through collectively analysing data that were obtained with the respective instrument; make accessible and comprehensible these instruments to the wider PCK research community; and reach consensus on a model of PCK that is strongly connected with empirical data of varying nature and can be used as a framework for the design of future PCK studies. The Summit consisted of sessions where participants worked in small groups with a focused task, alternated with whole-group sessions. The focused tasks were determined by the two facilitators of the Summit (Loughran and Cooper), in consultation with the Summit organisers. The tasks included interrogating data sets from participants' research projects, comparing and contrasting data collected using similar instruments, and analysing processes for inferring PCK for multiple data sets. These tasks were strongly driven by the discussions and outcomes of previous sessions to ensure that progress was made over the course of the Summit. The whole-group sessions were moderated by the two facilitators during the first half of the Summit; however, these sessions evolved and followed a more open format during the second half of the Summit. The whole-group sessions were an opportunity to discuss what had happened in the small groups sessions and focused more on the outcomes of these sessions. The Summit concluded with a model-building session that included all participants. One of the small groups was focused on working towards building the consensus model, and the whole group was given the opportunity later to continue their work. The whole-group model-building session was powerful in that it provided an opportunity for collective thought on a model to unify PCK research in science education and offer the beginning of a shared language for portraying PCK.

Immediately after the model-building session, a post-Summit meeting took place. A group of around 20 local researchers, most of who were doing a Ph.D. with a focus on PCK in science education or other disciplines, met with the Summit participants. In mixed groups of six to eight people, the local researchers presented their studies and received feedback from the Summit participants. This feedback led to lively discussions in all groups, and at the end of the session, there was a consensus that the presentations had been a great opportunity for both parties to share ideas and learn about each other's research.

In the following personal narrative, the other author of this chapter [Van Driel] shares some of his PCK research journey, including the influence of the PCK Summits.

Intermezzo 2: The PCK Journey of a Senior Researcher

In my first year as a chemistry teacher, I (Van Driel) was very lucky to be supervised by a senior colleague who generously shared his expertise on the teaching and learning of chemistry. He was able to explain in much detail how students would respond to certain teaching approaches and the conceptual struggles that students often would experience. This mentoring took place in the mid-1980s, and when I read Shulman's seminal PCK papers years later, I immediately recognised the expert knowledge my colleague had developed as PCK. That is, knowledge about student learning of particular subject matter and knowledge of specific teaching strategies that potentially help students to develop their knowledge and skills about this subject matter. In my Ph.D. (1985–1990), I was mostly focused on developing students' conceptual understanding of chemistry topics through specific lesson materials; however, I became increasingly interested in the different ways teachers implemented these materials in their practice. PCK provided a powerful framework to analyse the practical knowledge that teachers drew upon for this implementation. This interest resulted in a publication that helped to establish my reputation as a PCK researcher (Van Driel, Verloop, & De Vos, 1998).

In the next decade, I was involved in several PCK projects, collaborating with colleagues, Ph.D. students, and post-docs. During this period, I experienced the importance and benefits of communicating with international colleagues through a variety of modes. In particular, conference presentations (followed by direct interactions with colleagues) and publications, often with the same colleagues (followed by reactions and questions via email), have been extremely important to get feedback and inspiration for future research. Although some of these interactions led to ongoing collaborations (and personal friendships), in most cases, interactions were brief and limited in terms of depth. Mostly, these were fleeting interactions due to limitations of email and the length and frequency of conferences. I was therefore very happy when Julie Gess-Newsome introduced the idea of a PCK Summit to me. Participating in the 2012 Summit in Colorado was an incredible experience: the opportunity to talk and think for 5 days with a group of very committed and open colleagues about basically "everything you always wanted to know about PCK" will stay with me as a career highlight.

It was only natural for me to stay involved in the following developments (presenting at conferences and contributing to the "blue book"), and as soon as the idea of a 2nd Summit was proposed, I was keen to be involved in its organisation. Having direct access to the facility in Leiden (the Lorentz Center; http://www.lorentzcenter.nl/) made it logical for me to take the lead in the logistics of this Summit. After roughly a year of preparation (together with Kirschner, Borowski, Berry and Carlson), it was wonderful to see the actual meeting happen. Although organisational issues had to be attended to, I was able to concentrate on the discussions with the whole group and in the smaller working groups. I feel strongly that we made progress during this Summit, compared to the first one. In my view, elements that contributed to the success were (1) most of the participants knew each other's research quite well, and for some time, whereas (2) new participants brought new perspectives, and (3) the facilitators did a wonderful job, sensing very well where discussions were going and deciding, often on the spot, how progress could best be fostered. In addition, the physical layout of the facility and the support of its staff helped to keep everyone focused and distractions to a minimum.

After the Second (2nd) PCK Summit

Moving with the Momentum

The focus of the ongoing discussions for the participants post the 2nd PCK Summit concentrated on a revised consensus model of PCK, to be published in this book! As mentioned earlier, the final whole-group session at the 2nd PCK Summit was a model-building session. While the participants reached a somewhat final point, it was decided that it would be helpful to have a graphic designer to turn our rough sketches into a more coherent visual representation. Two participants (Carlson and Daehler) graciously took responsibility for this task. The visual representation, along with a comprehensive explanation, was shared with all participants of the 2nd PCK Summit using Google Docs, which allowed for the conversation around the model development to continue. In addition to electronic communication, an ad hoc meeting took place during NARST 2017 to discuss the revisions of the model and its visual representation. Fourteen participants of the 2nd Summit were present during this meeting. This ongoing development also fostered the preparation of more conference presentations (ESERA 2017, ASERA 2017, NARST 2018).

Sharing the Outcomes

Sharing the outcomes of the PCK Summit in relation to the progressing of PCK research is really important, and the time immediately post to the 2nd Summit was focused on opening up the revised consensus model for discussion. This discussion opportunity was also about sharing the data collection tools and processes for analysis and inferring PCK. In addition to sharing the research-focused outcomes, it became apparent that we had a broader story to tell that focused on the process of the Summit and the development of the research community. The existence of this story became evident at the end of a conference presentation (ASERA 2017) where the questions asked by the audience were delving further into the processes behind the planning and happenings at the Summits.

Impact and Next Steps

Sharing with the Broader PCK Research Community

Participation in both PCK Summits was by invitation only. Thus, while initially access to the discussions and offering of the summits was only provided to a small number of participants, these invited participants have a responsibility to provide access for the broader PCK research community. It has always been a priority of the organisers and the participants at large to share the outcomes and as much of the discussions from the Summit as possible. The 1st Summit managed this dissemination of ideas effectively through the creation and maintenance of a publicly available website, as mentioned earlier, but the website does not provide an indication of how what happened at the conference has influenced further research. Thus, it is a priority of Summit participants to present regularly at a variety of conferences to ensure that the Summit ideas are shared with and questioned by the broader PCK research community. Taking these ideas and questions further, there also are publications (i.e. the "blue book" and this current volume and an upcoming special issue of the *International Journal of Science Education*) that may not specifically address the proceedings during the Summits but do offer readers some insights into how the Summit has re-directed, influenced, or forwarded continuing PCK research. Speaking to the broader PCK research community has ensured that ideas have been articulated beyond that of the participants in attendance so that these ideas are shared, opened for discussion, and explored by more than those who were present. It has meant that other PCK researchers have had the opportunity to take the ideas and issues raised at the Summits and apply them to their own work, should they wish to do so, to progress their PCK research.

New Connections for Research and Writing

The Summits have been influential in generating new ideas to progress both individual and community research. It has provided the opportunity for robust discussion around individual researcher's plans and assisted them to further their research in more ambitious yet coherent ways. This expansion of their ambition is possible because their research has been more thoroughly critiqued before it even started, simply because they have been able to hear from other researchers at the Summits. There has been more cohesion to the research, in that the resulting research is better aligned in terms of the theory and models used. These alignments are important in building a genuine research community that can have conversations based on shared understanding of the foundations of their research to strengthen both individual and community research agendas and to forge new understandings.

Sharing the Evolution of PCK Research

The Summits have made it a priority to bring new researchers into the science PCK community to work alongside more experienced researchers and offer them networking and mentoring opportunities. The Summits have helped new researchers to better appreciate the genealogy of the PCK research field and become acclimatised to the

PCK research environment. These new researchers can therefore progress the field in science education with a genuine pursuit of new knowledge because they are well attuned to where progress is needed and why.

Sustaining a PCK Research Community

Further benefits include the significance of the processes involved in planning and carrying out the Summits. These processes are valuable in relation to the sustainability and cohesion of a research community so that there is consistency around the quality, validity, and reliability of research in the field. Keeping the number of participants at the Summit small is part of what made the Summits work so well, so it is never going to be a big event. However, the processes and understandings should be shared widely, as outlined previously, with all those in the PCK research community and beyond.

This chapter opens discussion around the processes underpinning the planning and implementing of PCK Summits and the communication prior, between, and after these Summits, in relation to the contribution the processes and ideas around continual communication between participants can make to the wider PCK research community. This contribution offers not only greater cohesion, but clarity around future thinking related to research in the PCK field. This approach may serve as a model or example for research in other fields/domains, especially when researchers are using a variety of models and methods to explore the same territory.

As part of their aims, both Summits included experienced and early career PCK researchers as participants, offering an opportunity for early career PCK researchers to be introduced to more experienced members of the PCK research community. As an international research community, the PCK research community in science education is forward thinking about the future of research in this area and assisting early career researchers to better plan and appreciate the trajectory of research in this field. Now, the science PCK research community needs to think about how to continue this conversation with Summit participants and the broader research community. This ongoing conversation will help to decide whether a third PCK Summit, in a couple of years from now, is necessary or useful to further research in the field and sustain the PCK research community.

References

Berry, A., Depaepe, F., & Van Driel, J. H. (2016). Pedagogical content knowledge in teacher education. In J. Loughran & M. L. Hamilton (Eds.), *International handbook of teacher education* (pp. 347–386). Dordrecht: Springer.

Berry, A., Friedrichsen, P., & Loughran, J. J. (2015). *Re-examining pedagogical content knowledge in science education*. New York, London: Routledge.

Berry, A., Loughran, J. J., & Van Driel, J. H. (2008). Revisiting the roots of pedagogical content knowledge. *International Journal of Science Education, 30,* 1271–1279.
Carlson, J., Stokes, L., Helms, J., Gess-Newsome, J., & Gardner, A. (2015). The PCK summit: A process and structure for challenging current ideas, provoking future work, and considering new directions. In A. Berry, P. Friedrichsen, & J. Loughran (Eds.), *Re-examining pedagogical content knowledge in science education* (pp. 14–27). New York, London: Routledge.
Friedrichsen, P., Van Driel, J. H., & Abell, S. K. (2011). Taking a closer look at science teaching orientations. *Science Education, 95,* 358–376.
Gess-Newsome, J., & Lederman, N. G. (1999). *Examining pedagogical content knowledge. The construct and its implications for science education*. Dordrecht, The Netherlands: Kluwer.
Loughran, J. J., Berry, A., & Mulhall, P. (2012). *Understanding and developing science teachers' pedagogical content knowledge* (2nd ed.). Rotterdam, The Netherlands: Sense Publishers.
Magnusson, S., Krajcik, J., & Borko, H. (1999). Nature, sources and development of pedagogical content knowledge. In J. Gess-Newsome & N. G. Lederman (Eds.), *Examining pedagogical content knowledge* (pp. 95–132). Dordrecht, The Netherlands: Kluwer.
Morine-Dershimer, G., & Kent, T. (1999). The complex nature and sources of teachers' pedagogical knowledge. In J. Gess-Newsome & N. G. Lederman (Eds.), *Examining pedagogical content knowledge* (pp. 21–50). Dordrecht, The Netherlands: Kluwer.
Van Driel, J. H., Verloop, N., & De Vos, W. (1998). Developing science teachers' pedagogical content knowledge. *Journal of Research in Science Teaching, 35,* 673–695.

Correction to: Repositioning Pedagogical Content Knowledge in Teachers' Knowledge for Teaching Science

Anne Hume, Rebecca Cooper, and Andreas Borowski

Correction to: A. Hume et al. (eds.), *Repositioning Pedagogical Content Knowledge in Teachers' Knowledge for Teaching Science*, https://doi.org/10.1007/978-981-13-5898-2

The original version of the book was inadvertently published with incorrect information in chapter 2, chapter 12 and in backmatter. The erratum book has been updated with the changes.

The updated version of this book can be found at
https://doi.org/10.1007/978-981-13-5898-2
https://doi.org/10.1007/978-981-13-5898-2_2
https://doi.org/10.1007/978-981-13-5898-2_6
https://doi.org/10.1007/978-981-13-5898-2_12

A. Hume et al. (eds.), *Repositioning Pedagogical Content Knowledge in Teachers' Knowledge for Teaching Science*, https://doi.org/10.1007/978-981-13-5898-2_15

Postscript: Considerations from an External Perspective

Oliver Tepner and Elke Sumfleth

Introduction

On many levels, this book is a unique systematic analysis of pedagogical content knowledge (PCK) concepts in science education. Both, the comprehensive review of recent literature on PCK research and especially the Refined Consensus Model (RCM) of PCK in science and its use with regard to different fields of research, show multiple perspectives on this complex concept in science education. Most of the chapters are used for referring the RCM to research projects that have been planned and executed without the RCM in order to show how fruitful and relevant the new model could be for concrete research projects. The book also offers perspectives on the future of PCK research. In addition to the retrospective linking of recent studies to the RCM that are presented in Part II, this postscript chapter is intended to give a personal view on the whole concept and on some selected aspects that seem to be especially important for understanding the ongoing debate on the nature of PCK.

In the first section, we highlight the most relevant aspects of each chapter that refer to the RCM and which can be used for capturing the nature of PCK. In the second section, we value the RCM from an external perspective, emphasise the potential of the RCM for prospective research, and make suggestions for its further development within the next years.

The original version of this chapter was revised: Reference citation and list were updated. The correction to this chapter is available at https://doi.org/10.1007/978-981-13-5898-2_15

O. Tepner
Department of Chemistry and Pharmacy, University of Regensburg, Regensburg, Germany
e-mail: oliver.tepner@ur.de

E. Sumfleth
Department of Chemistry and Pharmacy, University of Duisburg-Essen, Duisburg, Germany
e-mail: elke.sumfleth@uni-due.de

A. Hume et al. (eds.), *Repositioning Pedagogical Content Knowledge in Teachers' Knowledge for Teaching Science*, https://doi.org/10.1007/978-981-13-5898-2

Highlights

Chapter 1

The first chapter closes a gap in the literature by identifying methodologies used in empirical studies on science teachers' PCK. Besides the aims and statistical data of studies, the systematic review provides an overview of different conceptualisations and approaches to determine PCK. A substantive outcome of this review is a theory-driven system for categorising studies on PCK. The presented model of "different approaches for determining science teachers' PCK and their relationship with different forms of PCK" (Chan & Hume, 2019, Fig. 1.1) takes relevant aspects of the RCM like knowledge, skills, and reasoning into account. On this basis, the reviewed studies were organised into five major strands, including investigating the nature of science teachers' PCK, the development of teachers' PCK, the relationship between PCK and other variables, changes in teachers' PCK, and reporting the development and validation of a PCK test instrument. The review also reveals a lack of studies on the development of PCK over more than three years. One prominent question that has not been investigated broadly so far is whether a single PCK profile changes over different phases of a teaching cycle (i.e., lesson planning, enactment, and reflection) or not. Several points of accordance and divergence in thinking and wording around the PCK concepts in usage (including the Consensus Model (CM)) have been found within the PCK research community in the field of science education.

Conclusion: Chapter 1 gives a comprehensive, critical, and outstanding systematic review of 99 articles dealing with PCK in different ways. For anybody who wants to get a concise historical overview and a notion of research in the field of PCK, it is a must to read this chapter.

Chapter 2

The second book chapter gives insights both into the consensus model CM of teacher professional knowledge and skills including PCK developed in the context of the First (1st) PCK summit and into the process of refinement of the extant model that lead to the Refined Consensus Model (RCM) of PCK in science in the Second (2nd) PCK summit. Origins and backgrounds of the model development process are shown. Based on the researchers' experiences with working with the CM from 2012, advantages and limitations of that model are pointed out and communicated honestly. As a consequence, the key limitation of minimal information on PCK in the model itself has been removed (Carlson, Daehler, Alonzo, Barendsen, Berry, Borowski, & Wilson 2019). Therefore, the focus of PCK as a part of teachers' professional knowledge and skills in the CM from the 1st Summit has been changed to PCK itself. The 2016 Summit's debate on the nature of PCK revealed *collective PCK (cPCK), personal PCK (pPCK),* and *enacted PCK (ePCK)* as three distinct realms of PCK (Carlson, Cooper, Daehler, Friedrichsen, Heller, Kirschner, … Wong, 2019). The combination of knowledge and skills as an aspect of PCK is taken into account, both by the new model being centred on the practice of teaching and the new realm of ePCK. Further details of the RCM, its components, and their interactions are described in Chap. 2 as well and be discussed later in this chapter.

Conclusion: Chapter 2 allows for understanding the development of the RCM of PCK during the course of the second summit and the reasons for changes in thought compared to the 1st Summit. It gives deep insight into the thinking of the world's leading researchers on the nature of PCK for science teaching.

Chapter 3

Chapter 3 aims at combining theory in terms of the RCM and practice as revealed in a selection of research and education projects in science. Five short vignettes are presented to show how the RCM can be used for the purpose of teacher education and research and this post hoc perspective give hints at the applicability of the model in practice. One key result of this chapter is that the RCM offers diverse possibilities to connect it with empirical projects (Carlson, Daehler, et al., 2019). For example, it can bring together different ideas of personal development in a pre-service science teacher education programme and how cPCK, pPCK, and ePCK can be fostered in a biology methods course. Or the visual representation of the RCM could be used for orientation and for situating the range of activities occurring during a professional learning experience, maybe similar to an advance organiser. Taking two examples of existing instruments (paper-and-pencil tests and interviews) for measuring physics/science teachers' PCK, the vignettes also show that the RCM can be applied to test instruments, which have been developed independently from the RCM. The model could show which realms a test instrument can cover and which aspects cannot be addressed. Additionally, the RCM can link to existing and established tools like CoRes and PaP-sRs (Loughran, Berry, & Mulhall, 2006).

Conclusion: An important result of presenting five different vignettes is the range of personal views of researchers on the integration of the RCM into their own projects and on the implications of the RCM for future work. The model seems to be versatile and complex enough to meet the theoretical and practical needs of researchers and teacher educators.

Chapter 4

In Chap. 4, Soonhye Park describes how the pentagon model of PCK and the indispensable and idiosyncratic PCK model from the science education research field are situated within the RCM. On the one hand, congruency between different models is pointed out: For example, the enactment dimension of PCK in the pentagon model corresponds to ePCK and the understanding dimension of PCK meets the criteria of pPCK. On the other hand, differences between the two models are stated in terms of the relationship between ePCK and pPCK and between the understanding dimension and the enactment dimension (Park, 2019). Implications for research designs are given.

One outstanding result of Chap. 4 is a critical view on the usability of the RCM for measuring pPCK and ePCK. The authors question how ePCK can be measured when it is given that ePCK is an expression of pPCK? (Park, 2019). Additionally, the knowledge exchange between pPCK and ePCK in the RCM and the meaning of cPCK should be reconsidered.

Conclusion: While the chapter provides comparative insights into recent existing models on PCK, it also offers methodological approaches for how to measure RCM's PCK empirically through connected realms like ePCK, pPCK, and cPCK.

Chapter 5

Chapter 5 links the concept of topic-specific PCK (Davidowitz & Potgieter, 2016) to the RCM. Topic-specific PCK is seen as a "grain size in the continuum of PCK found within the three realms of PCK, namely ePCK, pPCK, and ePCK", and is based on content specific components like curricular saliency, learner prior knowledge, and representations (Mavhunga, 2019). The notion of topic-specific characteristics can be located at different aspects of the RCM and confirmed the usability of the RCM for research projects. On the basis of an intervention study, the development of pre-service teachers' pPCK and ePCK in chemistry is shown. The intervention refers to cPCK with a topic-specific focus on electrochemistry that helps student teachers enhance their pPCK in this topic. It was assessed by a paper-and-pencil test. A subsequent case study proves the development of ePCK in planning a lesson when using components of PCK at the topic level.

Conclusion: This chapter shows comprehensibly how the existing concept of a topic-specific characteristic of PCK is considered in the RCM.

Chapter 6

The sixth chapter provides a view on RCM through the lens of competence and focuses in the subject of physics on a broader knowledge base that informs both cPCK and pPCK/ePCK indirectly (Sorge, Stender, & Neumann, 2019). Two studies are presented to examine the exchange between the knowledge base CK and cPCK and the influence of cPCK on the development of pPCK and ePCK. While cPCK was more strongly related to PK in the beginning of university education, cPCK was more strongly related to CK for advanced pre-service physics teachers. The second study reveals an exchange between teachers' cPCK and pPCK/ePCK, moderated largely by teaching experience and by motivational aspects and self-regulatory skills, but not by beliefs.

Conclusion: The findings confirm the RCM and its postulated relationship between cPCK and a broader professional knowledge base (CK, PK). Additionally, the stated relationship between cPCK and pPCK/ePCK in the RCM is shown and the RCM can be used for anchoring processes of development in terms of science teachers' PCK.

Chapter 7

This chapter links the RCM and the learning study approach. Planning, teaching, and reflecting as a part of ePCK include anticipating students' ideas and analysing learning, which is typical of pedagogical reasoning within a learning study cycle (Schneider, 2019). Importantly, the learning study approach is used to illustrate teachers' ideas and thinking and to promote realms of the RCM at the same time. While cPCK can be a knowledge base for designing educative tasks for teachers, the learning study process itself informs pPCK development. Enacted PCK is shown, and influenced, by a learning study in terms of planning, teaching, and reflecting. Perhaps the most significant aspect of the learning study approach is its potential to map "the PCK construct and teachers' learning trajectories" (Schneider, 2019).

Conclusion: The RCM, especially the cPCK and ePCK realms, can be applied to learning studies that can be used as tools to foster teachers' development in PCK.

Chapter 8

Chapter 8 discusses how the PCK map approach of Park and Chen (2012) in science education can be situated in the context of the RCM (Park & Suh, 2019). The PCK map approach makes connections between different components of PCK visible when referring to the pentagon model of PCK (Park & Chen, 2012). Basic steps of the PCK map approach—in-depth analysis, e.g. via interviews, the enumerative approach referring to the number of connections, and the constant comparative method to find common aspects in different data sources—are applied to the RCM.

An outstanding aspect of the PCK map approach is its ability to reveal "coherent and synergistic interactions among the PCK components" (Park & Suh, 2019). This chapter shows how different PCK components can be integrated, for example, into ePCK of the RCM. The PCK map approach can be used to explicate a process of pedagogical reasoning that can foster knowledge exchange between pPCK and ePCK.

Conclusion: Using a previous study, Chap. 8 shows that the RCM can be supplemented by the PCK map approach by making the abstract and complex structure of the PCK more visible. On the other hand, the RCM influences and informs the PCK map approach so that in the future it takes into account contextual factors associated with a teacher's PCK.

Chapter 9

This chapter illustrates how a study on student science teachers' development of pPCK can be situated in the RCM. The process of planning, enacting, and reflecting on pedagogical constructions is used to monitor and analyse the pPCK development, while connections to three other theoretical models, that is, the Magnusson et al. Model (Magnusson, Krajcik, & Borko, 1999), the model of professional growth (Clarke & Hollingsworth, 2002) and a model of emotion and personal factors (Hong, 2010), that may influence the development of pPCK, are made. Moreover, the notion of ePCK "as a dynamic construct defined by the part of pPCK that is 'active' at a certain moment during teaching practice" (Henze & Barendsen, 2019) is explicitly suggested and expresses the connection between these two PCK realms of the RCM. One interesting differentiation the authors make is that long-term decision-making (e.g. decisions made in the process of planning) is based on pPCK, while short-term decision-making (e.g. in the process of instruction) involves ePCK.

Conclusion: Chap. 9 added theoretical value by connecting the RCM, especially the pPCK realm, to three alternative theoretical models and shows ways of scaffolding the development of student teachers' pPCK.

Chapter 10

This chapter elucidates how content representations (CoRes, Loughran et al., 2006) can explore pPCK and ePCK and especially how CoRes can be used to develop these two realms of PCK in a collaborative CoRe design workshop (Carpendale & Hume, 2019; Hume & Berry, 2011). The CoRe documents can be seen as a representation of cPCK, when it has been developed as a collaborative process and can form a basis to develop individual teachers' pPCK. Learning context as a factor influencing the knowledge exchange between different knowledge bases, including cPCK, pPCK, and ePCK, has been identified. That is particularly important "as knowledge that was shared within the cPCK realm was transferred into and enhanced the pPCK and ePCK of an individual teacher" (Carpendale & Hume, 2019). Additionally, a rubric system and several indicators for analysing ePCK (e.g. use of strategies that allow for metacognition) are presented.

Conclusion: The particular value of Chap. 10 is its account of a very recently completed study into science teachers' PCK development that was informed by the RCM. The research demonstrates links between the established qualitative research approach of CoRes and the RCM and how CoRes can be used as a tool to promote the RCM's realms of pPCK and ePCK.

Chapter 11

Chapter 11 provides a grand PCK rubric for differentiating the quality of science teachers' PCK. Its generic nature allows measurement of different realms of PCK referred to in the RCM, and it should be "valid, ubiquitously accepted, and support clear and unambiguous communication across researchers" (Chan, Rollnick, & Gess-Newsome, 2019). The authors of Chap. 11 make it clear that they have proceeded very systematically and with the involvement of many researchers in the field of science PCK. The result is a grand rubric that can be adopted for different quantitative and qualitative studies on science PCK. Suggestions for methodological approaches like interviews and lesson observations to explore five PCK components are combined with the RCM realms of cPCK, pPCK, and ePCK. Additionally, considerations on the static and dynamic character of different PCK realms are presented, where cPCK, pPCK, and ePCK in the planning and reflection phases encompass a more static form of PCK, while ePCK in the interactive phase of teaching corresponds to dynamic PCK.

Conclusion: The outstanding outcome of Chap. 11 is a template for a PCK rubric that can facilitate aggregation, communication, and comparison of data from different PCK studies in science education. The rubric complements the RCM from both a theoretical and a methodological perspective.

Chapter 12

The focus of Chap. 12 is on science teachers' ePCK and its relationship to pPCK. Via the process of pedagogical reasoning, a "transformation from pPCK to ePCK" and vice versa is achieved (Alonzo, Berry, & Nilsson, 2019). The aspect of explicit versus tacit knowledge is directly addressed. While ePCK can be transformed into pPCK as articulable knowledge (e.g., when explicitly reflecting on a lesson), the transformation of ePCK into a tacit form of pPCK is proposed as well. Furthermore, macro- and micro-perspectives on plan–teach–reflect cycles are introduced for distinguishing lesson level and the level of instructional moves as parts of a lesson. The significant role of teaching experience in ePCK and the development process of and interactions between pPCK and different ePCK stages are elucidated. Suggestions for capturing pPCK and especially ePCK via CoRes and pedagogical and professional-experience repertoires (PaP-eRs, Loughran, Milroy, Berry, Gunstone, & Mulhall, 2001) are made.

Conclusion: This chapter gives additional theoretical and empirical value to the RCM by its concrete conceptualisation of ePCK and the process of its transformation into pPCK and vice versa.

Chapter 13

The merit of Chap. 13 lies in its systematic overview of possible research priorities in science education and potentially other domains that can be situated in the RCM. The chapter proposes "A Framework for Future PCK Research" and describes "possible studies on the structure of PCK, the development of PCK, the measurement of PCK, and the broader impacts of the refined model itself" (Wilson, Borowski, & van Driel, 2019). Potential research on the structure of PCK comprises exploring the relationships between components like cPCK, pPCK, and ePCK and how these components correlate with student learning. The development of PCK as part of professional development processes can be examined via intervention and longitudinal studies including multiple variables like pPCK, cPCK, ePCK, CK, and student outcomes. These studies can or should include statistical methods to investigate contexts of PCK via mediation and moderation effects. In order to examine the impact of the RCM, researchers can, for example, focus on "the extent to which it becomes pervasive in studies of PCK, or in practice and policy" (Wilson et al., 2019). Research on the measurement of PCK is strongly connected to all of the above-mentioned aspects, while it can be a research goal in itself.

Conclusion: Within the suggested framework for future PCK research, four conceivable directions are presented and suggestions for research questions that could guide PCK research during the next years are given.

Chapter 14

The last chapter of this book provides a concise historical overview of the development of research on PCK in science education. The focus is on communication between scholars about this concept where different ways of communication like presentations at conferences, articles, books, and, primarily, the PCK summits are presented with meaningful examples. The significance of a debate on the nature of PCK in science and on PCK research is explained both from a general ("collective") and a personal perspective via literature review and personal reflective narratives of the authors. It may be by chance that the chosen format corresponds to aspects of cPCK, pPCK, and ePCK. The significance of research to the authors, their personal enthusiasm for research, and their great appreciation of communication and cooperation with colleagues become very clear. The two PCK summits provided the means for meeting these aims and the outcomes provide an excellent theoretical basis that future research can refer to.

Conclusion: Chapter 14 elucidates the importance and the process of developing the construct of PCK in science and PCK research in science education as a community.

External Perspectives on the Refined Consensus Model

In addition to the previous book chapters, the authors of this chapter take up aspects of the RCM, which seem to them to be particularly important. These are supplemented by their own views.

When considering the RCM, it might be meaningful to refer to the historical debate on the nature of PCK. For example, this debate was summarised by Julie Gess-Newsome almost 20 years ago, as an introduction to a special volume with N.G. Ledermann on examining PCK (Gess-Newsome, 1999). It is noticeable that the questions she asked on a model of PCK in 1999 are still valid in 2018 and a good reason for developing the CM and the RCM: "Is PCK with its related domains a more precise model, explaining better the teacher cognition data than other models? Or, could the knowledge divisions as offered in PCK be overly refined? Could three integrated categories of teacher knowledge—subject matter, pedagogy and context - offer a more precise and powerful organization?" (Gess-Newsome, 1999, p. 10). These three questions give hints of the complex nature of PCK, while the RCM could be seen as a consequence of a debate that was started many years ago. Today, different forms of PCK (cPCK, pPCK, ePCK) are stated, and in addition to content knowledge, pedagogical knowledge, and curricular knowledge, further knowledge categories like assessment knowledge and knowledge of students have been taken up.

Historically, both transformative and integrative views of the nature of PCK have sought recognition. Aspects of the transformative nature of PCK are included in the RCM as well as in the CM. A knowledge base (shown in outer circles) transformed into different forms of PCK is one prominent characteristic of RCM (Magnusson et al., 1999). By placing the act of teaching (ePCK) in the centre of the RCM, the notion of an integrative model of PCK is also taken into account: "Knowledge of subject matter, pedagogy, and context [are] developed separately and integrated in the act of teaching" (Gess-Newsome, 1999, p. 13). So the RCM takes into account the character of PCK and different (historical) theoretical frameworks.

It is remarkable that the RCM obviously provides multiple ways of interpretation. The statement that RCM "...does not specify the mechanisms and pathways by which science teachers strengthen their PCK, change their teaching, or connect various knowledge bases" (Carlson, Daehler, et al., 2019) is based on a focus on the nature of PCK, including static and dynamic aspects. At the same time, the "...model also offers a way to think about how to support teacher development over a career trajectory from pre-service to expert leadership..." (Carlson, Daehler, et al., 2019). This premise is perceived by Park as "... shifted focus from what PCK is towards how PCK develops". (Park, 2019).

The comparison of the two consensus models reveals other interesting aspects: "Whereas the former consensus model aimed to drive agreement in defining PCK, the refined consensus model intends to help researchers identify areas to study for advancing PCK research through situating their studies within the context of teaching practices (Carlson, Daehler, et al., 2019)" (Park, 2019). The RCM is no longer just the subject of research and debate, but a medium to foster research on PCK.

The authors of this postscript chapter are happy about reconsidering the CM developed in the context of the 1st Summit as they had similar difficulties to those Carlson, Daehler, et al. (2019) mentioned in Chap. 2: "While the 2012 Consensus Model differentiated between topic-specific professional knowledge (TSPK) and personal PCK&S, the distinction in terms of PCK was not sufficiently clear and researchers struggled with the specifics, such as where to put knowledge about instructional strategies. Was that PK or TSPK or personal PCK&S? When and under which conditions?" (Carlson, Daehler, et al., 2019). Additionally, Carlson, Daehler, et al. (2019) struggled with the role of personal PCK and PCK&S in the older model. The closeness to the term "competence" is obvious, but one could question whether both PCK terms were aiming primarily at the context of knowledge or at competency and use of knowledge? The distinction between pPCK and ePCK seems to be a fruitful way for orientating future research.

The main progress of the RCM in comparison to the CM is the focus on science teachers' practice of teaching based on knowledge referring to the processes of planning, teaching, and reflecting. The ePCK at the centre of the new model is surrounded by (and influenced by?) other categories like pPCK, learning context, cPCK, and a non-PCK knowledge base. By extending the understanding of PCK to include the knowledge and actions of teachers, an ongoing debate about the nature of PCK and possible deficits in its conception is addressed (Settlage, 2013). With the introduction of ePCK, the RCM takes into account both a general cognitive (knowledge) perspective and a situational teaching perspective (Depaepe, Verschaffel, & Kelchtermans, 2013) as: "a key feature of ePCK is the way a science teachers' actions draw on their knowledge to meet the unique needs of students in the classroom during a given instructional period". (Carlson, Daehler, et al., 2019). Additionally, the RCM attaches greater importance to the "…learning context as an amplifier and a filter of teacher PCK, as well as, a mediator for teacher actions" (Park, 2019), compared to the original CM.

The category ePCK widens the focus on knowledge as a basis for science teachers' action to knowledge in action and so to teachers' competence. Referring to Baumert and Kunter (2013), knowledge could be seen as a part of competency distinct to beliefs, motivational orientations, and self-regulation. Whereas the term competency is not uniformly defined, it can be reasonable to introduce the new knowledge category ePCK to explore knowledge apart from beliefs, motivational orientations, and self-regulation.

Furthermore, the RCM and CM are similar in that the cPCK realm is influenced by the broader teacher professional knowledge bases, i.e., content knowledge, pedagogical knowledge, knowledge of the students, curricular knowledge, and assessment knowledge.. Especially for science teacher educators, it seems reasonable to consider that they have to widen their university perspective of assessment of university students to a perspective that is applicable in schools and teachable at university. One could ask if assessment knowledge is a knowledge category that influences PCK or if it even is a part of cPCK. This question can be asked for other categories like knowledge of students and curricular knowledge as well. Both consensus models allow for proving hypothesised relations between postulated categories.

Putting pedagogical reasoning in the very centre of the model symbolises its great importance in teaching. The combination of reflection, planning, and teaching (ePCK) is a basis for reasoning and decision-making that are core activities of teachers. Science teachers, who are able to give explicit reason for their doing, are supposed to be more successful than teachers who hold implicit knowledge only. In order to test this supposition, research using group comparisons on the impact of implicit and explicit knowledge is needed. Constructs like understanding and knowledge need an observable and measurable action. Reasoning on the teacher (and on student) level is a significant indicator for understanding and knowledge. The (initially unspecified) wording "pedagogical reasoning" in the centre of the model is remarkable, as it suggests that it is a key component of subject-specific PCK. Perhaps a revised model might consider including the term "subject-specific pedagogical reasoning" in order to prevent misunderstandings at first glance.

The importance of knowledge-based reasoning becomes very clear in the RCM. One could ask, however, if there are other important applications of knowledge components of PCK like perception, interpretation, and decision-making in classroom situations (Blömeke, Gustafsson, & Shavelson, 2015)? The approach "modelling competence as a continuum" describes the effect of cognitive and affective motivation (cPCK/pPCK) on the processes of perception, interpretation, and decision-making (ePCK) that lead to observable actions (ePCK) (Blömeke et al., 2015). Since the model of Blömeke et al. (2015) is generally formulated, the RCM in science can be linked to this model in mathematics. While the RCM allows for investigation of many possible impact categories that probably influence teaching and learning, no direct chain of action is postulated in the RCM as it is in the model "competence as a continuum" (Blömeke et al., 2015).

At a first glance, one could argue that ePCK is less a knowledge category and more a description of competency, as Baumert and Kunter suggest in their model of professional competence of mathematics teachers (Baumert & Kunter, 2013). As a consequence, the RCM of PCK in science education could be said to focus not only on knowledge, but also on competency that includes knowledge. The notion that the RCM includes aspects of competence is made explicit by Sorge et al. (2019) in Chap. 6. Knowledge like PCK can be seen as a precondition for a competence visible in ePCK as planning, teaching, and reflecting on a lesson occurs. Maybe this distinction reflects a national (German) view on competence, whereas many international researchers include aspects of competence in the construct of PCK (Berry, Friedrichsen, & Loughran, 2015). Using the term ePCK instead of competence could be more meaningful and helpful because the even more complex construct of competence is avoided and the focus is on knowledge categories including the enactment of PCK.

The idea of knowledge exchanges as pedagogical reasoning processes, involving amplification and filtering by influences such as attitudes and beliefs, is included implicitly in the RCM and symbolised via two-way arrows between all concentric circles/layers in the diagrammatic model. For the future, it could be meaningful to make filters like teachers' perception, attitudes, and beliefs more explicit in the diagrammatic form of the model, because of their significant impact on teachers' enactment in classrooms and their students' learning achievement (Lumpe, Czerniak, Haney, & Beltyukova, 2012). Some researchers state that "beliefs and PCK were inextricably intertwined" (Veal, 2004, p. 346), so the meaning of non-knowledge aspects could be included in a revised RCM of PCK. The above-mentioned aspects show that the ideas of the PCK research community could not be represented entirely in the RCM, despite many science educators who have extensive experience in doing research on PCK being part of the development process of CM and RCM. That said, the model accomplishes an excellent base for the ongoing debate on PCK.

All in all, the RCM seems to accomplish substantive improvements on the older CM. Placing ePCK at the centre of the model gives consideration to a practitioner perspective on the nature of PCK as a significant consequence. A science teacher's knowledge can now be seen as a precondition for planning, teaching, and reflection, and in turn, planning, teaching, and reflection are seen as PCK itself. The RCM allows for different views on the nature of PCK and fosters many different approaches to PCK research. These affordances might prove to be the most important outcomes of bringing together some of the world's leading PCK researchers in science education.

Summary

The book provides a holistic picture of the conceptualisation of PCK in science education research and practice and its application in empirical research on science teachers' PCK. In addition, the reader gets an excellent review of recent literature on that field of research. The chapters offer insights into the process of developing a widely discussed model of PCK that includes the thinking of many outstanding researchers and their perspectives on PCK. The RCM and this book itself will both serve to foster research on PCK and continue the debate on the nature of PCK, which is exactly what research is all about!

References

Alonzo, A. C., Berry, A., & Nilsson, P. (2019). Unpacking the complexity of PCK in action: Enacted and personal PCK. In A. Hume, R. Cooper, & A. Borowski (Eds.), *Repositioning pedagogical content knowledge in teachers' knowledge for teaching science* (pp. 271–288). Singapore: Springer.

Baumert, J., & Kunter, M. (2013). The COACTIV model of teachers' professional competence. In M. Kunter, J. Baumert, W. Blum, U. Klusmann, S. Krauss, & M. Neubrand (Eds.), *Mathematics teacher education* (Vol. 8, pp. 25–48)., Cognitive activation in the mathematics classroom and professional competence of teachers: Results from the COACTIV project. New York: Springer.

Berry, A., Friedrichsen, P., & Loughran, J. (Eds.). (2015). *Re-examining pedagogical content knowledge in science education. Teaching and learning in science series.* Hoboken: Taylor and Francis. Retrieved from http://gbv.eblib.com/patron/FullRecord.aspx?p=2001988.

Blömeke, S., Gustafsson, J.-E., & Shavelson, R. J. (2015). Beyond Dichotomies. *Zeitschrift für Psychologie, 223*(1), 3–13. https://doi.org/10.1027/2151-2604/a000194.

Carlson, J., Cooper, R., Daehler, K. R., Friedrichsen, P., Heller, J., Kirschner, S., … Wong, N. (2019). Vignettes illustrating practitioners' and researchers' applications of the refined consensus model of pedagogical content knowledge. In A. Hume, R. Cooper, & A. Borowski (Eds.), *Repositioning pedagogical content knowledge in teachers' knowledge for teaching science* (pp. 93–116). Singapore: Springer.

Carlson, J., Daehler, K.R., Alonzo, A.C., Barendsen, E., Berry, A., Borowski, A., … & Wilson, C.D. (2019). The Refined Consensus Model of pedagogical content knowledge in science education. In A. Hume, R. Cooper, & A. Borowski (Eds.), *Repositioning pedagogical content knowledge in teachers' knowledge for teaching science* (pp. 77–92). Singapore: Springer.

Carpendale, J., & Hume, A. (2019). Investigating practising science teachers' pPCK and ePCK development as a result of collaborative CoRe design. In A. Hume, R. Cooper, & A. Borowski (Eds.), *Repositioning pedagogical content knowledge in teachers' knowledge for teaching science* (pp. 223–250).Singapore: Springer.

Chan, K. H., & Hume, A. (2019). Towards a consensus model: Literature review of how science teachers' pedagogical content knowledge is investigated in empirical studies. In A. Hume, R. Cooper, & A. Borowski (Eds.), *Repositioning pedagogical content knowledge in teachers' knowledge for teaching science* (pp. 3–76). Singapore: Springer.

Chan, K. H., Rollnick, M., & Gess-Newsome, J. (2019). A grand rubric for measuring science teachers' pedagogical content knowledge. In A. Hume, R. Cooper, & A. Borowski (Eds.), *Repositioning pedagogical content knowledge in teachers' knowledge for teaching science* (pp. 251–270). Singapore: Springer.

Clarke, D., & Hollingsworth, H. (2002). Elaborating a model of teacher professional growth. *Teaching and Teacher Education, 18*(8), 947–967. https://doi.org/10.1016/S0742-051X(02)00053-7.

Davidowitz, B., & Potgieter, M. (2016). *Use of the Rasch measurement model to explore the relationship between content knowledge and topic-specific pedagogical content knowledge for organic chemistry.* (Keine Angabe). Retrieved from https://open.uct.ac.za/bitstream/11427/20320/1/Davidowitz2016.pdf.

Depaepe, F., Verschaffel, L., & Kelchtermans, G. (2013). Pedagogical content knowledge: A systematic review of the way in which the concept has pervaded mathematics educational research. *Teaching and Teacher Education, 34,* 12–25. https://doi.org/10.1016/j.tate.2013.03.001.

Gess-Newsome, J. (1999). Pedagogical content knowledge: An introduction and orientation. In J. Gess-Newsome & N. G. Lederman (Eds.), *Examining pedagogical content knowledge* (pp. 3–17). Dordrecht: Kluwer.

Henze, I., & Barendsen, E. (2019). Unraveling student teachers' pPCK development and the influence of psychological factors using authentic data sources. In A. Hume, R. Cooper, & A. Borowski (Eds.), *Repositioning pedagogical content knowledge in teachers' knowledge for teaching science* (pp. 201–221). Singapore: Springer.

Hong, J. Y. (2010). Pre-service and beginning teachers' professional identity and its relation to dropping out of the profession. *Teaching and Teacher Education, 26*(8), 1530–1543. https://doi.org/10.1016/j.tate.2010.06.003.

Hume, A., & Berry, A. (2011). Constructing CoRes—A strategy for building PCK in pre-service science teacher education. *Research in Science Education, 41*(3), 341–355. https://doi.org/10.1007/s11165-010-9168-3.

Loughran, J., Berry, A., & Mulhall, P. (2006). *Understanding and developing science teachers' pedagogical content knowledge.* Rotterdam: Sense Publishers.

Loughran, J., Milroy, P., Berry, A., Gunstone, R., & Mulhall, P. (2001). Documenting science teachers' pedagogical content knowledge through PaP-eRs. *Research in Science Education, 31,* 289–307.

Lumpe, A., Czerniak, C., Haney, J., & Beltyukova, S. (2012). Beliefs about Teaching Science: The relationship between elementary teachers' participation in professional development and student achievement. *International Journal of Science Education, 34*(2), 153–166. https://doi.org/10.1080/09500693.2010.551222.

Magnusson, S., Krajcik, J., & Borko, H. (1999). Nature, sources, and development of pedagogical content knowledge. In J. Gess-Newsome & N. G. Lederman (Eds.), *Examining pedagogical content knowledge* (pp. 95–132). Dordrecht: Kluwer.

Mavhunga, E. (2019). Exposing pathways for developing teacher pedagogical content knowledge at the topic level. In A. Hume, R. Cooper, & A. Borowski (Eds.), *Repositioning pedagogical content knowledge in teachers' knowledge for teaching science* (pp. 129–148). Singapore: Springer.

Park, S. (2019). Reconciliation between the refined consensus model of PCK and extant PCK models for advancing PCK research. In A. Hume, R. Cooper, & A. Borowski (Eds.), *Repositioning pedagogical content knowledge in teachers' knowledge for teaching science* (pp. 117–128). Singapore: Springer.

Park, S., & Chen, Y.-C. (2012). Mapping out the integration of the components of pedagogical content knowledge (PCK): Examples from high school biology classrooms. *Journal of Research in Science Teaching, 49*(7), 922–941. https://doi.org/10.1002/tea.21022.

Park, S., & Suh, J. J. (2019). The PCK map approach to capturing the complexity of enacted PCK (ePCK) and pedagogical reasoning. In A. Hume, R. Cooper, & A. Borowski (Eds.), *Repositioning pedagogical content knowledge in teachers' knowledge for teaching science* (pp. 185–197). Singapore: Springer.

Schneider, R. M. (2019). Illustrating and developing teachers' pedagogical content knowledge through learning study. In A. Hume, R. Cooper, & A. Borowski (Eds.), *Repositioning pedagogical content knowledge in teachers' knowledge for teaching science* (pp. 165–184). Singapore: Springer.

Settlage, J. (2013). On acknowledging PCK's shortcomings. *Journal of Science Teacher Education, 24*(1), 1–12. https://doi.org/10.1007/s10972-012-9332-x.

Sorge, S., Stender, A., & Neumann, K. (2019). The development of teachers' professional competence. In A. Hume, R. Cooper, & A. Borowski (Eds.), *Repositioning pedagogical content knowledge in teachers' knowledge for teaching science* (pp. 149–164). Singapore: Springer.

Veal, W. R. (2004). Beliefs and knowledge in chemistry teacher development. *International Journal of Science Education, 26*(3), 329–351.

Wilson, C. D., Borowski, A., & van Driel, J. H. (2019). Perspectives on the Future of PCK Research. In A. Hume, R. Cooper, & A. Borowski (Eds.), *Repositioning Pedagogical content knowledge in teachers' knowledge for teaching science* (pp. 289–300). Singapore: Springer.